WILEY SERIES IN PURE AND APPLIED OPTICS

Founded by Stanley S. Ballard, University of Florida

EDITOR: Bahaa E. A. Saleh

T0203090

Optical Waves in
Layered Media

Optical Waves in Layered Media

POCHI YEH

University of California
Santa Barbara

A JOHN WILEY & SONS, INC., PUBLICATION

For general information on our other products and services please contact our Customer Care Department within the U.S. at 877-762-2974, outside the U.S. at 317-572-3993 or fax 317-572-4002.

Wiley also publishes its books in a variety of electronic formats. Some content that appears in print, however, may not be available in electronic format.

Library of Congress Cataloging-in-Publication is available.

ISBN 0-471-73192-7

10 9 8 7 6 5

Preface

This book treats the theory of electromagnetic propagation in layered media. It is intended as a text for a course in modern optics for electrical engineering or applied physics students. Students are assumed to have general knowledge in electromagnetism and elementary matrix algebra. Some mathematical background in Fourier expansion and elementary differential equations would be helpful. The primary objectives of this book are to present a clear picture of the propagation of optical waves in layered media and to teach the reader how to analyze and design optical devices using such layered media. Although there exist a number of books on similar subjects, most are either too narrow in scope or are in the form of monographs, which are not suitable as textbooks.

Layered media play a very important role in many applications of modern optics. To fully utilize these media for transmission optics, we must understand the propagation of electromagnetic waves in these media. In addition, we must also be familiar with the systematic approaches used in the design of the layered structures. The emphasis is therefore on the theory of the propagation of optical waves in these media. An effort is made to bridge the gap between theory and practice through the use of numerical examples based on real situations. Only classical electrodynamics is used in dealing with the interaction of light with matter, except in the last chapter, where the subject of quantum wells is treated. Layered media that consist of isotropic and anisotropic materials are considered. Transmission and reflection of optical waves as well as the propagation of confined electromagnetic radiation are covered. A very wide range of topics is included, as may be seen from the table of contents.

I am deeply indebted to Professor Amnon Yariv for introducing the optics of layered media to me during the years when I was a graduate student at Caltech, and for his enlightening teaching. Portions of chapters 6, 9, and 11 first appeared, in different form, in *Optical Waves in Crystals*, co-authored by Professor Yariv and me. These materials are included here for completeness. I thank John Wiley & Sons and Professor Amnon Yariv for permission to reproduce these materials. My grateful thanks are also due to Drs. Joseph Longo, Derek Cheung, and Monte Khoshnevisan for their constant support and encouragement. I am also indebted to Drs. William Southwell and Kuo-Liang Chen and Mr. Paul Beckwith for their patient reading of the manuscript and helpful suggestions and to Sandy Nestor for her patient and competent typing of the manuscript. Finally, I am deeply grateful to my wife, Linda. Her love and devotion as a mother and wife have made the task at hand palatable and worthwhile.

POCHI YEH

Thousand Oaks, California
May 1988

Contents

1

The Electromagnetic Field

This book deals with the propagation of optical waves in layered media. In the first chapter, we review some of the basic properties of the propagation of electromagnetic radiation. These background materials are used frequently throughout the book and are included for completeness and as a ready source of reference.

We begin by briefly reviewing Maxwell's equations and the material equations. We then discuss the boundary conditions and the energy flow associated with electromagnetic radiation. These are followed by a derivation of the wave equations and an analysis of the propagation of monochromatic plane waves and some of their properties. Finally, we discuss the polarization state as well as the coherence of electromagnetic radiation.

1.1 MAXWELL'S EQUATIONS AND BOUNDARY CONDITIONS

1.1.1 Maxwell's Equations

The most fundamental equations in electrodynamics are Maxwell's equations, which are given in the following in rationalized MKS units:

$$\nabla \times \mathbf{E} + \frac{\partial \mathbf{B}}{\partial t} = 0, \tag{1.1-1}$$

$$\nabla \times \mathbf{H} - \frac{\partial \mathbf{D}}{\partial t} = \mathbf{J}, \tag{1.1-2}$$

$$\nabla \cdot \mathbf{D} = \varrho, \tag{1.1-3}$$

$$\nabla \cdot \mathbf{B} = 0. \tag{1.1-4}$$

In these equations, $\mathbf{E}$ and $\mathbf{H}$ are the electric field vector (in volts per meter) and magnetic field vector (in amperes per meter), respectively. These two field vectors are often used to describe an electromagnetic field. The quantities $\mathbf{D}$ and $\mathbf{B}$ are called the electric displacement (in coulombs per square meter) and

1

the magnetic induction (in webers per square meter), respectively. These two quantities are introduced to include the effect of the field on matter. The quantities ϱ and $\mathbf{J}$ are the electric charge (in coulombs per cubic meter) and current (in amperes per square meter) densities, respectively, and may be considered as the sources of the fields $\mathbf{E}$ and $\mathbf{H}$. These four Maxwell equations completely determine the electromagnetic field and are the fundamental equations of the theory of such fields, that is, of electrodynamics.

In optics, one often deals with propagation of electromagnetic radiation in regions of space where both charge density and current density are zero. In fact, if we set $\varrho = 0$ and $\mathbf{J} = 0$ in Maxwell's equations, we find that nonzero solutions exist. This means that an electromagnetic field can exist even in the absence of any charges and currents. Electromagnetic fields occurring in media in the absence of charges are called electromagnetic waves.

Maxwell's Equations (Eq. 1.1-1 to 1.1-4) consist of 8 scalar equations that relate a total of 12 variables, 3 for each of the 4 vectors $\mathbf{E}$, $\mathbf{H}$, $\mathbf{D}$, and $\mathbf{B}$. They cannot be solved uniquely unless the relationship between $\mathbf{B}$ and $\mathbf{H}$ and that between $\mathbf{E}$ and $\mathbf{D}$ are known. To obtain a unique determination of the field vectors, Maxwell's equations must be supplemented by the so-called constitutive equations (or material equations),

$$\mathbf{D} = \varepsilon\mathbf{E} = \varepsilon_0\mathbf{E} + \mathbf{P}, \qquad (1.1\text{-}5)$$

$$\mathbf{B} = \mu\mathbf{H} = \mu_0\mathbf{H} + \mathbf{M}, \qquad (1.1\text{-}6)$$

where the constitutive parameters ε and μ are tensors of rank 2 and are known as the dielectric tensor (or permittivity tensor) and the permeability tensor, respectively; $\mathbf{P}$ and $\mathbf{M}$ are electric and magnetic polarizations, respectively. When an electromagnetic field is present in matter, the electric field can perturb the motion of electrons and produce a dipole polarization $\mathbf{P}$ per unit volume. Analogously, the magnetic field can also induce a magnetization $\mathbf{M}$ in materials having a permeability that is different from μ_0. The constant ε_0 is called the permittivity of a vacuum and has a value of 8.854×10^{-12} F/m. The constant μ_0 is known as the permeability of a vacuum. It has, by definition, the exact value of $4\pi \times 10^{-7}$ H/m. If the material medium is isotropic, both ε and μ tensors reduce to scalars. In many cases, the quantities ε and μ can be assumed to be independent of the field strengths. However, if the fields are sufficiently strong, such as obtained, for example, by focusing a laser beam or applying a strong dc electric field to an electro-optic crystal, the dependence of these quantities on $\mathbf{E}$ and $\mathbf{H}$ must be considered. These nonlinear optical effects are beyond the scope of this book.

1.1.2 Boundary Conditions

One of the most important problems in determining the reflection and transmission of electromagnetic radiation through a layered medium is the continuity of some components of the field vectors at the dielectric interfaces between the layers. Although the physical properties (characterized by ε and μ) may change abruptly across the dielectric interfaces, there exist continuity relationships of some of the components of the field vectors at the dielectric boundary. These continuity conditions can be derived directly from Maxwell's equations.

Consider a boundary surface separating two media with different dielectric permittivity and permeability (medium 1 and medium 2). To obtain the boundary conditions for $\mathbf{B}$ and $\mathbf{D}$, we construct a thin cylinder over a unit area of the surface, as shown in Fig. 1.1(a). The end faces of the cylinder are parallel to the surface. We now apply the Gauss divergence theorem

$$\int \nabla \cdot \mathbf{F} \, dV = \int \mathbf{F} \cdot d\mathbf{S} \tag{1.1-7}$$

to both sides of Eqs. (1.1-3) and (1.1-4). The surface integral reduces, in the limit as the height of the cylinder approaches zero, to an integral over the end surfaces only. This leads to

$$\mathbf{n} \cdot (\mathbf{B}_2 - \mathbf{B}_1) = 0, \qquad \mathbf{n} \cdot (\mathbf{D}_2 - \mathbf{D}_1) = \sigma, \tag{1.1-8}$$

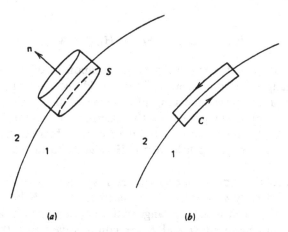

(a)   (b)

Figure 1.1 (a) A short cylinder about the interface between two media: S is the surface of this cylinder. (b) A narrow rectangle about the interface between two media; C is the boundary of this rectangle. (Adapted from A. Yariv and P. Yeh, *Optical Waves in Crystals*, Wiley, New York, 1984, p. 3. Copyright © 1984. By permission of John Wiley & Sons, Inc.)

where $\mathbf{n}$ is the unit normal to the surface directed from medium 1 into medium 2, σ is the surface charge density (in coulombs per square meter), and the subscripts refer to values at the surfaces in the two media. The boundary conditions (1.1-8) are often written as

$$B_{2n} = B_{1n}, \qquad D_{2n} - D_{1n} = \sigma, \qquad (1.1\text{-}9)$$

where $B_{2n} = \mathbf{B}_2 \cdot \mathbf{n}$, $B_{1n} = \mathbf{B}_1 \cdot \mathbf{n}$, $D_{2n} = \mathbf{D}_2 \cdot \mathbf{n}$, and $D_{1n} = \mathbf{D}_1 \cdot \mathbf{n}$. In other words, the normal component of the magnetic induction $\mathbf{B}$ is always continuous, and the difference between the normal components of the electric displacement $\mathbf{D}$ is equal in magnitude to the surface charge density σ.

For the field vectors $\mathbf{E}$ and $\mathbf{H}$, we draw a rectangular contour with two long sides parallel to the surface of discontinuity, as shown in Fig. 1.1(b). We now apply the Stokes theorem

$$\int \nabla \times \mathbf{F} \cdot d\mathbf{S} = \int \mathbf{F} \cdot d\mathbf{l} \qquad (1.1\text{-}10)$$

to both sides of Eqs. (1.1-1) and (1.1-2). The contour integral reduced, in the limit as the width of the rectangle approaches zero, to an integral over the two long sides only. This leads to

$$\mathbf{n} \times (\mathbf{E}_2 - \mathbf{E}_1) = 0, \qquad \mathbf{n} \times (\mathbf{H}_2 - \mathbf{H}_1) = \mathbf{K}, \qquad (1.1\text{-}11)$$

where $\mathbf{K}$ is the surface current density (in amperes per meter). Again, the boundary conditions for the electric and magnetic field vectors (1.1-11) are often written as

$$E_{2t} = E_{1t}, \qquad H_{2t} - H_{1t} = K, \qquad (1.1\text{-}12)$$

where the subscript t means the tangential component of the field vector. (*Note:* The tangential components of these field vectors to the boundary surface are still vectors in the tangential plane of the surface.) In other words, the tangential component of the electric field vector $\mathbf{E}$ is always continuous at the boundary surface, and the difference between the tangential components of the magnetic field vector $\mathbf{H}$ is equal to the surface current density $\mathbf{K}$.

In many areas of optics, one often deals with situations in which the surface charge density σ and the surface current density $\mathbf{K}$ both vanish. It follows that, in such a case, the tangential components of $\mathbf{E}$ and $\mathbf{H}$ and the normal components of $\mathbf{D}$ and $\mathbf{B}$ are continuous across the interface separating media 1 and 2. These boundary conditions are important in solving many wave propagation problems in optics, such as guided-wave optics (Chapter 11) and wave propagation in layered media.

1.2 ENERGY DENSITY AND ENERGY FLUX

It has been known for some time that light carries energy with it and is a form of electromagnetic radiation. The first and most conspicuous success of Maxwell's theory was the prediction of the existence of electromagnetic waves and the transport of energy. We now consider two of the most important aspects of electrodynamics: the energy density stored with an electromagnetic wave and the energy flux associated with an electromagnetic wave. To derive the energy density and the energy flux, we consider the conservation of energy in a small volume. The work done per unit volume by an electromagnetic field is $\mathbf{J} \cdot \mathbf{E}$, which may also be considered as the energy dissipation per unit volume. This energy dissipation must be connected with the net decrease in the energy density and the energy flow out of the volume. According to Eq. (1.1-2), the work done by the electromagnetic field can be written as

$$\mathbf{J} \cdot \mathbf{E} \;=\; \mathbf{E} \cdot (\nabla \times \mathbf{H}) - \mathbf{E} \cdot \frac{\partial \mathbf{D}}{\partial t}. \qquad (1.2\text{-}1)$$

If we now employ the vector identity

$$\nabla \cdot (\mathbf{E} \times \mathbf{H}) \;=\; \mathbf{H} \cdot (\nabla \times \mathbf{E}) - \mathbf{E} \cdot (\nabla \times \mathbf{H}) \qquad (1.2\text{-}2)$$

and use Eq. (1.1-1), the right side of (1.2-1) becomes

$$\mathbf{J} \cdot \mathbf{E} \;=\; -\nabla \cdot (\mathbf{E} \times \mathbf{H}) - \mathbf{H} \cdot \frac{\partial \mathbf{B}}{\partial t} - \mathbf{E} \cdot \frac{\partial \mathbf{D}}{\partial t}. \qquad (1.2\text{-}3)$$

If we now further assume that the material medium involved is linear in its electromagnetic properties (i.e., ε and μ are independent of the field strengths), Eq. (1.2-3) can be written as

$$\frac{\partial U}{\partial t} + \nabla \cdot \mathbf{S} \;=\; -\mathbf{J} \cdot \mathbf{E}, \qquad (1.2\text{-}4)$$

where U and $\mathbf{S}$ are defined as

$$U \;=\; \tfrac{1}{2}(\mathbf{E} \cdot \mathbf{D} + \mathbf{B} \cdot \mathbf{H}), \qquad (1.2\text{-}5)$$

$$\mathbf{S} \;=\; \mathbf{E} \times \mathbf{H}. \qquad (1.2\text{-}6)$$

The scalar U represents the energy density of the electromagnetic fields and has the dimensions of joules per cubic meter. The vector $\mathbf{S}$, representing the energy flux, is called Poynting's vector and has the dimensions of joules per square meter per second. It is consistent to view $|S|$ as the power per unit area (watts per square meter) carried by the field in the direction of S. The quantity $\nabla \cdot S$ thus represents the net electromagnetic power flowing out of a unit volume. Equation (1.2-4) is known as the continuity equation or Poynting's theorem. It represents conservation of energy for the electromagnetic field. The conservation laws for the linear momentum of electromagnetic fields can be obtained in a similar way. This is left as a problem for the student (Problem 1.4).

1.3 COMPLEX NUMBERS AND MONOCHROMATIC FIELDS

It is known that monochromatic light has a unique angular frequency of oscillation. Although most light sources consist of a distribution of the angular frequencies, we will investigate the properties of layered media using monochromatic light. This is legitimate because throughout the book we assume that the materials involved in layered media are linear and that each frequency component of light interacts independently with the media. For monochromatic light, the field vectors are sinusoidal functions of time, and it is convenient to represent each field vector as a complex exponential function. The reason we do this is that it is easier to work with exponential functions than with cosine or sine. As an example, consider some component of the field vectors:

$$a(t) = |A| \cos(\omega t + \alpha), \tag{1.3-1}$$

where ω is the angular frequency and α is the phase. If we define a complex amplitude of $a(t)$ by

$$A = |A| e^{i\alpha}, \tag{1.3-2}$$

Eq. (1.3-1) can be written as

$$a(t) = \mathrm{Re}[A e^{i\omega t}]. \tag{1.3-3}$$

We will often represent $a(t)$ by

$$a(t) = A e^{i\omega t} \tag{1.3-4}$$

instead of by Eq. (1.3-1) or (1.3-3). This is sometimes referred to as the analytic representation. We must understand that the complex number so defined is not a real physical component because no electromagnetic field in physics is complex; actually, field vectors have no imaginary parts, only real parts. We shall, however, speak of the "field" $A \exp(i\omega t)$, but of course, the actual field is the real part of that expression. Using complex representations, the monochromatic fields are written as exponential functions. This leads to great mathematical simplification. For example, differentiation can now be replaced by a simple multiplication. The exceptions are cases that involve the product (or powers) of field vectors such as energy density and Poynting's vector. In these cases, one must use the real form of the physical quantities.

As an example, consider the product of two sinusoidal functions $a(t)$ and $b(t)$, where

$$
\begin{aligned}
a(t) &= |A| \cos(\omega t + \alpha) \\
&= \mathrm{Re}[Ae^{i\omega t}]
\end{aligned}
\tag{1.3-5}
$$

and

$$
\begin{aligned}
b(t) &= |B| \cos(\omega t + \beta) \\
&= \mathrm{Re}[Be^{i\omega t}],
\end{aligned}
\tag{1.3-6}
$$

with $A = |A| \exp(i\alpha)$ and $B = |B| \exp(i\beta)$. Using the real functions, we get

$$
a(t)b(t) = \tfrac{1}{2}|AB|[\cos(2\omega t + \alpha + \beta) + \cos(\alpha - \beta)].
\tag{1.3-7}
$$

But if we were to evaluate the product $a(t)b(t)$ with the complex form of the functions, we would get

$$
a(t)b(t) = ABe^{i2\omega t} = |AB|e^{i(2\omega t + \alpha + \beta)}.
\tag{1.3-8}
$$

A comparison of the last result to Eq. (1.3-7) shows that the time-independent (dc) term $\tfrac{1}{2}|AB| \cos(\alpha - \beta)$ is missing, and thus the use of the complex form led to an error. Generally speaking, the product of the real part of two complex numbers may not be equal to the real part of the product of these two complex numbers. In other words, if x and y are two arbitrary complex numbers, the following is generally true:

$$
\mathrm{Re}[x] \cdot \mathrm{Re}[y] \neq \mathrm{Re}[xy].
\tag{1.3-9}
$$

1.3.1 Time Averaging of Sinusoidal Products

In optical fields, the field vectors are rapidly varying functions of time. For example, the period of a time-varying field with a wavelength $\lambda = 1 \, \mu\mathrm{m}$ is $T = \lambda/c = 0.33 \times 10^{-14} \, \mathrm{s}$. One often considers the time-averaged values rather than the instantaneous values of many physical quantities such as Poynting's vector and the energy density. It is frequently necessary to find the time average of the product of two sinusoidal functions of the same frequency:

$$\langle a(t)b(t)\rangle = \frac{1}{T}\int_0^T |A|\cos(\omega t + \alpha)|B|\cos(\omega t + \beta)\,dt, \quad (1.3\text{-}10)$$

where $a(t)$ and $b(t)$ are given by Eqs. (1.3-5) and (1.3-6) and the angle brackets denote time averaging; $T = 2\pi/\omega$ is the period of oscillation. Since the integral in Eq. (1.3-10) is periodic in T, the averaging can be performed over a time T. By using Eq. (1.3-7), we obtain directly

$$\langle a(t)b(t)\rangle = \tfrac{1}{2}|AB|\cos(\alpha - \beta) \qquad (1.3\text{-}11)$$

since the average of T of the term involving $\cos(2\omega t + \alpha + \beta)$ is zero. This last result can be written in terms of the complex amplitudes A and B, defined immediately following Eq. (1.3-6) as

$$\langle a(t)b(t)\rangle = \tfrac{1}{2}\text{Re}[AB^*] \qquad (1.3\text{-}12)$$

or in terms of the analytic form of $a(t)$ and $b(t)$ directly as

$$\langle \text{Re}[a(t)]\,\text{Re}[b(t)]\rangle = \tfrac{1}{2}\text{Re}[a(t)b^*(t)]. \qquad (1.3\text{-}13)$$

where the superscript asterisk indicates the complex conjugate. The time dependence on the right side of Eq. (1.3-13) disappears because both $a(t)$ and $b(t)$ have the same sinusoidal time dependence $\exp(i\omega t)$. These two results, Eqs. (1.3-12) and (1.3-13), are important and will find frequent use throughout the book.

By using the complex formalism (or analytic representation) for the field vectors $\mathbf{E}$, $\mathbf{H}$, $\mathbf{D}$, and $\mathbf{B}$, the time-averaged Poynting's vector (1.2-6) and the energy density (1.2-5) for sinusoidally varying fields are given by

$$\mathbf{S} = \tfrac{1}{2}\text{Re}[\mathbf{E}\times\mathbf{H}^*] \qquad (1.3\text{-}14)$$

and

$$U = \tfrac{1}{4}\text{Re}[\mathbf{E}\cdot\mathbf{D}^* + \mathbf{B}\cdot\mathbf{H}^*], \qquad (1.3\text{-}15)$$

respectively.

1.4 WAVE EQUATIONS AND MONOCHROMATIC PLANE WAVES

Two of the most important results of Maxwell's equations are the wave equations and the existence of electromagnetic waves that are solutions to them. We now derive the wave equations in material media. This is achieved by mathematical elimination so that each of the field vectors satisfies a differential equation. We limit our attention to regions where both charge density ϱ and current density $\mathbf{J}$ vanish. We also assume in this section that the medium is isotropic, so that ε and μ are scalars.

If we use the constitutive relation (1.1-6) for $\mathbf{B}$ in Eq. (1.1-1), divide both sides by μ, and apply the curl operator, we obtain

$$\nabla \times \left(\frac{1}{\mu} \nabla \times \mathbf{E} \right) + \frac{\partial}{\partial t} \nabla \times \mathbf{H} = 0. \qquad (1.4\text{-}1)$$

If we now differentiate Eq. (1.1-2) with respect to time, combine it with Eq. (1.4-1), and use the material Eq. (1.1-5), we obtain

$$\nabla \times \left(\frac{1}{\mu} \nabla \times \mathbf{E} \right) + \varepsilon \frac{\partial^2 \mathbf{E}}{\partial t^2} = 0. \qquad (1.4\text{-}2)$$

We now employ the vector identities

$$\nabla \times \left(\frac{1}{\mu} \nabla \times \mathbf{E} \right) = \frac{1}{\mu} \nabla \times (\nabla \times \mathbf{E}) + \left(\nabla \frac{1}{\mu} \right) \times (\nabla \times \mathbf{E}) \qquad (1.4\text{-}3)$$

and

$$\nabla \times (\nabla \times \mathbf{E}) = \nabla(\nabla \cdot \mathbf{E}) - \nabla^2 \mathbf{E}, \qquad (1.4\text{-}4)$$

and Eq. (1.4-2) becomes

$$\nabla^2 \mathbf{E} - \mu\varepsilon \frac{\partial^2 \mathbf{E}}{\partial t^2} + (\nabla \ln \mu) \times (\nabla \times \mathbf{E}) - \nabla(\nabla \cdot \mathbf{E}) = 0. \qquad (1.4\text{-}5)$$

By substituting for $\mathbf{D}$ from Eq. (1.1-5) into Eq. (1.1-3) and applying the vector identity

$$\nabla \cdot (\varepsilon \mathbf{E}) = \varepsilon \nabla \cdot \mathbf{E} + \mathbf{E} \cdot \nabla \varepsilon, \qquad (1.4\text{-}6)$$

we obtain, from Eq. (1.4-5),

$$\nabla^2 \mathbf{E} - \mu\varepsilon \frac{\partial^2 \mathbf{E}}{\partial t^2} + (\nabla \ln \mu) \times (\nabla \times \mathbf{E}) + \nabla(\mathbf{E} \cdot \nabla \ln \varepsilon) = 0. \qquad (1.4\text{-}7)$$

This is the wave equation for the field vector $\mathbf{E}$. The wave equation for the magnetic field vector $\mathbf{H}$ can be obtained in a similar way and is given by

$$\nabla^2 \mathbf{H} - \mu\varepsilon \frac{\partial^2 \mathbf{H}}{\partial t^2} + (\nabla \ln \varepsilon) \times (\nabla \times \mathbf{H}) + \nabla(\mathbf{H} \cdot \nabla \ln \mu) = 0. \qquad (1.4\text{-}8)$$

Inside a homogeneous and isotropic medium, the gradient of the logarithm of ε and μ vanishes, and the wave Eqs. (1.4-7) and (1.4-8) reduce to

$$\nabla^2 \mathbf{E} - \mu\varepsilon \frac{\partial^2 \mathbf{E}}{\partial t^2} = 0, \quad \nabla^2 \mathbf{H} - \mu\varepsilon \frac{\partial^2 \mathbf{H}}{\partial t^2} = 0. \qquad (1.4\text{-}9)$$

These are the standard electromagnetic wave equations. They are satisfied by the well-known monochromatic plane wave

$$\psi = A e^{i(\omega t - \mathbf{k}\cdot\mathbf{r})}, \qquad (1.4\text{-}10)$$

where A is a constant and is the amplitude. In Eq. (1.4-10), the angular frequency ω and the magnitude of the wave vector $\mathbf{k}$ are related by

$$|\mathbf{k}| = \omega\sqrt{\mu\varepsilon} \qquad (1.4\text{-}11)$$

and ψ can be any Cartesian component of $\mathbf{E}$ and $\mathbf{H}$.

Let us now examine the meaning of this solution. In each plane, $\mathbf{k} \cdot \mathbf{r} =$ constant (const), the field is a sinusoidal function of time. At each given moment, the field is a sinusoidal function of space. It is clear that the field has the same value for coordinates $\mathbf{r}$ and times t, which satisfy

$$\omega t - \mathbf{k}\cdot\mathbf{r} = \text{const}, \qquad (1.4\text{-}12)$$

where the constant is arbitrary and determines the field value. Equation (1.4-12) determines a plane normal to the wave vector $\mathbf{k}$ at any instant t. This plane is called a *surface of constant phase*. The surfaces of constant phases are often referred to as *wavefronts*. The electromagnetic wave represented by Eq. (1.4-10) is called a plane wave because all the wavefronts are planar. It is easily seen that the surfaces of constant phase travel in the direction of $\mathbf{k}$ with a velocity whose magnitude is

$$v = \frac{\omega}{k}. \qquad (1.4\text{-}13)$$

This is the phase velocity of the wave. We let $t = 0$ and examine the spatial variation, the separation between two neighboring field peaks, that is, the wavelength is

$$\lambda' = \frac{2\pi}{k} = 2\pi \frac{v}{\omega}. \qquad (1.4\text{-}14)$$

where the prime indicates the wavelength of light inside the medium. In optics, λ is reserved for the wavelength of light in a vacuum.

The value of the phase velocity is a characteristic of the medium and can be expressed in terms of the dielectric constant ε and the magnetic permeability μ. From Eqs. (1.4-13) and (1.4-11), we obtain

$$v = \frac{1}{\sqrt{\mu\varepsilon}}. \tag{1.4-15}$$

The phase velocity of the electromagnetic radiation in a vacuum is

$$c = \frac{1}{\sqrt{\mu_0\varepsilon_0}} = 2.997\,930 \times 10^8 \text{ m/s}$$

whereas in material media it has the value

$$v = \frac{c}{n},$$

where $n = \sqrt{\mu\varepsilon/\mu_0\varepsilon_0}$. Most transparent media are nonmagnetic and have a magnetic permeability μ_0. In that case, $n = \sqrt{\varepsilon/\varepsilon_0}$, and it is called the *index of refraction* of the medium. Table 1.1 lists the index of refraction of some common optical materials. We must keep in mind, however, that ε, and therefore n, of a nonmagnetic material ($\mu = \mu_0$) are functions of the frequency. The variation of n with frequency gives rise to the well-known phenomenon of *dispersion* in optics. In a dispersive medium, the phase velocity of a light wave depends on the frequency. The physical origin of dispersion and refractive index is discussed in some detail in Chapter 2.

We now turn our attention to the vector nature of the electromagnetic field and the requirements of satisfying Maxwell's equations. Using the complex-function formalism, we write the electromagnetic fields of the plane wave in the form

$$\mathbf{E} = \mathbf{u}_1 E_0 e^{i(\omega t - \mathbf{k}\cdot\mathbf{r})}, \tag{1.4-16}$$

$$\mathbf{H} = \mathbf{u}_2 H_0 e^{i(\omega t - \mathbf{k}\cdot\mathbf{r})}, \tag{1.4-17}$$

where $\mathbf{u}_1$ and $\mathbf{u}_2$ are two constant unit vectors, and E_0 and H_0 are complex amplitudes that are constant in space and time. In a homogeneous charge-free medium, the divergent Maxwell equations are $\nabla \cdot \mathbf{E} = 0$ and $\nabla \cdot \mathbf{H} = 0$, which, when applied to (1.4-16) and (1.4-7), result in

$$\mathbf{u}_1 \cdot \mathbf{k} = \mathbf{u}_2 \cdot \mathbf{k} = 0. \tag{1.4-18}$$

This means that $\mathbf{E}$ and $\mathbf{H}$ are both perpendicular to the direction of propagation. For this reason, electromagnetic waves are said to be transverse

Table 1.1. Index of Refraction of Selected Materials by Wavelength (λ)

Material	0.488 μm	0.5 μm	0.5145 μm	0.6328 μm	1.0 μm	3.0 μm	5.0 μm	10.6 μm
As–S glass	2.786	2.77	2.75	2.606	2.478	2.41	2.407	2.378
BaF$_2$	1.478	1.4779	1.477	1.473	1.4686	1.4611	1.451	1.3928
Bi$_{12}$GeO$_{20}$	—	—	2.55	2.54	—	—	—	—
Bi$_{12}$SiO$_{20}$	—	—	—	2.54	—	—	—	—
Bi$_{12}$TiO$_{20}$	—	—	—	2.55	—	—	—	—
CaF$_2$	1.437	1.4366	1.4362	1.433	1.429	1.418	1.399	1.2803
CdS n_o	—	—	2.747	2.46	2.334	2.279	2.266	2.226
CdSe n_o	—	—	—	—	2.55	2.454	2.446	2.43
CdTe	—	—	—	—	2.84	2.81	2.77	2.69
CuBr	2.201	2.184	2.167	2.102	—	—	—	—
CuCl	2.019	2.01	2.002	1.9613	1.924	1.92	1.92	1.93
CuI	2.437	2.422	2.405	2.321	—	—	—	—
GaAs	—	—	—	—	3.5	3.35	3.29	3.135
GaP	—	—	3.66	3.38	3.17	2.97	2.94	2.90
GaSb	—	—	—	—	—	3.898	3.824	3.843
Ge	—	—	—	—	—	4.045	4.0163	4.0029
InAs	—	—	—	—	—	—	3.46	3.42
InP	—	—	—	—	3.327	—	3.08	3.05
InSb	—	—	—	—	—	—	—	3.95
Irtran 1 (MgF$_2$)	—	—	—	—	1.378	1.364	1.337	—
Irtran 2 (ZnS)	—	—	—	—	2.291	2.256	2.245	2.1902
Irtran 3 (CdF$_2$)	—	—	—	—	1.429	1.418	1.399	1.2817
Irtran 4 (ZnSe)	—	—	—	—	2.485	2.44	2.432	2.4034
Irtran 5 (MgO)	—	—	—	—	1.723	1.692	1.637	—
KBr	1.572	1.57	1.568	1.558	1.5443	1.536	1.534	1.525
KCl	1.498	1.497	1.496	1.488	1.4799	1.474	1.471	1.454

KI	1.686	1.684	1.68	1.661	1.64	1.6284	1.626	1.6191
LiF	1.395	1.394	1.394	1.392	1.387	1.367	1.327	1.05
MgF$_2$	—	1.3703	—	—	1.363	—	—	—
MgF$_2$ crystal n_o	1.392	1.3917	1.391	1.389	1.385	—	—	—
MgO	—	1.76	—	—	1.7237	1.6922	1.6262	1.625
PbF$_2$	1.786	1.782	1.779	1.761	1.742	1.724	1.708	—
Quartz, crystal n_o	1.55	1.549	1.548	1.507	1.535	1.5	1.417	—
Ruby, Al$_2$O$_3$ n_o	1.776	1.775	1.774	1.766	—	—	—	—
Sapphire, Al$_2$O$_3$ n_o	1.776	1.775	1.774	1.766	1.756	1.7115	1.6239	—
Si	—	—	—	—	—	3.432	3.422	3.418
SiC, crystal	2.698	2.691	2.682	2.64	2.583	—	—	—
SiO$_2$, fused silica	—	1.462	—	—	1.45	1.4185	—	—
SrTiO$_3$, crystal	2.489	2.477	2.461	2.389	2.316	2.231	2.122	—
TiO$_2$, rutile n_o	2.731	2.712	2.694	2.59	2.484	—	—	—
ZnO, crystal n_o	2.064	2.051	2.044	1.99	1.944	1.9072	—	—
ZnS	—	2.42	—	—	2.32	—	—	—
ZnS, sphalerite	2.433	2.421	2.406	2.354	2.293	—	—	—
ZnS, wurtzite n_o	2.43	2.421	2.41	2.352	2.301	—	—	—
ZnSe	—	2.7	—	—	2.48	—	—	2.392
ZnTe	—	—	—	2.984	2.79	—	—	2.70

waves. The transverse condition (1.4-18) holds for all the four field vectors for plane-wave propagation in homogeneous and isotropic media. In a general anisotropic medium, only the field vectors $\mathbf{D}$ and $\mathbf{B}$ of a plane wave are perpendicular to the direction of propagation.

The curl Maxwell equations provide further restrictions on the field vectors. These are obtained by substituting Eqs. (1.4-16) and (1.4-17) into Eq. (1.1-1) and are given by

$$\mathbf{u}_2 = \frac{\mathbf{k} \times \mathbf{u}_1}{|\mathbf{k}|} \tag{1.4-19}$$

and

$$H_0 = \frac{E_0}{\eta}, \qquad \eta = \sqrt{\frac{\mu}{\varepsilon}}. \tag{1.4-20}$$

This shows that the triad ($\mathbf{u}_1$, $\mathbf{u}_2$, and $\mathbf{k}$) forms a set of orthogonal vectors and that $\mathbf{E}$ and $\mathbf{H}$ are in phase and in constant ratio provided ε and μ are real. The η in Eq. (1.4-20) has the dimension of resistance and is called the impedance of space. In a vacuum, $\eta_0 = \sqrt{\mu_0/\varepsilon_0} \simeq 377\,\Omega$.

The plane wave we have just described is a transverse wave propagating in the direction $\mathbf{k}$. It represents a time-averaged flux of energy given, according to Poynting's theorem (1.3-4), by

$$\mathbf{S} = \frac{|E_0|^2 \mathbf{u}_3}{2\eta} = \frac{\mathbf{k}}{2\omega\mu} |E_0|^2 = \frac{\mathbf{k}}{2\omega\mu} \mathbf{E}^* \cdot \mathbf{E}, \tag{1.4-21}$$

where $\mathbf{u}_3$ is a unit vector in the direction of $\mathbf{k}$. The time-averaged energy density is

$$U = \tfrac{1}{2}\varepsilon|E_0|^2 = \tfrac{1}{2}\varepsilon\mathbf{E}^* \cdot \mathbf{E}. \tag{1.4-22}$$

We note from Eqs. (1.4-21) and (1.4-22) that the energy flux is directed along the direction of propagation. We also note from Eqs. (1.4-13) and (1.4-15) that we have the relation

$$\text{Energy flux} = v \times \text{energy density.} \tag{1.4-23}$$

This equation simply says that the energy is flowing at a speed v along the direction of propagation.

1.5 POLARIZATION STATES OF LIGHT

An electromagnetic wave is specified by its frequency and direction of propagation as well as by the direction of oscillation of the field vector. The latter is known as the polarization state of the electromagnetic wave. It is usually specified by the electric field vector **E**. For a monochromatic plane wave, the time variation of the electric field vector **E** is exactly sinusoidal. In other words, the electric field must oscillate at a definite frequency. By virtue of the transverse nature of the electromagnetic wave, the variation of the electric field vector is confined in a plane perpendicular to the direction of propagation. If we assume that the light is propagating in the z direction, the electric field vector will lie on the xy plane. Since the x and y components of the field vector can oscillate independently at a definite frequency, one must first consider the resultant effect produced by the vector addition of these two oscillating orthogonal components. The problem of superposing two independent oscillations at right angles to each other and with the same frequency is well known and is completely analogous to the classical motion of a two-dimensional harmonic oscillator. The general motion of the oscillator is an ellipse, which corresponds to oscillations in which the x and y components are not in the same phase. There are, of course, many special cases that are of great importance in optics. We will start with a discussion of the general properties of elliptic polarization and follow with a consideration of some special cases.

In the complex-function representation, the electric field vector of a monochromatic plane wave propagating in the z direction is given by

$$\mathbf{E}(z,\ t)\ =\ \mathrm{Re}\,[\mathbf{A}e^{i(\omega t - kz)}], \tag{1.5-1}$$

where **A** is a complex vector that lies in the xy plane. We shall now consider the nature of the curve that the end point of the electric field vector **E** describes at a typical point in space. This curve is the time evolution locus of the points whose coordinates $(E_x,\ E_y)$ are

$$E_x\ =\ A_x\cos(\omega t - kz + \delta_x), \qquad E_y\ =\ A_y\cos(\omega t - kz + \delta_y), \tag{1.5-2}$$

where we have defined the complex vector **A** as

$$\mathbf{A}\ =\ \mathbf{x}A_x e^{i\delta_x} + \mathbf{y}A_y e^{i\delta_y}, \tag{1.5-3}$$

where A_x and A_y are positive numbers and **x** and **y** are unit vectors. The curve described by the end point of the electric vector as time evolves can be obtained by eliminating $\omega t - kz$ in Eqs. (1.5-2). After several steps of elementary algebra, we obtain

$$\left(\frac{E_x}{A_x}\right)^2 + \left(\frac{E_y}{A_y}\right)^2 - 2\,\frac{\cos\delta}{A_x A_y}\,E_x E_y\ =\ \sin^2\delta, \tag{1.5-4}$$

where

$$\delta = \delta_y - \delta_x. \qquad (1.5\text{-}5)$$

All the phase angles are defined in the range $-\pi < \delta \leqslant \pi$.

Equation (1.5-4) is the equation of a conic. From (1.5-2) it is obvious that this conic is confined in a rectangular region with sides parallel to the coordinate axes and whose lengths are $2A_x$ and $2A_y$. Therefore, the curve must be an ellipse. The wave (1.5-1) is then said to be elliptically polarized. A complete description of an elliptical polarization requires the orientation of the ellipse with respect to the coordinate axes, the shape, and the sense of revolution of **E**. In general, the principal axes of the ellipse are not in the x and y directions. By using a transformation (rotation) of the coordinate system, we are able to diagonalize (1.5-4). Let x' and y' be a new set of axes along the principal axes of the ellipse. Then the equation of the ellipse in this new coordinate system becomes

$$\left(\frac{E_{x'}}{a}\right)^2 + \left(\frac{E_{y'}}{b}\right)^2 = 1, \qquad (1.5\text{-}6)$$

where a and b are the principal axes of the ellipse and $E_{x'}$ and $E_{y'}$ are the components of the electric field vector in this principal coordinate system.

Let ϕ $(0 \leqslant \phi < \pi)$ be the angle between the direction of the major axis x' and the x axis (see Fig. 1.2). Then the lengths of the principal axes are given by

$$
\begin{aligned}
a^2 &= A_x^2 \cos^2\phi + A_y^2 \sin^2\phi + 2A_x A_y \cos\delta \cos\phi \sin\phi, \\
b^2 &= A_x^2 \sin^2\phi + A_y^2 \cos^2\phi - 2A_x A_y \cos\delta \cos\phi \sin\phi.
\end{aligned}
\qquad (1.5\text{-}7)
$$

The angle ϕ can be expressed in terms of A_x, A_y, and $\cos\delta$ as

$$\tan 2\phi = \frac{2A_x A_y}{A_x^2 - A_y^2} \cos\delta. \qquad (1.5\text{-}8)$$

The sense of revolution of an elliptical polarization is determined by the sign of $\sin\delta$. The end point of the electric vector will revolve in a clockwise direction if $\sin\delta > 0$ and in a counterclockwise direction if $\sin\delta < 0$.

Before discussing some special cases of polarization, it is important to familiarize ourselves with the terminology. Light is linearly polarized when the tip of the electric field vector **E** moves along a straight line. When the end point of the electric field vector **E** describes an ellipse, the light is elliptically polarized. When the end point of the electric field vector travels around a

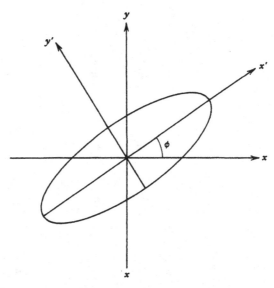

Figure 1.2 A typical polarization ellipse. (Adapted from A. Yariv and P. Yeh, *Optical Waves in Crystals*, Wiley, New York, 1984, p. 57. Copyright © 1984. By permission of John Wiley & Sons, Inc.)

circle, the light is circularly polarized. If the end point of the electric field vector is seen to move in a counterclockwise direction by an observer facing the approaching wave, the field is said to possess right-handed polarization. Our convention for labeling right-hand and left-hand polarization is consistent with the terminology of modern physics in which a photon with a right-hand circular polarization has a positive angular momentum along a direction of propagation (see Table 1.2 and Problem 1.10).

1.5.1 Linear and Circular Polarizations

Two special cases are of significant importance, namely, when the polarization ellipse degenerates into a straight line or becomes a circle. According to (1.5-4), the ellipse will reduce to a straight line when

$$\delta = \delta_y - \delta_x = m\pi \quad (m = 0, 1). \quad (1.5-9)$$

Here we recall that all the phase angles are defined in the range $-\pi < \delta \leqslant \pi$. In this case, the ratio of the components of the electric field

Table 1.2. Relations between Angular Momentum and Handedness of Circularly Polarized Light

Field of Single Photon ($\hbar\omega$)		L_z	Helicity	Handedness
$E_x = A \cos(\omega t - kz)$ $E_y = A \cos(\omega t - kz - \frac{1}{2}\pi)$	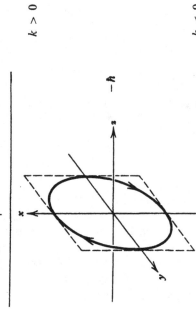	$k > 0$	$\hbar$	Right-handed
		$k < 0$	$-\hbar$	Left-handed
$E_x = A \cos(\omega t - kz)$ $E_y = A \cos(\omega t - kz + \frac{1}{2}\pi)$		$k > 0$	$-\hbar$	Left-handed
		$k < 0$	$\hbar$	Right-handed

Source: A. Yariv and P. Yeh, *Optical Waves in Crystals*, Wiley, New York, 1984, p. 59. Copyright © 1984. Reprinted by permission of John Wiley & Sons, Inc.

vector is a constant,

$$\frac{E_y}{E_x} = (-1)^m \frac{A_y}{A_x}, \tag{1.5-10}$$

and the light is linearly polarized.

The other special case of importance is that of a circularly polarized wave. According to (1.5-4) and (1.5-7), the ellipse will reduce to a circle when

$$\delta = \delta_y - \delta_x = \pm\tfrac{1}{2}\pi \tag{1.5-11}$$

and

$$A_y = A_x. \tag{1.5-12}$$

According to our convention, the light is right-hand circularly polarized when $\delta = -\tfrac{1}{2}\pi$, which corresponds to a counterclockwise rotation of the electric field vector. The light is left-hand circularly polarized when $\delta = \tfrac{1}{2}\pi$, which corresponds to a clockwise rotation of the electric field vector (see Table 1.2).

The ellipticity of a polarization ellipse is defined as

$$e = \pm\frac{b}{a}, \tag{1.5-13}$$

where a and b are the lengths of the principal axes. The ellipticity is taken as positive when the rotation of the electric field vector is right handed and is negative otherwise.

The polarization state of a monochromatic plane wave is usually represented mathematically by a complex number or a column vector.

1.5.2 Complex-Number Representation

From the preceding discussion, we found how the polarization state of a monochromatic plane wave can be described in terms of the amplitudes and the phase angles of the x and y components of the electric field vector. In fact, all the information about the polarization state is contained in the complex amplitude A of the plane wave (1.5-1). Therefore, a complex number χ defined as

$$\chi = e^{i\delta} \tan\psi = \frac{A_y}{A_x} e^{i(\delta_y - \delta_x)} \tag{1.5-14}$$

is sufficient enough to describe the polarization states. The angle ψ is defined between 0 and $\frac{1}{2}\pi$. A complete description of the ellipse of polarization that includes the orientation, sense of revolution, and ellipticity [see (1.5-13)] can all be expressed in terms of δ and ψ. The right-handed elliptical polarization states are in the lower half of the plane, whereas the left-handed elliptical polarization states are in the upper half of the complex plane. The origin corresponds to a linear polarization state with direction of oscillation parallel to the x axis. Thus, each point on the complex plane represents a unique polarization state. Each point on the x axis represents a linearly polarized state with different azimuth angles of oscillation. Only two points $(0, \pm 1)$ correspond to circularly polarized states. Each point of the rest of the complex plane corresponds to a unique elliptical polarization state.

The inclination angle ϕ and the ellipticity angle θ ($\theta \equiv \tan^{-1} e$) of the polarization ellipse correspond to a given number χ are given by

$$\tan 2\phi = \frac{2\,\mathrm{Re}\,[\chi]}{1 - |\chi|^2} \tag{1.5-15}$$

and

$$\sin 2\theta = -\frac{2\,\mathrm{Im}\,[\chi]}{1 + |\chi|^2}. \tag{1.5-16}$$

1.5.3 Jones Vector Representation

The Jones vector, introduced in 1941 by R. C. Jones [1], describes efficiently the polarization state of a plane wave. In this representation, the plane wave (1.5-1) is expressed in terms of its complex amplitudes as a column vector

$$\mathbf{J} = \begin{pmatrix} A_x e^{i\delta_x} \\ A_y e^{i\delta_y} \end{pmatrix}. \tag{1.5-17}$$

Notice that the Jones vector is a complex vector because its elements are complex numbers. Here $\mathbf{J}$ is not a vector in the real physical space; rather it is a vector in an abstract mathematical space. To obtain, as an example, the real x component of the electric field at $z = 0$, we must perform the operation $E_x(t) = \mathrm{Re}\,[J_x e^{i\omega t}] = \mathrm{Re}\,[A_x e^{i(\omega t + \delta_x)}]$.

The Jones vector contains complete information about the amplitudes and the phases of the electric field vector components. It thus specifies the wave uniquely. If we are only interested in the polarization state of the wave, it is convenient to use the normalized Jones vector, which satisfies the condition that

$$\mathbf{J}^* \cdot \mathbf{J} = 1, \tag{1.5-18}$$

where the asterisk (*) denotes complex conjugation. Thus, a linearly polarized light wave with the electric field vector oscillating along a given direction can be represented by the Jones vector

$$\begin{pmatrix} \cos \psi \\ \sin \psi \end{pmatrix},$$ (1.5-19)

where ψ is the azimuth angle of the oscillation direction with respect to the x axis. The state of polarization that is orthogonal to the state represented by (1.5-19) can be obtained by the substitution of ψ by $\psi + \frac{1}{2}\pi$, leading to a Jones vector

$$\begin{pmatrix} -\sin \psi \\ \cos \psi \end{pmatrix}.$$ (1.5-20)

For the special case when $\psi = 0$ represents linearly polarized waves whose electric field vector oscillate along the coordinate axes, the Jones vectors are given by

$$\mathbf{x} = \begin{pmatrix} 1 \\ 0 \end{pmatrix}, \quad \mathbf{y} = \begin{pmatrix} 0 \\ 1 \end{pmatrix}.$$ (1.5-21)

Jones vectors for the right- and left-hand circularly polarized light waves are given by

$$\mathbf{R} = \frac{1}{\sqrt{2}} \begin{pmatrix} 1 \\ -i \end{pmatrix},$$ (1.5-22)

$$\mathbf{L} = \frac{1}{\sqrt{2}} \begin{pmatrix} 1 \\ i \end{pmatrix}.$$ (1.5-23)

These two circular polarizations are mutually orthogonal in the sense that

$$\mathbf{R}^* \cdot \mathbf{L} = 0.$$ (1.5-24)

Since the Jones vector is a column matrix of rank 2, any pair of orthogonal Jones vectors can be used as a basis of the mathematical space spanned by all the Jones vectors. Any polarization can be represented as a superposition of two mutually orthogonal polarizations $\mathbf{x}$ and $\mathbf{y}$, or $\mathbf{R}$ and $\mathbf{L}$. In particular, we can resolve the basic linear polarization $\mathbf{x}$ and $\mathbf{y}$ into two circular polarizations $\mathbf{R}$ and $\mathbf{L}$, and vice versa. These relations are given by

$$\mathbf{R} = \frac{1}{\sqrt{2}} (\mathbf{x} - i\mathbf{y}),$$ (1.5-25)

$$\mathbf{L} = \frac{1}{\sqrt{2}} (\mathbf{x} + i\mathbf{y}), \qquad (1.5\text{-}26)$$

$$\mathbf{x} = \frac{1}{\sqrt{2}} (\mathbf{R} + \mathbf{L}), \qquad (1.5\text{-}27)$$

$$\mathbf{y} = \frac{i}{\sqrt{2}} (\mathbf{R} - \mathbf{L}). \qquad (1.5\text{-}28)$$

Circular polarizations are seen to consist of linear oscillations along the x and y directions with equal amplitude $1/\sqrt{2}$ but with a phase difference $\pi/2$. Similarly, a linear polarization can be viewed as a superposition of two oppositely sensed circular polarizations.

We have so far discussed only the Jones vectors of some simple special cases of polarization. It is easy to show that a general elliptic polarization can be represented by the Jones vector

$$\mathbf{J}(\psi, \delta) = \begin{pmatrix} \cos\psi \\ e^{i\delta}\sin\psi \end{pmatrix}. \qquad (1.5\text{-}29)$$

This Jones vector represents the same polarization state as the one represented by the complex number $\chi = e^{i\delta}\tan\psi$ (1.5-14). Table 1.3 shows the Jones vectors of some typical polarization states.

The most important application of Jones vectors is in conjunction with the Jones calculus. This is a powerful technique used to study the propagation of plane waves with arbitrary states of polarizations through an arbitrary sequence of birefringent elements and polarizers. This topic will be considered in some detail in Chapter 9.

1.6 PARTIALLY POLARIZED AND UNPOLARIZED LIGHT

By virtue of its nature, a monochromatic plane wave must be polarized; that is, the end point of its electric field vector at each point in space must trace out periodically an ellipse or one of its special forms, such as a circle or a straight line. However, if the light is not absolutely monochromatic, the relative phase δ and amplitudes between the x and y components can both vary with time, and the electric field vector will first vibrate in one ellipse and then in another. As a result, the polarization state of a polychromatic plane wave may be constantly changing. If the polarization state changes more rapidly on the scale of observation, we say the light is partially polarized or unpolarized depending on the time-averaged behavior of the polarization

Table 1.3. Jones Vectors of Some Typical Polarization States

Polarization Ellipse	Jones Vector
	$\begin{pmatrix} 1 \\ 0 \end{pmatrix}$
	$\begin{pmatrix} 0 \\ 1 \end{pmatrix}$
	$\dfrac{1}{\sqrt{2}} \begin{pmatrix} 1 \\ 1 \end{pmatrix}$
	$\dfrac{1}{\sqrt{2}} \begin{pmatrix} 1 \\ -1 \end{pmatrix}$
	$\dfrac{1}{\sqrt{2}} \begin{pmatrix} 1 \\ -i \end{pmatrix}$
	$\dfrac{1}{\sqrt{2}} \begin{pmatrix} 1 \\ i \end{pmatrix}$
	$\dfrac{1}{\sqrt{5}} \begin{pmatrix} 1 \\ 2i \end{pmatrix}$
	$\dfrac{1}{\sqrt{5}} \begin{pmatrix} 2 \\ -i \end{pmatrix}$

Source: A. Yariv and P. Yeh, *Optical Waves in Crystals*, Wiley, New York, p. 65. Copyright © 1984. Reprinted by permission of John Wiley & Sons, Inc.

state. In optics, one often deals with the light with oscillation frequencies of about 10^{14} s^{-1}, whereas the polarization state may change in a time period of 10^{-8} s depending on the nature of the light source.

We will limit ourselves to the case of the quasi-monochromatic wave, whose frequency spectrum is confined to a narrow band of width $\Delta\omega$ (i.e., $\Delta\omega \ll \omega$). Such a wave can still be described by Eq. (1.5-1), provided we relax the constancy condition of the amplitude A. Now ω denotes the center frequency, and the complex amplitude A is a function of time. Because the bandwidth is narrow A(t) may change only by a relatively small amount in a time interval $1/\Delta\omega$, and in this sense it is a slowly varying function of time. However, if the time constant of the detector, τ_D, is greater than $1/\Delta\omega$, A(t) may change significantly in a time interval τ_D. Although the amplitudes and phases are irregularly varying functions of time, certain correlations may exist among them.

To describe the polarization state of this type of radiation, we introduce the following time-averaged quantities:

$$
\begin{aligned}
S_0 &= \langle\!\langle A_x^2 + A_y^2 \rangle\!\rangle, \\
S_1 &= \langle\!\langle A_x^2 - A_y^2 \rangle\!\rangle, \\
S_2 &= 2\langle\!\langle A_x A_y \cos\delta \rangle\!\rangle, \\
S_3 &= 2\langle\!\langle A_x A_y \sin\delta \rangle\!\rangle,
\end{aligned}
\tag{1.6-1}
$$

where the amplitudes A_x, A_y and the relative phase δ are assumed to be time dependent, and the double brackets denote averages performed over a time interval τ_D that is the characteristic time constant of the detection processs. These four quantities are known as the *Stokes parameters* of a quasi-monochromatic plane wave. Notice that all four quantities have the same dimension of intensity. It can be shown that the Stokes parameters satisfy the relation

$$
S_0^2 \geqslant S_1^2 + S_2^2 + S_3^2,
\tag{1.6-2}
$$

where the equality sign holds only for polarized waves.

It is a simple exercise to compute, from the definitions, various Stokes parameters of principal interest. Consider, for example, unpolarized light. There is no preference between A_x and A_y; consequently, $\langle A_x^2 + A_y^2 \rangle$ reduces to $2\langle A_x^2 \rangle$, and $\langle A_x^2 - A_y^2 \rangle$ reduces to zero. The other two quantities also reduce to zero because δ is a random function of time. If the field is normalized such that $S_0 = 1$, the Stokes vector representation of an unpolarized light wave is (1, 0, 0, 0). Similar reasoning shows that a horizontally polarized beam can be represented by the Stokes vector (1, 1, 0, 0), and a vertically polarized beam can be represented by $(1, -1, 0, 0)$. Right-hand circularly polarized light $(\delta = -\frac{1}{2}\pi)$ is represented by $(1, 0, 0, -1)$, and the left-hand circularly polarized light $(\delta = \frac{1}{2}\pi)$ is represented by $(1, 0, 0, 1)$. From the definition, none of the parameters can be greater than the first S_0, which is normalized to 1. Therefore, each of the others lies in the range from -1 to 1. If the beam is entirely unpolarized, $S_1 = S_2 = S_3 = 0$. If it is completely polarized, $S_1^2 + S_2^2 + S_3^3 = 1$. The degree of polarization is therefore defined as

$$\gamma = \frac{(S_1^2 + S_2^2 + S_3^2)^{1/2}}{S_0}. \qquad (1.6\text{-}3)$$

The parameter γ is thus very useful in describing the partially polarized light. The polarization preference of a partially polarized light can be seen directly from the sign of the parameters S_1, S_2, and S_3.

The parameter S_1 describes the linear polarization along the x or y axis; the probability that the light is linearly polarized along the x axis is, $\frac{1}{2}(1 + S_1)$ and along the y axis, $\frac{1}{2}(1 - S_1)$. The values $S_1 = 1, -1$ therefore correspond to complete polarization in these directions. The parameter S_2 describes the linear polarization along directions at angles $\phi = \pm 45°$ to the x axis. The probability that the light is linearly polarized along these directions is, respectively, $\frac{1}{2}(1 + S_2)$ and $\frac{1}{2}(1 - S_2)$. The values $S_2 = 1, -1$ correspond to complete polarization in these directions. Finally, the parameter S_3 represents the degree of circular polarization; the probability that the light wave had right-hand or left-hand circular polarization is, $\frac{1}{2}(1 - S_3)$ and left-hand circular polarization, $\frac{1}{2}(1 + S_3)$.

The Stokes parameters for a polarized light with a complex representation $e^{i\delta} \tan \psi$ (1.5-14) are given by [according to (1.6-1)]

$$
\begin{aligned}
S_0 &= 1, \\
S_1 &= \cos 2\psi, \\
S_2 &= \sin 2\psi \cos \delta, \\
S_3 &= \sin 2\psi \sin \delta.
\end{aligned}
\qquad (1.6\text{-}4)
$$

According to our convention, a positive S_3 corresponds to left-hand elliptical polarization ($\sin \delta > 0$, clockwise revolution).

1.7 ELEMENTARY THEORY OF COHERENCE

It is known that the principle of superposition applies to the electromagnetic field. This principle states that the total field due to all sources is the sum of the fields due to each source. The reason why this is true is that Maxwell's equations, which determine the electromagnetic field, are linear differential equations, provided both ε and μ are independent of the field strength. Consider now two linearly polarized plane waves of the same frequency ω. Let the electric fields be

$$
\begin{aligned}
\mathbf{E}_1 &= \mathbf{A}_1 \exp[i(\omega t - \mathbf{k}_1 \cdot \mathbf{r} + \phi_1)], \\
\mathbf{E}_2 &= \mathbf{A}_2 \exp[i(\omega t - \mathbf{k}_2 \cdot \mathbf{r} + \phi_2)],
\end{aligned}
\tag{1.7-1}
$$

where $\mathbf{A}_1$ and $\mathbf{A}_2$ are the amplitudes and are real vectors because of the linear polarization states. The real quantities ϕ_1 and ϕ_2 are the phases, and $\mathbf{k}_1$ and $\mathbf{k}_2$ are the wave vectors. The sources of these two plane waves are said to be mutually coherent if the phase difference $\phi_2 - \phi_1$ is constant. On the other hand, if the quantity $\phi_2 - \phi_1$ varies with time in a random fashion, the sources of these two waves are said to be mutually incoherent. To define the degree of mutual coherence of the two waves (1.7-1), we need to examine the interference pattern formed by these two waves. From Eqs. (1.4-21) and (1.4-22), we know that the time-averaged intensity of radiation at any point in space is proportional to the square of the amplitude of the electric field. Thus, aside from a constant factor, the time-averaged intensity distribution of the interference pattern formed by the two waves is given by

$$
I = |\mathbf{E}_1 + \mathbf{E}_2|^2 = I_1 + I_2 + 2\mathbf{A}_1 \cdot \mathbf{A}_2 \cos(\mathbf{K} \cdot \mathbf{r} - \phi), \quad (1.7\text{-}2)
$$

where $I_1 = |\mathbf{A}_1|^2$, $I_2 = |\mathbf{A}_2|^2$, $\mathbf{K} = \mathbf{k}_2 - \mathbf{k}_1$, and $\phi = \phi_2 - \phi_1$. The first two terms are the time-averaged intensities of the two waves, respectively. The third term is the interference term and contains information about the degree of mutual coherence. Note that Eq. (1.7-2) is a time-averaged result, averaging over a period of $2\pi/\omega$. If we limit ourselves to the case of quasi-monochromatic plane waves, all the phases ϕ_1, ϕ_2, and ϕ are considered constant during such a small time interval $2\pi/\omega$, which is on the order of 10^{-15} s for visible light.

If the two waves are mutually coherent (i.e., $\phi = $ constant), the interference pattern is stationary with a spatial period

$$
\Lambda = \frac{2\pi}{|\mathbf{K}|} = \frac{\lambda}{2\sin(\theta/2)},
\tag{1.7-3}
$$

where θ is the angle between the two wave vectors and λ is the wavelength of light (see Fig. 1.3). If the two sources are not mutually coherent, the phase ϕ varies with time. As a result, the time-averaged intensity (1.7-2) fluctuates rapidly with time. If ϕ varies significantly over a time interval τ_D, the characteristic time constant associated with the detection process, the fluctuation is

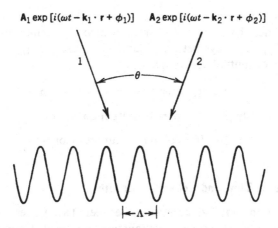

$$\mathbf{A}_1 \exp\left[i(\omega t - \mathbf{k}_1 \cdot \mathbf{r} + \phi_1)\right] \qquad \mathbf{A}_2 \exp\left[i(\omega t - \mathbf{k}_2 \cdot \mathbf{r} + \phi_2)\right]$$

Figure 1.3 Intensity distribution of a fringe pattern formed by the interference of two waves.

too fast for the detector to respond, and the detected intensity is averaged over a time interval τ_D. As a result of this time averaging, the visibilities of the fringes decreases. For mutually incoherent sources, the phase ϕ varies with time in a random fashion between 0 and 2π, and the averaging of the third term in Eq. (1.7-2) is obviously zero. This leads to the disappearance of the fringe pattern. Thus, for many practical purposes, the degree of mutual coherence between the two waves (1.7-1) can be defined as

$$\gamma_{12} \equiv \langle\!\langle e^{i(\phi_2 - \phi_1)} \rangle\!\rangle \equiv \frac{1}{\tau_D} \int_0^{\tau_D} e^{i(\phi_2 - \phi_1)} \, dt, \tag{1.7-4}$$

where the double brackets indicate time averaging over a time interval τ_D, the characteristic time constant of the detection system. The parameter γ_{12} is, in general, complex and can be written as

$$\gamma_{12} = |\gamma_{12}| \exp(i\alpha), \tag{1.7-5}$$

where α is a real number.

We now evaluate the detection of the time-averaged intensity. From Eq. (1.7-2), we have

$$\langle\!\langle I \rangle\!\rangle = I_1 + I_2 + 2\mathbf{A}_1 \cdot \mathbf{A}_2 \langle\!\langle \cos(\mathbf{K} \cdot \mathbf{r} - \phi) \rangle\!\rangle, \tag{1.7-6}$$

where we assumed $\mathbf{A}_1$ and $\mathbf{A}_2$ are constants.

Using $\cos(\mathbf{K} \cdot \mathbf{r} - \phi) = \cos(\mathbf{K} \cdot \mathbf{r})\cos\phi + \sin(\mathbf{K} \cdot \mathbf{r})\sin\phi$ and Eq. (1.7-5), we obtain

$$\langle\!\langle I \rangle\!\rangle = I_1 + I_2 + 2\mathbf{A}_1 \cdot \mathbf{A}_2 |\gamma_{12}| \cos(\mathbf{K} \cdot \mathbf{r} - \alpha). \tag{1.7-7}$$

Thus, an interference fringe pattern still results if $|\gamma_{12}|$, called the degree of mutual coherence, has a value other than zero. By definition (1.7-4), the magnitude of γ_{12} is always between 0 and 1. In terms of $|\gamma_{12}|$, we have the following type of mutual coherence:

$$|\gamma_{12}| = 1, \quad \text{complete mutual coherence,}$$

$$0 < |\gamma_{12}| < 1, \quad \text{partial mutual coherence,}$$

$$|\gamma_{12}| = 0, \quad \text{complete mutual incoherence.}$$

1.7.1   Coherence Time and Coherence Length

In addition to the mutual coherence between two waves, we can also define the coherence of an electromagnetic wave itself. Consider a quasi-monochromatic wave of the form

$$\mathbf{E}(\mathbf{r}, t) = \mathbf{A}(t)\exp[i(\omega t - \mathbf{k} \cdot \mathbf{r})], \qquad (1.7\text{-}8)$$

where $\mathbf{A}(t)$ is a time-varying complex amplitude. In the two-wave interference, the two electric fields $\mathbf{E}_1$ and $\mathbf{E}_2$ may originate from the same electric field given in Eq. (1.7-8). This happens in many interferometers, such as the Michelson interferometer, the Mach–Zehnder interferometer, and Young's two-slit interferometer. In these interference experiments, the two fields are only different in their optical paths. Referring to Fig. 1.3, we assume that the two waves are

$$\begin{aligned} \mathbf{E}_1 &= \mathbf{A}(t)\exp[i(\omega t - \mathbf{k}_1 \cdot \mathbf{r})], \\ \mathbf{E}_2 &= \mathbf{A}(t + \tau)\exp[i(\omega t - \omega\tau - \mathbf{k}_2 \cdot \mathbf{r})], \end{aligned} \qquad (1.7\text{-}9)$$

where τ is the time delay. We can see that these two waves originate from the same wave (1.7-8). The degree of mutual coherence between the two waves (1.7-9) is a measure of the self-coherence of the wave (1.7-8). Thus, a self-coherence function $\gamma(\tau)$ is defined as

$$\gamma(\tau) = \frac{\langle\!\langle \mathbf{E}^*(t)\mathbf{E}(t + \tau)\rangle\!\rangle}{\langle\!\langle \mathbf{E}^*(t)\mathbf{E}(t)\rangle\!\rangle^{1/2}\langle\!\langle \mathbf{E}^*(t + \tau)\mathbf{E}(t + \tau)\rangle\!\rangle^{1/2}}. \qquad (1.7\text{-}10)$$

Assume that all the quantities are stationary, that is, the time average is independent of the choice of the origin. Then the self-coherence function $\gamma(\tau)$ becomes

$$\gamma(\tau) = \frac{\langle\!\langle \mathbf{E}^*(t)\mathbf{E}(t + \tau)\rangle\!\rangle}{\langle\!\langle \mathbf{E}^*(t)\mathbf{E}(t)\rangle\!\rangle}. \qquad (1.7\text{-}11)$$

The self-coherence function defined in the preceding also satisfies the condition

$$0 \leqslant |\gamma(\tau)| \leqslant 1. \tag{1.7-12}$$

For monochromatic plane waves, $\mathbf{A}(t)$ is a constant vector, and the self-coherence function is unity for any τ. Such a wave is called a coherent radiation. For most polychromatic waves, $|\gamma(\tau)| = 1$ only when $\tau = 0$, and $|\gamma(\tau)|$ approaches zero when τ becomes large. Such waves are called partially coherent radiation. If $|\gamma(\tau)|$ is a monotonically decreasing function of τ and, in addition, if $|\gamma(\tau)|$ drops to virtually zero for $\tau > \tau_c$, we say that τ_c is the coherent time of the wave. Thus, in the two-beam interference experiment, if the path difference between the two beams does not exceed the value

$$l_c = c\tau_c, \tag{1.7-13}$$

interference fringe appears. The quantity l_c is the coherence length of the wave (1.7-8).

For many practical purposes, the time interval for averaging in Eq. (1.7-10) or (1.7-11) can be taken as infinity. Thus, the self-coherence function can be considered as the normalized autocorrelation function of the electromagnetic field. A basic theorem of the stochastic theory, the so-called *Wiener–Khintchine theorem*, states that the power spectrum (or spectral density) of the electromagnetic field and the autocorrelation function form a Fourier transform pair. Therefore, $\gamma(\tau)$, as given by Eq. (1.7-11), is the normalized Fourier transform of the power spectrum of the electromagnetic field. Let $\Delta\omega$ be the width of the frequency distribution of the field. Then, according to the theory of Fourier transform, we have

$$\tau_c = \frac{2\pi}{\Delta\omega} \equiv \frac{1}{\Delta f}, \tag{1.7-14}$$

which states that the coherence time of a quasi-monochromatic radiation is given by the inverse of its spectral line width Δf.

REFERENCES AND SUGGESTED READINGS

1. R. C. Jones, New calculus for the treatment of optical systems, *J. Opt. Soc. Am.* **31**, 488 (1941).
2. M. Born and E. Wolf, *Principles of Optics*, Macmillan, New York, 1964.
3. J. D. Jackson, *Classical Electrodynamics*, Wiley, New York, 1962.

4. L. D. Landau and E. M. Lifshitz, *Electrodynamics of Continuous Media*, Addison-Wesley, Reading, MA, 1965.

5. G. J. Troup, *Optical Coherence Theory*, Methuen, London, 1967.

6. A. Yariv and P. Yeh, *Optical Waves in Crystals*, Wiley, New York, 1984.

PROBLEMS

1.1   Conservation of charge requires that the charge density at any point in space be related to the current density in that neighborhood by a continuity equation

$$\frac{\partial \varrho}{\partial t} + \nabla \cdot \mathbf{J} = 0.$$

Derive this equation from Maxwell's equation.

1.2   Derive the boundary conditions for the normal components of **B** and **D**, Eq. (1.1-8), by using an application of Gauss's divergence theorem to Maxwell's equations.

1.3   Derive the boundary conditions that apply to the tangential components of **E** and **H**, Eq. (1.1-11), by using an application of Stokes theorem to Maxwell's equations.

1.4   The electromagnetic field momentum density and Maxwell's stress tensor are given, respectively, by

$$\mathbf{P} = \mu\varepsilon(\mathbf{E} \times \mathbf{H}),$$
$$T_{ij} = \varepsilon E_i E_j + \mu H_i H_j - \tfrac{1}{2}\delta_{ij}(\varepsilon E^2 + \mu H^2).$$

Derive the hydrodynamical equation of motion,

$$\frac{\partial \mathbf{P}}{\partial t} = \nabla \cdot T - \mathbf{F},$$

where **F** is the Lorentz force exerted on a distribution of charges and currents by the electromagnetic field

$$\mathbf{F} = \varrho\mathbf{E} + \mathbf{J} \times \mathbf{B}.$$

In a source-free region, the time rate of the field momentum density is equal to the force exerted by Maxwell's stresses on the region considered.

1.5   If the magnetic monopole exists, the static magnetic field produced by
       the magnetic charges obeys the equation

$$\nabla \cdot \mathbf{B} = \varrho_m,$$

where ϱ_m is the magnetic charge density.

(a) Show that the magnetic field produced by a point charge m in
    vacuum is

$$\mathbf{H} = \frac{m}{4\pi\mu_0 r^3} \mathbf{r},$$

where $\mathbf{r}$ is measured from the magnetic point charge.

(b) Consider the electromagnetic field produced by a combination of
    an electric point charge e located at $(0, 0, \frac{1}{2}d)$ and a magnetic point
    charge m located at $(0, 0, -\frac{1}{2}d)$. If the angular momentum
    density of the field is

$$\mathbf{L} = \mathbf{r} \times \mathbf{P},$$

where $\mathbf{P}$ is the linear momentum density given in Problem 1.4, find
the total angular momentum of the field by integrating $\mathbf{L}$ over the
space. *Answer:* $-(em/4\pi)\mathbf{z}$. Note that the angular momentum is
independent of the separation d.

(c) Let the origin of the coordinate system be on the magnetic point
    charge so that the electric charge is located at $(0, 0, d)$. Move the
    electric charge e from $(0, 0, d)$ to $(0, 0, -d)$ on a semicircular
    path. Find the torque required and then integrate the torque with
    respect to time. Show that the angular momentum required to
    move the electric charge is $-em/2\pi$. This method offers an easier
    way to calculate the angular momentum than does the direct
    approach in (b).

1.6   Let $\mathbf{E} = \mathbf{E}_0(\mathbf{r})\exp(i\omega t)$ and $\mathbf{H} = \mathbf{H}_0(\mathbf{r})\exp(i\omega t)$ be solutions to
       Maxwell's equations.

(a) Show that $\mathbf{E}^*$ and $\mathbf{H}^*$ also satisfy Maxwell's equations. Note that
    $\mathbf{E}^*, \mathbf{H}^*$ and $\mathbf{E}, \mathbf{H}$ are, in fact, the same field since only the real parts
    have physical meaning.

(b) Show that the conjugate waves

$$\mathbf{E}^c = \mathbf{E}_0^*(\mathbf{r})\exp(i\omega t) \quad \text{and} \quad \mathbf{H}^c = \mathbf{H}_0^*(\mathbf{r})\exp(i\omega t)$$

also satisfy the wave equation, provided the medium is lossless
(i.e., ε and μ are the real tensors).

1.7   Derive Eq. (1.5-4).

1.8   Derive Eqs. (1.5-7) and (1.5-8), and show that for $A_x = A_y$

$$e = -\tan \tfrac{1}{2}\delta.$$

1.9   Show that the end point of the electric vector of an elliptically
       polarized light will revolve in a clockwise direction if $\sin \delta > 0$ and in
       a counterclockwise direction if $\sin \delta < 0$.

1.10   Consider a beam of right-hand circularly polarized wave ($\sin \delta < 0$)
       which propagates along the z direction. Let the beam crosssection be
       several wavelengths broad and that the amplitude modulation be
       slowly varying.

       (a) Show that the electric and magnetic fields can be given approxi-
           mately by

$$E(x, y, z, t) \simeq \left[E_0(x, y)(\mathbf{x} - i\mathbf{y}) + \frac{-i}{k}\left(\frac{\partial E_0}{\partial x} - i\frac{\partial E_0}{\partial y} \right)\mathbf{z} \right] e^{i(\omega t - kz)}$$

$$H(x, y, z, t) \simeq i\frac{k}{\omega\mu} E(x, y, z, t).$$

           where $\mathbf{x}$, $\mathbf{y}$ and $\mathbf{z}$ are unit vectors.

       (b) Calculate the time-averaged component of angular momentum
           along the direction of propagation ($+z$). Show that this com-
           ponent of angular momentum is $\hbar$ provided the energy of the wave
           is normalized to $\hbar\omega$. This shows that a right-hand circularly
           polarized photon carries a positive angular momentum $\hbar$ along the
           direction of its momentum vector (helicity).

       (c) Show that the transverse components of the angular momentum
           vanish.

1.11   *Orthogonal polarization states.*

       (a) Find a polarization state that is orthogonal to the polarization
           state

$$J(\psi, \delta) = \begin{pmatrix} \cos\psi \\ e^{i\delta}\sin\psi \end{pmatrix}.$$

       Answer:

$$\begin{pmatrix} \sin\psi \\ e^{i(\pi+\delta)}\cos\psi \end{pmatrix}.$$

(b) Show that the major axes of the ellipse of two mutually orthogonal polarization states are perpendicular to each other and the sense of revolution is opposite.

1.12 (a) An elliptically polarized beam propagating in the z direction has a finite extent in the x and y directions,

$$\mathbf{E}(x, y, z, t) \sim E_0(x, y)(\alpha\mathbf{x} + \beta\mathbf{y})e^{i(\omega t - kz)},$$

where $\alpha = \cos\psi$ and $\beta = \sin\psi e^{i\delta}$. Show that the electric field must have a component in the z direction [see Problem 1.10(a)] and derive the expressions for the electric field and the magnetic field.

(b) Calculate the z component of the angular momentum assuming that the total energy of the wave is $\hbar\omega$. *Answer:* $L_z = -\hbar\sin 2\psi \sin\delta$.

(c) Decompose the elliptically polarized light into a linear superposition of right- and left-hand polarized states $\mathbf{R}$ and $\mathbf{L}$; that is, if $\mathbf{J}$ is the Jones vector of the polarization state shown in (a), find r and l such that

$$\mathbf{J} = r\mathbf{R} + l\mathbf{L}.$$

(d) Here r and l are the probability amplitudes that the photon is right- and left-hand circularly polarized, respectively. The angular momentum can be obtained by evaluating $|r|^2 - |l|^2$; that is,

$$L_z = \hbar(|r|^2 - |l|^2).$$

(e) Express the angular momentum in terms of the ellipticity angle θ and the Stokes parameter S_3.

1.13 Derive (1.5-15) and (1.5-16).

1.14 *Orthogonal polarization states.* Consider two monochromatic plane waves of the form

$$\mathbf{E}_a(z, t) = \text{Re}[\mathbf{A}e^{i(\omega t - kz)}],$$

$$\mathbf{E}_b(z, t) = \text{Re}[\mathbf{B}e^{i(\omega t - kz)}].$$

The polarization states of these two waves are orthogonal; that is, $\mathbf{A}^* \cdot \mathbf{B} = 0$.

(a) Let δ_a, δ_b be the phase angles defined in (1.5-5). Show that $\delta_a - \delta_b = \pm\pi$.

(b) Since δ_a, δ_b are in the range $-\pi < \delta \leqslant \pi$, show that $\delta_a \delta_b \leqslant 0$.

(c) Let χ_a, χ_b be the complex numbers representing the polarization states of these two waves. Show that $\chi_a \chi_b = -1$.

(d) Show that the major axes of the polarization ellipses are mutually orthogonal and the ellipticities are of the same magnitude with opposite signs.

1.15 Show that, for polarized light, the Stokes parameters can be expressed in terms of the orientation of the ellipse (angle ϕ with respect to the x axis) and the algebraic ellipticity angle θ as

$$
\begin{aligned}
S_0 &= 1, \\
S_1 &= \cos 2\theta \cos 2\phi, \\
S_2 &= \cos 2\theta \sin 2\phi, \\
S_3 &= -\sin 2\theta.
\end{aligned}
$$

1.16 Show that a polychromatic wave is polarized when the ratio of the amplitudes, A_y/A_x, and the phase differences δ are absolute constants (i.e., show that $S_0^2 = S_1^2 + S_2^2 + S_3^2$).

1.17 *Amplitude-stabilized classical oscillator.* Laser radiations such as continuous-wave gas He–Ne and GaAs lasers are good approximations to the "classical" amplitude-stabilized oscillators. Such radiations are quasi-monochromatic and can be expressed as

$$
E(t) = A e^{i\phi} e^{i\omega t},
$$

where A is a constant amplitude and ϕ is a random phase uniformly distributed over 2π. We may assume that ϕ changes to a new random phase in every time interval τ_c. Derive the self-coherence function of such a wave.

1.18 (a) Show that for unpolarized light, the x and y components of the electric field vector are mutually incoherent.

(b) Show that for partially polarized light, the x and y components of the electric vector are partially coherent.

(c) Try to derive a relation between the degree of polarization and the degree of mutual coherence between the x and y components.

1.19 Show that:

(a) The Poynting vector and the energy density of a monochromatic plane wave are related by $S = vU$, where v is the velocity of the wave.

(b) The intensity (magnitude of the Poynting vector) of a monochromatic plane wave in a vacuum is

$$I = \tfrac{1}{2} c \varepsilon_0 |E|^2.$$

(c) The electric field of a laser beam with an intensity of $1\,\text{W/cm}^2$ is
$E = 27.45\,\text{V/cm}$ regardless of the wavelength of the beam.

(d) The electric field of sunlight at the surface of earth is approximately $E = 10\,\text{V/cm}$. *Hint:* Use an intensity of $135.3\,\text{mW/cm}^2$ for
the sunlight.

2

Interaction of Electromagnetic Radiation with Matter

In the previous chapter, we defined the index of refraction of a medium in terms of its dielectric constant. We now investigate the physical origin of refractive index and dispersion as well as absorption. We first discuss the classical electron model and derive the atomic polarizability. We then discuss absorption and gain as well as the Kramers–Kronig relations. Finally, we take up the subject of pulse propagation in dispersive media.

2.1 DIELECTRIC CONSTANT AND ATOMIC POLARIZABILITY

Consider the presence of an electric field in an isotropic medium. Under the action of the electric field, the positive and negative charges inside each atom are displaced from their equilibrium positions. In most dielectrics, this charge separation is directly proportional to and in the same direction as the electric field. Thus, the induced dipole moment is

$$\mathbf{p} = \alpha \mathbf{E} \tag{2.1-1}$$

where the constant α is known as the atomic polarizability. This atomic polarizability α is derived using a simple classical electron model in the next section. The dielectric constant of a medium will depend on the manner in which the atoms are assembled. Let N be the number of atoms per unit volume. Then the polarization can be written approximately as

$$\mathbf{P} = N\mathbf{p} = N\alpha \mathbf{E} = \varepsilon_0 \chi \mathbf{E}, \tag{2.1-2}$$

where χ is electric susceptibility. In Eq. (2.1-2), we ignore the difference between the local field and the macroscopic field. This is legitimate for gas media when $N\alpha$ is $\ll 1$.

The dielectric constant ε of the medium is thus given by

$$\varepsilon = \varepsilon_0(1 + \chi) = \varepsilon_0\left(1 + \frac{N\alpha}{\varepsilon_0}\right). \tag{2.1-3}$$

If the medium is nonmagnetic (i.e., $\mu = \mu_0$), the index of refraction is given by

$$n^2 = 1 + \chi = 1 + \frac{N\alpha}{\varepsilon_0}. \tag{2.1-4}$$

2.2 CLASSICAL ELECTRON MODEL

In order to find the atomic polarizability, we need to solve for the dipole moment of an atom induced by the electric field of an incident optical wave. Since the atoms are so small compared to the wavelength of light, we can assume that the electric field is uniform in each atom. Let the electric field of the optical wave in the atom be

$$E = E_0 e^{i\omega t}, \tag{2.2-1}$$

where E_0 is a constant amplitude and ω is the frequency. Each of the charges in the atom will feel this electric field and will be driven up and down (or back and forth) by the electric force. Because of the great difference in their masses, we expect that electrons contribute dominantly to the induced dipole moment. In classical physics, we further assume that the electrons are fastened elastically to the atoms and obey the following equation of motion:

$$m\frac{d^2}{dt^2}X + m\gamma\frac{d}{dt}X + m\omega_0^2 X = -eE, \tag{2.2-2}$$

where X is the position of the electron relative to the atom and m is the mass of the electron. The parameter ω_0 is the resonant frequency of the electron motion, and γ is the damping coefficient. Here E is the electric field, and $-e$ is the electronic charge. For harmonic driving field (2.2-1), this equation has the following steady-state solution:

$$X = \frac{-eE_0}{m(\omega_0^2 - \omega^2 + i\omega\gamma)} e^{i\omega t}. \tag{2.2-3}$$

The induced dipole moment is thus given by

$$p = -eX = \frac{e^2}{m(\omega_0^2 - \omega^2 + i\gamma\omega)} E. \tag{2.2-4}$$

Using definition (2.1-2), we obtain the following expression for atomic polarizability:

$$\alpha = \frac{e^2}{m(\omega_0^2 - \omega^2 + i\gamma\omega)}. \tag{2.2-5}$$

If there are N atoms per unit volume, the index of refraction of such a medium is given by, according to Eqs. (2.2-5) and (2.1-4),

$$n^2 = 1 + \frac{Ne^2}{\varepsilon_0 m(\omega_0^2 - \omega^2 + i\gamma\omega)} . \qquad (2.2\text{-}6)$$

If the second term in this equation is small compared to 1, the index of refraction can be given approximately as

$$n = 1 + \frac{Ne^2}{2\varepsilon_0 m(\omega_0^2 - \omega^2 + i\gamma\omega)} . \qquad (2.2\text{-}7)$$

Equation (2.2-7) not only gives the index of refraction in terms of the basic atomic quantities, but also shows the variation with the frequency ω of light. However, it is not very practical to compute the index of refraction because both N and ω_0 vary from material to material.

In addition, information about the nature of resonance frequency ω_0 requires a calculation using quantum mechanics. Therefore, we cannot expect to get a simple formula for the index of refraction that applies to all substances. However, Eq. (2.2-7) is still very useful for many practical purposes. First, consider gas media such as air, helium and hydrogen. The natural resonance frequencies of the electron oscillations correspond to ultraviolet light. These frequencies are higher than the frequencies of visible light; that is, ω_0 is much larger than the ω of visible light. To a first approximation, we can neglect ω^2 and $i\gamma\omega$ in the denominator in comparison with ω_0^2. Then we find that the index of refraction is nearly constant. If we take $N = 2.69 \times 10^{25} \, \text{m}^{-3}$ for air at standard temperature ($0\,^\circ$C) and pressure ($760\,\text{mm Hg}$) and take $\omega_0 = 4\pi \times 10^{15}\,\text{s}^{-1}$, Eq. (2.2-7) yields $n = 1.0003$ for air.

For most transparent substances such as glass and water, ω_0 is also in the ultraviolet region. However, for solids and liquid, N is three orders of magnitude larger than that of air; thus, the electric susceptibility is accordingly higher.

2.3   DISPERSION AND COMPLEX REFRACTIVE INDEX

If we examine closely the expression for the index of refraction, Eq. (2.2-7), we notice that as ω rises and becomes closer to ω_0, the index of refraction also rises. So n rises slowly with ω. This is true for almost all the transparent materials. Thus the index of refraction is higher for blue light than for red light. This is the reason a prism bends the light more in the blue than in the red. The phenomenon that the refractive index depends on frequency is called the phenomenon of dispersion. Equation (2.2-7) is called the dispersion equation.

So far, we have been ignoring the imaginary term $i\gamma\omega$ in the denominator of Eq. (2.2-7). This term accounts for the damping of electron motion (for $\gamma > 0$) and gives rise to the phenomenon of *optical absorption*. Because of this damping term, $i\gamma\omega$, the index of refraction is a complex number according to Eq. (2.2-7). The imaginary part of the index of refraction is significant only when ω is close to ω_0. By working out the real and imaginary parts of Eq. (2.2-7), the complex refractive index is

$$n \rightarrow n - ik = 1 + \frac{Ne^2(\omega_0^2 - \omega^2)}{2\varepsilon_0 m[(\omega_0^2 - \omega^2)^2 + \gamma^2\omega^2]}$$

$$- i\frac{Ne^2\gamma\omega}{2\varepsilon_0 m[(\omega_0^2 - \omega^2)^2 + \gamma^2\omega^2]}, \qquad (2.3-1)$$

where $-k$ is the imaginary part of the refractive index. This k is referred to as the *extinction coefficient* because it represents an absorption (or attenuation) of the electromagnetic wave. To see the effect of k on electromagnetic radiation, we consider the propagation of a monochromatic plane wave in a medium with a complex refractive index $n - ik$. According to the discussion in Section 1.4, such a wave can be written as

$$\mathbf{E} = \mathbf{A} \exp[i(\omega t - Kz)], \qquad (2.3-2)$$

where K denotes the wave number in the complex medium and is given by

$$K = \omega\sqrt{\mu\varepsilon} = \frac{\omega}{c}n = \frac{2\pi}{\lambda}n. \qquad (2.3-3)$$

Using the complex refractive index (2.3-1), the wave number K becomes complex and is

$$K = \frac{2\pi}{\lambda}(n - ik), \qquad (2.3-4)$$

where n is now the real part of the complex refractive index and k is the extinction coefficient. Substituting Eq. (2.3-4) for K in Eq. (2.3-2), the electric field of the wave becomes

$$\mathbf{E} = \mathbf{A} \exp\left[i\left(\omega t - \frac{2\pi n}{\lambda}z\right)\right]\exp\left(-\frac{2\pi k}{\lambda}z\right). \qquad (2.3-5)$$

We notice that the imaginary part of the complex refractive index leads to an attenuation of electromagnetic radiation along its direction of propagation. The *attenuation coefficient* is usually defined as

$$\alpha \equiv \frac{1}{I}\frac{d}{dz}I, \qquad (2.3\text{-}6)$$

where I is the intensity of electromagnetic radiation and, aside from a constant factor, is proportional to $|E|^2$. According to Eq. (2.3-6), intensity is

$$I(z) = |E|^2 = I(0)e^{-\alpha z}, \qquad (2.3\text{-}7)$$

where $I(0)$ is the intensity at $z = 0$. The attenuation coefficient α defined in Eq. (2.3-6) is related to the extinction coefficient k by

$$\alpha = \frac{4\pi}{\lambda}k \qquad (2.3\text{-}8)$$

according to Eqs. (2.3-7) and (2.3-5).

According to Eq. (2.3-8), even a small k will lead to a strong attenuation for visible light. For example, with $k = 0.0001$ and $\lambda = 0.5\,\mu m$, Eq. (2.3-8) yields attenuation coefficient $\alpha = 25\,cm^{-1}$.

Both the real part n and the imginary part k of the complex index of refraction are functions of ω. The dependence on ω is significant, especially when ω is close to ω_0. To examine the dispersion at frequencies near resonance, we put $\omega \simeq \omega_0$ in Eq. (2.3-1) and obtain

$$n - ik = 1 + \frac{Ne^2(\omega_0 - \omega)}{4\varepsilon_0 m\omega_0[(\omega_0 - \omega)^2 + (\gamma/2)^2]}$$

$$- i\frac{Ne^2\gamma}{8\varepsilon_0 m\omega_0[(\omega_0 - \omega)^2 + (\gamma/2)^2]}. \qquad (2.3\text{-}9)$$

Figure 2.1 shows a normalized plot of $(n - 1)$ and k as functions of ω. Note that the extinction coefficient k is maximum at $\omega = \omega_0$ and decreases like $|\omega_0 - \omega|^{-2}$ as $|\omega_0 - \omega|$ increases. In fact, the $k(\omega)$ in Eq. (2.3-9) has a Lorentzian shape. On the other hand, the real part of the index of refraction approaches 1 like $|\omega_0 - \omega|^{-1}$ as $|\omega_0 - \omega|$ increases. This explains the fact that dispersion exists in most transparent materials at frequencies where k is negligible.

According to Eq. (2.3-9), the index of refraction is greater than unity for low frequencies ($\omega < \omega_0$) and increases with ω as the resonant frequency ω_0

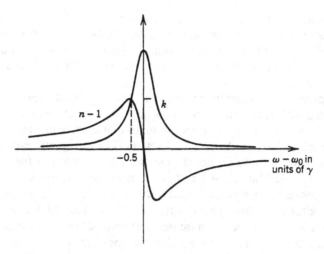

Figure 2.1 Normalized plot of $n - 1$ and k versus $\omega - \omega_0$.

is approached. Such an increasing behavior is the case of *normal dispersion* and occurs in most transparent media. Very near the resonance frequency, there is a small region of ω's for which the slope is negative. Such a negative slope is often referred to as *anomalous dispersion*. As we can see from Fig. 2.1, anomalous dispersion always occurs when absorption is significant.

In the classical electron model discussed in Section 2.2, we only assumed a single resonance frequency ω_0, which is somewhat simpler than in nature. In fact, there are several resonance frequencies for each atom. We can now modify the dispersion equation by assuming that not all the electrons are identically bound and all these electrons oscillate independently. Let f_j be the fraction of electrons whose natural resonance frequency is ω_j and whose damping factor is γ_j. Then by adding the contribution from all the electron oscillators, we obtain

$$n^2 = 1 + \frac{Ne^2}{\varepsilon_0 m} \sum_j \frac{f_j}{\omega_j^2 - \omega^2 + i\gamma_j\omega}. \qquad (2.3\text{-}10)$$

The fractions f_j are known as oscillator strengths and are dimensionless. By definition, the oscillator strengths obey the following sum rule:

$$\sum_j f_j = Z, \qquad (2.3\text{-}11)$$

where Z is the number of electrons per atom. Theoretical evaluation of oscillator strengths requires knowledge of quantum mechanics. To illustrate the magnitude of the oscillator strength and the rule (2.3-11), we note that the oscillator strength corresponding to the $1s \rightarrow 2p$ transition ($\omega_{1s \rightarrow 2p} = 1.55 \times 10^{-16} \text{s}^{-1}$) in the hydrogen atom is 0.4162. The sum of the oscillator strengths for transitions from the ground state $1s$ to all other states different from $2p$ is 0.5838.

In the preceding discussion, we assumed that there is a damping coefficient γ_j associated with each resonance frequency. These coefficients are all positive, and electron damping leads to attenuation of electromagnetic radiation. If some electrons are in their excited states, electromagnetic radiation will drive these electrons and induce the emission of radiation at the resonance frequency. For such an excited system (or inverted system), Eq. (2.3-10) is still valid, provided we replace the parameter N by ($N_g - N_j$), where N_g is the number of electrons in the ground state per unit volume and N_j is the number of electrons in the jth excited state per unit volume. In an inverted system, N_j is greater than N_g. Thus, we see that the complex refractive index of an excited system has a positive imaginary part in a small frequency range near the resonance frequency ω_j for which $N_g - N_j$ is negative. According to Eq. (2.3-5), the propagation of electromagnetic radiation in a medium with $k < 0$ will lead to amplification of the energy. Such a medium is called a *gain medium* and is described by a complex index of refraction with a positive imaginary part (i.e., $-k > 0$ or $k < 0$).

Figure 2.2 shows the dispersion of the index of refraction of several transparent media in the spectra range $\lambda = 0.3 \, \mu\text{m}$ to $\lambda = 30 \, \mu\text{m}$. Traditionally, the dispersion of n is usually expressed in terms of wavelength λ.

2.3.1 Complex Refractive Index of Metals

In a metal, there are free electrons that do not oscillate around atoms but are free to move under the action of the applied electric field. For these electrons, there is no restoring force. The equation of motion (2.2-2) is still valid, provided we set $\omega_0 = 0$. Thus, all the results obtained in Section 2.2 can still be used for metals. Putting $\omega_0 = 0$ in Eq. (2.2-6), we obtain the complex refractive index

$$n^2 = 1 - \frac{Ne^2}{\varepsilon_0 m(\omega^2 - i\gamma\omega)} \qquad (2.3\text{-}12)$$

where N is electron density. In fact, Eq. (2.3-12) becomes a good approximation to the actual complex refractive index, provided we can neglect the contribution from the deeply bound electrons. The damping constant γ is

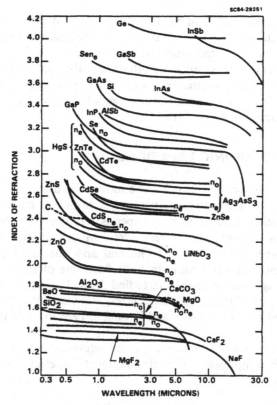

Figure 2.2 The n versus λ for some materials. For birefringent crystals, n_o is ordinary refractive index and n_e is the extraordinary refractive index.

some sort of average rate of collisions involving appreciable momentum transfer. If we further assume that $\gamma \ll \omega$, the index of refraction is given by

$$n^2 \simeq 1 - \frac{\omega_p^2}{\omega^2}, \tag{2.3-13}$$

where ω_p is *plasma angular frequency* given by

$$\omega_p^2 = \frac{Ne^2}{\varepsilon_0 m}. \tag{2.3-14}$$

For high-frequency radiation ($\omega > \omega_p$), the index of refraction is real and the waves propagate freely. For frequencies lower than the plasma frequency ω_p, n is purely imaginary. Thus, within the metal, the fields fall off exponentially with distance from the surface (see also Problem 2.3). Consequently, such electromagnetic radiation incident on a high-conductivity metal will be reflected from the surface. For metals such as aluminum, copper, gold, and silver, the density of the free electrons are on the order of $N \sim 10^{23}\,\text{cm}^{-3}$. This means that $\omega_p \sim 2 \times 10^{16}\,\text{s}^{-1}$, so that for visible and infrared radiation with $\omega < \omega_p$, the index of refraction is imaginary according to Eq. (2.3-13). The index of refraction, in general, is complex when γ is finite. Table 2.1 lists the index of refraction of several metals.

2.4 KRAMERS–KRONIG RELATIONS

The dielectric function $\varepsilon(\omega)$ or refractive index $n(\omega)$ describes the response of a material to electromagnetic radiation. According to the preceding discussion, the dielectric function $\varepsilon(\omega)$ depends sensitively on the electronic structure of the medium and is very useful in the determination of the electronic structure of a material. The real and imaginary parts of the complex dielectric function $\varepsilon(\omega)$ are both functions of frequency ω. The Kramers–Kronig relations enables us to find the real part of $\varepsilon(\omega)$ if we know the imaginary part of $\varepsilon(\omega)$ at all frequencies, and vice versa. Such relations were first introduced by Kramers and Kronig in 1926 to study the dielectric constant of a substance.

The dielectric function derived in the previous section relates the monochromatic components of the displacement and electric vectors in the following form:

$$\mathbf{D}(\omega) = \varepsilon(\omega)\mathbf{E}(\omega). \qquad (2.4\text{-}1)$$

As a consequence of the frequency dependence of $\varepsilon(\omega)$, there is a temporally nonlocal connection between the displacement $\mathbf{D}(t)$ and the electric field $\mathbf{E}(t)$. Such a connection can be easily constructed by Fourier transform. Let us put

$$\mathbf{D}(t) = \int_{-\infty}^{\infty} \mathbf{D}(\omega)e^{i\omega t}\,d\omega, \qquad (2.4\text{-}2)$$

$$\mathbf{E}(t) = \int_{-\infty}^{\infty} \mathbf{E}(\omega)e^{i\omega t}\,d\omega. \qquad (2.4\text{-}3)$$

The substitution of Eq. (2.4-1) for $\mathbf{D}(\omega)$ in Eq. (2.4-2) gives

$$\mathbf{D}(t) = \int_{-\infty}^{\infty} \varepsilon(\omega)\mathbf{E}(\omega)e^{i\omega t}\,d\omega. \qquad (2.4\text{-}4)$$

If we invert the Fourier integral in Eq. (2.4-3), we obtain

$$\mathbf{E}(\omega) = \frac{1}{2\pi}\int_{-\infty}^{\infty} \mathbf{E}(t)e^{-i\omega t}\,dt. \qquad (2.4\text{-}5)$$

Table 2.1. Complex Index of Refraction of Some Metals and Semiconductors by Wavelength

Substance		0.4 μm	0.5 μm	0.6 μm	0.7 μm	0.8 μm	1.0 μm	4.0 μm	10.0 μm
Aluminum	n	0.4	0.667	1.043	1.55	1.99	1.991	6.1	26
	k	4.45	5.573	6.568	7	7.05	9.3	30.4	67.3
Copper	n	—	0.88	0.186	0.15	0.17	0.197	1.94	10.52
	k	—	2.42	2.98	4.05	4.84	6.272	23.1	59.29
Germanium	n	2.3	3.4	4.5	5.15	5.27	5.1	4.35	4.3
	k	2.8	2.25	1.7	1.3	0.9	0.45	—	—
Gold	n	—	0.84	0.2	0.131	0.149	0.179	2.04	11.5
	k	—	1.84	2.897	3.842	4.654	6.044	27.9	67.5
Indium	n	—	1.02	1.29	1.39	1.5	1.76	7.6	23.8
	k	—	2.08	2.59	5.75	6.69	7.69	26.1	51.7
Lead	n	—	—	—	1.68	1.51	1.41	6.58	21
	k	—	—	—	3.67	4.24	5.4	20.8	37.4
Mercury	n	0.73	1.04	1.39	1.76	2.14	—	—	—
	k	3.01	3.7	4.32	4.83	5.33	—	—	—
Silver	n	0.075	0.05	0.06	0.075	0.09	0.25	2.34	10.8
	k	1.93	2.87	3.75	4.62	5.45	6.81	26.9	60.7
GaP	n	4.15	3.73	3.46	3.33	3.27	3.17	2.95	2.9
	k	0.78	0.0025	4×10^{-6}	—	—	1×10^{-4}	0.0026	0.038
GaAs	n	4.149	4.394	3.866	3.672	3.62	3.5	3.3	3.1
	k	2.137	0.4	0.25	0.15	0.068	—	—	—
GaSb	n	—	—	—	—	—	—	3.833	3.843
	k	—	—	—	—	—	—	0.0013	0.0093
InP	n	4.1	3.61	3.44	3.41	3.4	3.327	3.1	3.05
	k	1.41	0.47	0.32	0.198	0.15	—	—	—
InAs	n	3.2575	4.5	4.18	3.92	3.78	3.63	3.51	3.42
	k	1.667	1.37	0.42	0.22	0.15	0.083	—	—
InSb	n	3.14	3.65	4.24	5.09	4.65	4.27	4.02	3.95
	k	1.9	2.17	1.82	1.2	0.56	0.34	0.11	0.002

We now substitute Eq. (2.4-5) for $\mathbf{E}(\omega)$ in Eq. (2.4-4) and obtain

$$\mathbf{D}(t) = \frac{1}{2\pi} \int_{-\infty}^{\infty} \varepsilon(\omega) e^{i\omega t} \, d\omega \int_{-\infty}^{\infty} \mathbf{E}(t') e^{-i\omega t'} \, dt'. \qquad (2.4\text{-}6)$$

By rearranging the orders of integration and letting $\tau = t - t'$, the last equation can be written as

$$\mathbf{D}(t) = \frac{1}{2\pi} \int_{-\infty}^{\infty} d\tau \int_{-\infty}^{\infty} d\omega \, \varepsilon(\omega) e^{i\omega \tau} \, \mathbf{E}(t - \tau). \qquad (2.4\text{-}7)$$

We now define a function $F(\tau)$ as

$$F(\tau) = \frac{1}{2\pi} \int_{-\infty}^{\infty} [\varepsilon(\omega) - \varepsilon_0] e^{i\omega \tau} \, d\omega \qquad (2.4\text{-}8)$$

so that Eq. (2.4-7) can be written as

$$\mathbf{D}(t) = \varepsilon_0 \mathbf{E}(t) + \int_{-\infty}^{\infty} F(\tau) \mathbf{E}(t - \tau) \, d\tau. \qquad (2.4\text{-}9)$$

Equations (2.4-9) and (2.4-10) give a nonlocal relation between $\mathbf{D}(t)$ and $\mathbf{E}(t)$ in which the displacement $\mathbf{D}(t)$ at time t depends on the electric field $\mathbf{E}$ at times other than t. If the dielectric function $\varepsilon(\omega)$ is independent of ω [i.e., $\varepsilon(\omega) = $ const], Eq. (2.4-8) yields $F(\tau) = \varepsilon_0 \chi \delta(\tau)$, where χ is electric susceptibility, and the temporal nonlocal response disappears. But if $\varepsilon(\omega)$ varies with ω, $F(\tau)$ is no longer a delta function, and the nonlocal response results.

Using $[\varepsilon(\omega) - \varepsilon_0] = \varepsilon_0[n^2(\omega) - 1]$ and Eq. (2.3-10) for $n^2(\omega)$, a contour integration (see Problem 2.10) of Eq. (2.4-8) yields

$$F(\tau) = 0 \quad \text{for } \tau < 0. \qquad (2.4\text{-}10)$$

This result agrees with the *principle of causality*, which states that the displacement $\mathbf{D}(t)$ at time t depends on the value of electric field $\mathbf{E}$ at all earlier times. Using Eq. (2.4-10), Eq. (2.4-9) becomes

$$\mathbf{D}(t) = \varepsilon_0 \mathbf{E}(t) + \int_{0}^{\infty} F(\tau) \mathbf{E}(t - \tau) \, d\tau. \qquad (2.4\text{-}11)$$

If we allow $F(\tau)$ to be an arbitrary real function of τ such that the integral in Eq. (2.4-11) always converges, this equation can be viewed as the most general relation between $\mathbf{D}$ and $\mathbf{E}$ in a uniform isotropic medium since only the causality condition is assumed. We now start from Eq. (2.4-11) and substitute Eqs. (2.4-3) and (2.4-4) for $E(t)$ and $D(t)$, respectively, and we get

$$\varepsilon(\omega) = \varepsilon_0 + \int_0^\infty F(\tau)e^{-i\omega\tau}\,d\tau. \tag{2.4-12}$$

This formula can be used to determine the frequency dependence of the dielectric constant, provided $F(\tau)$ is known. In general, $\varepsilon(\omega)$ is complex. However, since $F(\tau)$ is a real function, $\varepsilon(\omega)$ satisfies the following symmetry relation:

$$\varepsilon^*(\omega) = \varepsilon(-\omega). \tag{2.4-13}$$

If we separate the real and imaginary parts through the relation

$$\varepsilon(\omega) = \varepsilon'(\omega) - i\varepsilon''(\omega), \tag{2.4-14}$$

we obtain

$$\varepsilon'(-\omega) = \varepsilon'(\omega), \tag{2.4-15}$$

$$\varepsilon''(-\omega) = -\varepsilon''(\omega), \tag{2.4-16}$$

which shows that the real part of the dielectric constant is an even function and the imaginary part an odd function of the frequency ω.

The relation (2.4-12) determines the dielectric constant as a function of frequency ω, which is a real variable. Let us now view ε as a function of complex variable, that is, let us put

$$\varepsilon(z) = \varepsilon_0 + \int_0^\infty F(\tau)e^{-iz\tau}\,d\tau, \tag{2.4-17}$$

where z is a complex variable. Such an $\varepsilon(z)$ is an analytic function of z in the lower half of the complex plane. In the upper half plane, where $z = \omega + i\gamma$ with $\gamma > 0$, the integral in Eq. (2.4-17) diverges.

We now consider the contour integral

$$J = \oint_C \frac{\varepsilon(z) - \varepsilon_0}{z - \omega}\,dz \tag{2.4-18}$$

in which the contour of integration is a counterclockwise path that consists of the real axis going around the point $z = \omega$ from below, along an infinitesimal semicircle, and being closed by an infinite semicircle lying in the lower half plane (see Fig. 2.3). Note that the integrand of Eq. (2.4-18) has no singularity

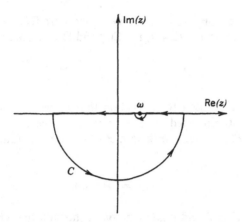

Figure 2.3   Contour of the integration in Eq. (2.4-18).

in the region surrounded by the contour C. According to Cauchy's theorem, the integral is zero (i.e., $J = 0$). Furthermore, the value of $\varepsilon(z) - \varepsilon_0$ tends to zero as $z \to \infty$. Therefore, the integration in J over the infinite semicircle gives no contribution, and Eq. (2.4-18) can be written as

$$0 = J = \lim_{a \to 0} \left(\int_{\infty}^{\omega+a} \frac{\varepsilon(z) - \varepsilon_0}{z - \omega} \, dz + \int_{\omega-a}^{-\infty} \frac{\varepsilon(z) - \varepsilon_0}{z - \omega} \, dz \right) - i\pi[\varepsilon(\omega) - \varepsilon_0],$$

$$(2.4\text{-}19)$$

where a is the radius of the infinitesimal semicircle and the term $-i\pi[\varepsilon(\omega) - \varepsilon_0]$ is a result of the integration over the infinitesimal semicircle around the point $z = \omega$.

The preceding equation can be rewritten as

$$\varepsilon(\omega) - \varepsilon_0 = \frac{i}{\pi} P \int_{-\infty}^{\infty} \frac{\varepsilon(z) - \varepsilon_0}{z - \omega} \, dz, \qquad (2.4\text{-}20)$$

where P indicates that the integral is a principal-value integral. Since the integration is over the real axis, we may replace z by ω' and Eq. (2.4-20) becomes

$$\varepsilon(\omega) - \varepsilon_0 = \frac{i}{\pi} P \int_{-\infty}^{\infty} \frac{\varepsilon(\omega') - \varepsilon_0}{\omega' - \omega} \, d\omega'. \qquad (2.4\text{-}21)$$

Using Eq. (2.4-14) and equating the real and imaginary parts of both sides of Eq. (2.4-21) yields the Kramers–Kronig relations:

$$\varepsilon'(\omega) - \varepsilon_0 = \frac{1}{\pi} P \int_{-\infty}^{\infty} \frac{\varepsilon''(\omega')}{\omega' - \omega} \, d\omega', \tag{2.4-22}$$

$$\varepsilon''(\omega) = \frac{1}{\pi} P \int_{-\infty}^{\infty} \frac{\varepsilon'(\omega') - \varepsilon_0}{\omega' - \omega} \, d\omega'. \tag{2.4-23}$$

The symmetry properties (2.4-15) and (2.4-16) can be used to rewrite Eqs. (2.4-22) and (2.4-23) into the following forms, which contain only positive frequencies:

$$\varepsilon'(\omega) - \varepsilon_0 = \frac{2}{\pi} P \int_{0}^{\infty} \frac{\varepsilon''(\omega')\omega'}{\omega'^2 - \omega^2} \, d\omega', \tag{2.4-24}$$

$$\varepsilon''(\omega) = \frac{2\omega}{\pi} P \int_{0}^{\infty} \frac{\varepsilon'(\omega') - \varepsilon_0}{\omega'^2 - \omega^2} \, d\omega'. \tag{2.4-25}$$

The Kramers–Kronig relations enable us to calculate the function $\varepsilon'(\omega)$ if we know the function $\varepsilon''(\omega)$, and vice versa. They are of very general validity and are derived directly from Eq. (2.4-11), which contains only the assumption of causality between the displacement $\mathbf{D}$ and electric field $\mathbf{E}$.

As an example of using the Kramers–Kronig relations, let us consider the attenuation of electromagnetic radiation determined by the imaginary part of the dielectric constant [i.e., $\varepsilon''(\omega)$]. Since the integral $P \int_{0}^{\infty} d\omega'/(\omega'^2 - \omega^2)$ vanishes identically, it follows directly from Eq. (2.4-25) that in a dispersionless medium, that is, when $\varepsilon'(\omega') - \varepsilon_0 = $ constant, the imaginary part of the dielectric constant vanishes. In other words, any dispersive medium is at the same time also an absorbing medium.

Consider now, as a second example, the real part of the dielectric constant of an atomic system for which the damping constant γ_j is small and negligible. From Eq. (2.3-10), we have

$$\varepsilon'(\omega) - \varepsilon_0 = \frac{Ne^2}{m} \sum_j \frac{f_j}{\omega_j^2 - \omega^2}. \tag{2.4-26}$$

This equation can be written as

$$\varepsilon'(\omega) - \varepsilon_0 = \frac{Ne^2}{m} P \int_{0}^{\infty} \sum \frac{f_j \delta(\omega' - \omega_j)}{\omega'^2 - \omega^2} \, d\omega', \tag{2.4-27}$$

where δ is the Dirac delta function. Comparing Eq. (2.4-27) with Eq. (2.4-24), we obtain an expression for the imaginary part of the dielectric constant in terms of the oscillator strengths,

$$\varepsilon''(\omega) = \frac{\pi Ne^2}{2\omega m} \sum_j f_j \delta(\omega - \omega_j). \tag{2.4-28}$$

Using the sum rule (2.3-11) and integrating Eq. (2.4-28), we obtain the sum rule

$$\int_0^\infty \varepsilon'(\omega)\omega \, d\omega = \frac{\pi e^2}{2m} NZ, \tag{2.4-29}$$

where Z is the number of electrons per atom and N is the number of atoms per unit volume.

2.5 OPTICAL PULSES AND GROUP VELOCITY

In Chapter 1, we discussed the propagation of monochromatic plane waves in homogeneous isotropic media and studied some basic properties. Laser radiation in a continuous-wave (CW) operation may be considered monochromatic and can often be represented by a plane wave. However, many important applications of lasers involve pulse operation. In the pulsed mode, the pump energy can be concentrated into extremely short time durations, thereby increasing the peak power. This is of key importance in numerous industrial applications such as machining and welding with lasers as well as in scientific applications. In the latter category, the use of extremely short pulses makes it possible to probe very short-lived ($< 10^{-12}$ s) transient phenomena.

The finite duration of a laser pulse results in a finite spread of frequencies or, equivalently, wavelengths. Laser pulse propagation in a linear medium can be described in terms of the propagation of an appropriate linear superposition of plane waves with different frequencies since Maxwell's equations are linear. However, there are new features associated with laser pulse propagation in a dispersive medium, one where the phase velocity depends on the frequency. As a result, different frequency components of the wave propagate with different speeds and tend to change phase with respect to one another. This usually leads to a spreading of the laser pulse as it propagates in a dispersive medium. Also, the velocity of energy flow of a laser pulse propagating in a dispersive medium may differ greatly from the phase velocity. This is a subtle subject and requires careful investigation.

These effects on the propagation of a laser pulse due to dispersion can be described by representing the pulse as a sum of many plane-wave solutions of Maxwell's equations. In the limit, we will replace the summation by an integral. For the sake of clarity in introducing the basic concepts, we will only consider the case of one-dimensional scalar waves. The scalar amplitude $\psi(z, t)$ can be thought of as one of the components of electromagnetic field vectors. If $A(k)$ denotes the amplitude of the plane-wave component with wave number k, the pulse $\psi(z, t)$ can be written as

$$\psi(z, t) = \int_{-\infty}^{\infty} A(k)e^{i(\omega t - kz)} \, dk. \qquad (2.5\text{-}1)$$

This integral satisfies Maxwell's equation since the integrand is a basic plane-wave solution to the same equations. Note that if we view $\psi(z, t)$ at some instance of time, say, $t = 0$, $A(k)$ is then formally the Fourier transform of $\psi(z, 0)$. We will then refer to $|A(k)|^2$ as the Fourier spectrum of the field $\psi(z, t)$. The relationship between ω and k (called the *dispersion relation*) is given by Eq. (1.4-11) for the electromagnetic field. In an isotropic medium, the dispersion properties cannot depend on the direction of propagation; therefore, ω must be an even function of k, $\omega(-k) = \omega(k)$. Here, we further assume that both k and $\omega(k)$ are real. A typical pulse and its Fourier spectrum are shown in Fig. 2.4.

A laser pulse is usually characterized by its center frequency ω_0 (or its corresponding wave number k_0) and the frequency spread $\Delta\omega$ around ω_0 (or the corresponding spread in wave number, Δk). In other words, $A(k)$ is sharply peaked around k_0. To study the evolution of such a pulse in time, we expand $\omega(k)$ around the value k_0 in terms of a Taylor series:

$$\omega(k) = \omega_0 + \left(\frac{d\omega}{dk}\right)_0 (k - k_0) + \cdots. \qquad (2.5\text{-}2)$$

Then we substitute for ω from (2.5-2) into (2.5-1). Thus, Eq. (1.5-1) becomes

$$\psi(z, t) \simeq e^{i(\omega_0 t - k_0 z)} \int_{-\infty}^{\infty} A(k)e^{i[(d\omega/dk)_0 t - z](k - k_0)} \, dk, \qquad (2.5\text{-}3)$$

where we neglect the higher order terms in $k - k_0$. The integral in Eq. (2.5-3) is a function of the composite variable $[z - (d\omega/dk)_0 t]$ only and is called the envelope function $E[z - (d\omega/dk)_0 t]$. Thus, the amplitude of the pulse can be written as

$$\psi(z, t) = e^{i(\omega_0 t - k_0 z)} E[z - (d\omega/dk)_0 t]. \qquad (2.5\text{-}4)$$

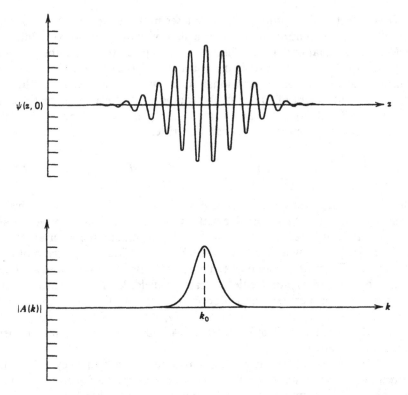

Figure 2.4 A laser pulse of finite extent and its Fourier spectrum in wave number space k. (From A. Yariv and P. Yeh, *Optical Waves in Crystals*, Wiley, New York, 1984, p. 15. Copyright © 1984. Reprinted by permission of John Wiley & Sons, Inc.)

This shows that, apart from an overall phase factor, the laser pulse travels along undistorted (see Fig. 2.5) in shape, with a velocity

$$v_g = \left(\frac{d\omega}{dk}\right)_0, \qquad (2.5\text{-}5)$$

which is called the *group velocity* of the pulse. This approximation is legitimate only when the distribution $A(k)$ is sharply peaked at k_0 and the frequency ω is a smoothly varying function of k around k_0. If the electromagnetic energy density of a laser pulse is associated with the absolute square of the amplitude, it is obvious that in this approximation the group velocity represents the transport of energy (see problem 2.5). We note that in the case of a pulse, the phase and group velocites are, in general, different, that is,

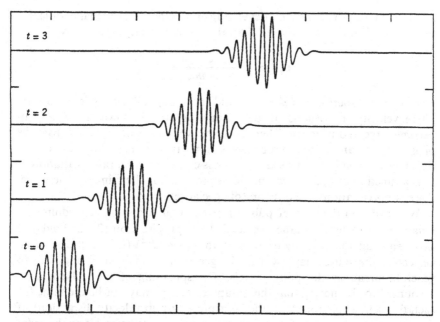

Figure 2.5 Schematic plot of the field amplitude, illustrating the propagation of the wave packet in the regime where Eq. (2.5-2) is valid and the pulse shape remains undistorted. (Adapted from A. Yariv and P. Yeh, *Optical Waves in Crystals*, Wiley, New York, 1984, p. 16. Copyright © 1984. By permission of John Wiley & Sons, Inc.)

$(d\omega/dk)_0 \neq \omega_0/k_0$. The phase velocity, which is usually greater than the group velocity, has the same significance as in the case of a plane wave. It is the velocity needed to stay on top of a given wave crest or valley.

In optics, the dispersion of a material medium is usually described by the index of refraction, $n(\omega)$, as a function of frequency (or wavelength), and the relation between ω and k is given by

$$k = n(\omega)\frac{\omega}{c},\qquad(2.5\text{-}6)$$

where c is the velocity of light in a vacuum. The phase velocity is

$$v_p = \frac{c}{n(\omega)}\qquad(2.5\text{-}7)$$

and is greater or smaller than c depending on whether $n(\omega)$ is smaller or larger than unity. From (2.5-5) and (2.5-6), the group velocity is given by

$$v_g = \frac{c}{n + \omega(dn/d\omega)}. \tag{2.5-8}$$

For normal dispersion $dn/d\omega > 0$, and the group velocity is less than the phase velocity. In regions of anomalous dispersion, however, $dn/d\omega$ can become large and negative. Then the group velocity differs greatly from the phase velocity and sometimes becomes larger than c. The latter case occurs only when $dn/d\omega$ is very negative, which is equivalent to a rapid variation of ω as a function of k, thus making the approximation illegitimate. Therefore, there is no violation of the theory of special relativity.

We have seen that a laser pulse propagating in a dispersive medium will remain undistorted in shape provided the approximation (2.5-2) holds. In the event that the next higher order term $\frac{1}{2}(d^2\omega/dk^2)_0(k - k_0)^2$ cannot be neglected, the pulse shape will no longer remain unchanged and will, in general, spread as the pulse travels. The spreading of the pulse can be accounted for by noting that the group velocity v_g may not be the same for each frequency component of the laser pulse (group velocity dispersion). If Δk is the spectral spread of the pulse, the group velocity spread is on the order of

$$\Delta v_g \sim \frac{d^2\omega}{dk^2} \Delta k. \tag{2.5-9}$$

As the pulse propagates, one expects a spread in position on the order of $\Delta v_g t$. The issue of pulse spreading is considered again in Problem 2.6.

PROBLEMS

2.1   *Clausius–Mossotti relation.* The electric susceptibility of a medium is defined through $\mathbf{P} = \varepsilon_0 \chi \mathbf{E}$ (2.1-2), where $\mathbf{E}$ is the macroscopic electric field. The local field, that is, the electric field acting on the atom, is given by

$$\mathbf{E}_{\text{local}} = \mathbf{E} + \frac{1}{3\varepsilon_0} \mathbf{P}.$$

Show that

$$\frac{\chi}{\chi + 3} = \frac{N\alpha}{3\varepsilon_0},$$

which is the Clausius–Mossotti relation.

2.2   *Free-electron gas.* Equation (2.2-2) can also be used to describe the classical motion of free electrons, provided we set $\omega_0 = 0$, which indicates the absence of a restoring force. The damping coefficient γ can be replaced by $1/\tau$, where τ is collision time.

(a) Show that the dielectric constant of a free-electron gas can be written as

$$\frac{\varepsilon(\omega)}{\varepsilon_0} = 1 - \frac{\omega_p^2}{\omega^2 - i\omega/\tau},$$

where ω_p is the plasma frequency,

$$\omega_p^2 = \frac{Ne^2}{\varepsilon_0 m}.$$

(b) Show that, for $\tau \to \infty$, the dielectric constant is positive and real if $\omega > \omega_p$.

(c) Using Eqs. (2.3-2) and (2.3-3), show that electromagnetic radiation cannot propagate in a free-electron gas when the frequency ω is below the plasma frequency.

(d) In the plasma of a gas discharge, N is typically on the order of 10^{18} electrons/m^3. Find the plasma frequency. In the ionosphere, N is on the order of 10^{11} electrons/m^3. Find the plasma frequency.

2.3   *Lineshape function.* Using Eqs. (2.3-9) and (2.3-8), the absorption coefficient as a function of ω can be written, when ω is near ω_0, as

$$\alpha(\omega) = \alpha_0 \frac{(\gamma/2)^2}{(\omega - \omega_0)^2 + (\gamma/2)^2},$$

where α_0 is the absorption coefficient at line center [i.e., $\alpha_0 = \alpha(\omega_0)$]. This is called a Lorentzian lineshape function.

(a) Using Eq. (2.3-9), show that

$$\alpha_0 = \frac{Ne^2}{c\varepsilon_0 m\gamma}.$$

Note that when $\gamma \to 0$, $\alpha_0 \to \infty$.

(b) Show that

$$I = \int_0^\infty \alpha(\omega)\, d\omega = \tfrac{1}{4}\pi\, \alpha_0\gamma.$$

(c) Show that the integral I in (b) tends to a constant when $\gamma \to 0$.

2.4   Equations (2.5-3) and (2.5-4) are approximate descriptions of a laser pulse. These equations become exact in a dispersionless medium. Given a distribution in wave number $A(k)$, an envelope function $E(\xi)$ can be obtained by carrying out the integral in (2.5-3). For each of the forms $A(k)$ that follows, calculate the envelope function $E(\xi)$, plot $|A(k)|^2$ and $|E(\xi)|^2$, evaluate the standard deviations from the means (Δk and $\Delta \xi$), and test the Heisenberg uncertainty relation $\Delta k \, \Delta \xi \geqslant \frac{1}{2}$.

(a) $A(k) = A(k_0)\exp[-(k-k_0)^2/4q^2]$. This spectral distribution corresponds to a Gaussian pulse and is the minimum wave packet, that, $\Delta k \, \Delta \xi = \frac{1}{2}$.

(b) $A(k) = A(k_0)\exp(-|k-k_0|/2q)$. This spectral distribution corresponds to a Lorentzian pulse.

(c) $A(k) = A(k_0)\sin[(k-k_0)/q]/[(k-k_0)/q]$. This spectral distribution corresponds to a square-wave pulse.

2.5   A one-dimensional laser pulse is centered at frequency ω_0 and the corresponding wave number k_0 with an envelope function $E(z - v_g t)$ for the electric field, that is,

$$E_x = E(z - v_g t)e^{i(\omega_0 t - k_0 z)}.$$

Find the envelope for the magnetic field, H_y, and calculate the Poynting power.

2.6   The envelope of a laser pulse will remain undistorted in shape provided the second- and higher order terms in the expansion of $\omega(k)$ around k_0 can be neglected (2.5-2). Now consider the case when the second-order terms is not small and cannot be neglected, that is,

$$\omega(k) = \omega_0 + \left(\frac{d\omega}{dk}\right)_0 (k-k_0) + \frac{1}{2}\left(\frac{d^2\omega}{dk^2}\right)_0 (k-k_0)^2.$$

Find the envelope function of a pulse with a spectral distribution $A(k)$ given in Problem 2.4(a) and the width of the envelope as a function of time t. The following integral may be useful:

$$\int_{-\infty}^{\infty} e^{-(\alpha x^2 + \beta x)} \, dx = \sqrt{\frac{\pi}{\alpha}} \exp\left(\frac{\beta^2}{4\alpha}\right),$$

where the real part of α must be positive.

2.7 If the index of refraction is expressed as a function of wavelength λ, show that the group velocity (2.5-8) can be written as

$$v_g = \frac{c}{n - \lambda(dn/d\lambda)} .$$

2.8 Show that the Taylor expansion coefficient $\frac{1}{2}(d^2\omega/dk^2)$ is proportional to the group velocity dispersion and that

$$\frac{dv_g}{d\lambda} = v_g^2 \frac{\lambda}{c} \frac{d^2n}{d\lambda^2} .$$

2.9 Using Eq. (2.3-9) for the index of refraction, derive an expression for the group velocity v_g.

2.10 *Principle of causality.* The principle of causality requires that $F(\tau) = 0$ for $\tau < 0$. Prove it by using Eqs. (2.4-8) and (2.3-10).

(a) Show that, according to Eq. (2.3-10), the poles of the integrand in Eq. (2.4-8) occur at

$$\omega = i\gamma_j \pm [\omega_j^2 - (\gamma_j/2)^2]^{1/2}.$$

Note that these poles are all in the upper half of the complex plane regardless of the plus or minus sign.

(b) The integrand in Eq. (2.4-8) is analytic in the lower half plane as well as on the real axis. Consider a contour integration from $-R$ to $+R$ and also along a semicircle in the lower half plane that is centered at the origin and has a radius R. Show that this contour integration is zero provided $\tau < 0$. This proves the principle of causality.

3

Reflection and Refraction
of Plane Waves

One of the most fundamental issues in the optics of layered media is the reflection and refraction of plane electromagnetic waves at a boundary of dielectric discontinuity. In this chapter, we consider in detail the problem of reflection and refraction of monochromatic plane waves at a plane boundary between two homogeneous isotropic media. We will define some symbols and conventions and lay down the groundwork for treating more complicated structures.

3.1 SNELL'S LAW AND FRESNEL'S FORMULAS

The reflection and refraction of a plane wave at a plane interface between two media with different dielectric properties are familiar phenomena. A plane wave incident on the interface will, in general, be split into two waves: a transmitted wave proceeding into the second medium and a reflected wave propagating back into the first medium. The existence of these two waves is a direct consequence of the boundary conditions on the field vectors (see Section 1.1). We will now derive some general properties for the directions of propagation of these two waves.

Let $E_i \exp[i(\omega t - \mathbf{k}_i \cdot \mathbf{r})]$ be a field amplitude of an incident plane wave with frequency ω and wave propagation vector $\mathbf{k}_i$. The reflected and transmitted plane-wave amplitudes are designated as $E_r \exp[i(\omega t - \mathbf{k}_r \cdot \mathbf{r})]$ and $E_t \exp[i(\omega t - \mathbf{k}_t \cdot \mathbf{r})]$, respectively. Here $\mathbf{k}_r$ and $\mathbf{k}_t$ are the corresponding propagation vectors. Any boundary condition that relates these three field amplitudes at the plane interface $x = 0$ will require that the spatial (and temporal) variation of all fields be the same. Consequently, the arguments of these field amplitudes at any point on the boundary $x = 0$ must satisfy the equation

$$(\mathbf{k}_i \cdot \mathbf{r})_{x=0} = (\mathbf{k}_r \cdot \mathbf{r})_{x=0} = (\mathbf{k}_t \cdot \mathbf{r})_{x=0}. \qquad (3.1\text{-}1)$$

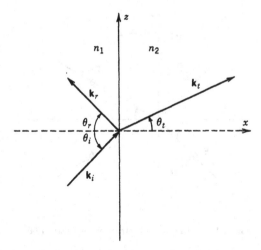

Figure 3.1 Reflection and refraction of plane wave at a boundary between two dielectric media.

Condition (3.1-1) must always be satisfied regardless of the nature of the boundary conditions.

Let n_1 and n_2 be the indices of refraction of medium 1 and 2, respectively. The wave numbers have the magnitudes

$$|\mathbf{k}_i| = |\mathbf{k}_r| = \frac{\omega}{c} n_1, \qquad |\mathbf{k}_t| = \frac{\omega}{c} n_2. \qquad (3.1\text{-}2)$$

Equations (3.1-1) and (3.1-2) yield the kinematic properties of the reflection and refraction. From (3.1-1) we see immediately that all three wave propagation vectors $\mathbf{k}_i$, $\mathbf{k}_r$, and $\mathbf{k}_t$ must lie in a plane. This plane is known as the *plane of incidence* (see Fig. 3.1). Furthermore the tangential components of all three wave vectors must be the same. If θ_i, θ_r, and θ_t are the incident, reflected, and transmitted angles, respectively, of the wave vectors with respect to the normal of the plane interface, the following relation must hold:

$$n_1 \sin \theta_i = n_1 \sin \theta_r = n_2 \sin \theta_t. \qquad (3.1\text{-}3)$$

This implies that the angle of reflection must equal the angle of incidence ($\theta_r = \theta_i$) as well as Snell's law,

$$\frac{\sin \theta_i}{\sin \theta_t} = \frac{n_2}{n_1}. \qquad (3.1\text{-}4)$$

Up to this point we considered the kinematic properties of reflection and refraction that are generally true for many types of wave propagation such

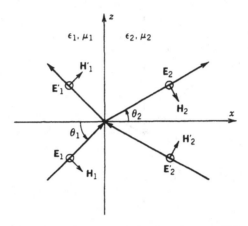

Figure 3.2 Reflection and refraction of s wave (TE).

as light waves, matter waves, and acoustic waves. Dynamical properties, such as the intensities of reflected and transmitted waves, phase changes, and the polarization effect depend entirely on the specific nature of the wave propagation and their boundary conditions. The amplitudes of the reflected and transmitted waves can be expressed in terms of the amplitude of the incident wave by imposing the boundary conditions at the interface $x = 0$.

The coordinate system and symbols appropriate to this problem are shown in Fig. 3.2. The media on both sides of the interface $x = 0$ have the permittivities and permeabilities μ_1, ε_1 and μ_2, ε_2, respectively. A plane wave with wave vector $\mathbf{k}_1$ and frequency ω is incident from medium 1. The transmitted and reflected waves have wave vectors $\mathbf{k}_2$ and $\mathbf{k}_1'$, respectively. The coordinate system is chosen such that the xz plane is the plane of incidence. A general solution of the wave equations (1.4-9) can be taken as the superposition of the incident and reflected waves in each medium:

$$\mathbf{E} = \begin{cases} (\mathbf{E}_1 e^{-i\mathbf{k}_1 \cdot \mathbf{r}} + \mathbf{E}_1' e^{-i\mathbf{k}_1' \cdot \mathbf{r}})e^{i\omega t}, & x < 0, \\ (\mathbf{E}_2 e^{-i\mathbf{k}_2 \cdot \mathbf{r}} + \mathbf{E}_2' e^{-i\mathbf{k}_2' \cdot \mathbf{r}})e^{i\omega t}, & x > 0. \end{cases} \quad (3.1\text{-}5)$$

where $\mathbf{E}_1$, $\mathbf{E}_2$, $\mathbf{E}_1'$, and $\mathbf{E}_2'$ are constant complex vectors; $\mathbf{k}_1$ is the incident wave vector, $\mathbf{k}_2$ is the transmitted wave vector; $\mathbf{k}_1'$ and $\mathbf{k}_2'$ are the mirror images of the wave vectors $\mathbf{k}_1$ and $\mathbf{k}_2$, respectively, with respect to the yz plane, that is, $k_{\alpha x}' = -k_{\alpha x}$, $\alpha = 1, 2$. The magnetic field vector $\mathbf{H}$ can be obtained from

(1.1-1) and (3.1-5):

$$\mathbf{H} = \frac{i}{\omega\mu} \nabla \times \mathbf{E}. \tag{3.1-6}$$

For the case of incidence from the left, $\mathbf{E}_1$ is the incident amplitude, $\mathbf{E}_1'$ is the reflected amplitude, and $\mathbf{E}_2$ is the transmitted amplitude. The amplitude $\mathbf{E}_2'$ is zero because there is no reflected light in the region $x > 0$. In Eq. (3.1-5), we include the term $\mathbf{E}_2'$ so that the result obtained may be generalized in order to treat multilayer structures.

The boundary conditions on the tangential components of the electric and magnetic field vectors require that E_z, E_y, H_z, and H_y be continuous at the interface $x = 0$. In applying these boundary conditions, it is convenient to resolve each vector into components parallel (denoted by subscript p) and perpendicular (denoted by subscript s) to the plane of incidence. These two components of waves (p and s waves) can be shown to be independent of each other, provided media 1 and 2 are homogeneous and isotropic. In other words, the s (or p) maintains its s (or p) character in reflection and refraction. The general case of arbitrary elliptic polarization can be obtained by appropriate linear combination of a p wave and an s wave.

3.1.1 Reflection and Transmission of s Wave (TE Wave)

The s wave is also known as a TE wave because the electric field vector $\mathbf{E}$ is transverse to the plane of incidence. Referring to Fig. 3.2, we consider the reflection and refraction of the s wave. All electric field vectors are perpendicular to the plane of incidence, and the magnetic field vectors are chosen to give a positive energy flow in the direction of the wave vectors. Imposing the continuity of E_y and H_z at the interface $x = 0$ leads to

$$E_{1s} + E_{1s}' = E_{2s} + E_{2s}'$$

$$\sqrt{\frac{\varepsilon_1}{\mu_1}} (E_{1s} - E_{1s}')\cos\theta_1 = \sqrt{\frac{\varepsilon_2}{\mu_2}} (E_{2s} - E_{2s}')\cos\theta_2, \tag{3.1-7}$$

where θ_1 and θ_2 are the angles of the wave vectors $\mathbf{k}_1$ and $\mathbf{k}_2$, respectively, with respect to the normal to the interface. These two equations can be rewritten as a matrix equation:

$$D_s(1) \begin{pmatrix} E_{1s} \\ E_{1s}' \end{pmatrix} = D_s(2) \begin{pmatrix} E_{2s} \\ E_{2s}' \end{pmatrix}, \tag{3.1-8}$$

where

$$D_s(i) = \begin{pmatrix} 1 & 1 \\ \sqrt{\dfrac{\varepsilon_i}{\mu_i}}\cos\theta_i & -\sqrt{\dfrac{\varepsilon_i}{\mu_i}}\cos\theta_i \end{pmatrix}, \qquad i = 1, 2. \quad (3.1\text{-}9)$$

The matrix $D_s(i)$ is called the dynamical matrix of the s wave for the medium i ($i = 1$, 2). If the light is incident from medium 1, the reflection and transmission coefficients are given for a single interface as

$$r_s = \left(\frac{E'_{1s}}{E_{1s}}\right)_{E'_{2s}=0}, \qquad t_s = \left(\frac{E_{2s}}{E_{1s}}\right)_{E'_{2s}=0}. \qquad (3.1\text{-}10)$$

The subscript $E'_{2s} = 0$ indicates that only the transmitted wave E_{2s} exists in medium 2 because the incident wave is launched from medium 1. From the definitions (3.1-10) and the boundary conditions (3.1-7), we obtain

$$r_s = \frac{n_1\cos\theta_1 - n_2\cos\theta_2}{n_1\cos\theta_1 + n_2\cos\theta_2}, \qquad (3.1\text{-}11)$$

$$t_s = \frac{2n_1\cos\theta_1}{n_1\cos\theta_1 + n_2\cos\theta_2}, \qquad (3.1\text{-}12)$$

where we assumed $\mu_2 = \mu_1$, which is generally true for most materials at optical frequencies. The indices of refraction of media 1 and 2 are n_1 and n_2, respectively.

3.1.2   Reflection and Transmission of p Wave (TM Wave)

The p wave is also known as a TM wave because the magnetic field vector **H** is perpendicular to the plane of incidence. Referring to Fig. 3.3, we consider the reflection and refraction of the p wave. All electric field vectors are in the plane of incidence, and the magnetic field vectors are chosen to give a positive energy flow in the direction of propagation. Imposing the continuity of E_z and H_y leads to

$$(E_{1p} + E'_{1p})\cos\theta_1 = (E_{2p} + E'_{2p})\cos\theta_2,$$

$$\sqrt{\frac{\varepsilon_1}{\mu_1}}(E_{1p} - E'_{1p}) = \sqrt{\frac{\varepsilon_2}{\mu_2}}(E_{2p} - E'_{2p}). \qquad (3.1\text{-}13)$$

Again these two equations can be written as

$$D_p(1)\begin{pmatrix} E_{1p} \\ E'_{1p} \end{pmatrix} = D_p(2)\begin{pmatrix} E_{2p} \\ E'_{2p} \end{pmatrix}, \qquad (3.1\text{-}14)$$

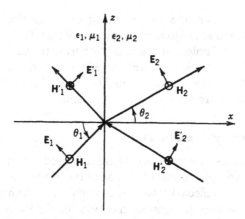

Figure 3.3 Reflection and refraction of p wave (TM).

where

$$D_p(i) = \begin{pmatrix} \cos\theta_i & \cos\theta_i \\ \sqrt{\dfrac{\varepsilon_i}{\mu_i}} & -\sqrt{\dfrac{\varepsilon_i}{\mu_i}} \end{pmatrix}, \qquad i = 1, 2. \qquad (3.1\text{-}15)$$

The matrix $D_p(i)$ is called the dynamical matrix of the p wave for medium i ($i = 1, 2$). If the light is incident from medium 1, the reflection and transmission coefficients for the p wave in the case of a single interface are defined as

$$r_p = \left(\frac{E'_{1p}}{E_{1p}}\right)_{E'_{2p}=0}, \qquad t_p = \left(\frac{E_{2p}}{E_{1p}}\right)_{E'_{2p}=0}. \qquad (3.1\text{-}16)$$

The subscript $E'_{2p} = 0$ indicates that only the transmitted wave E_{2p} exists in medium 2 because the incident p wave is launched from medium 1. From (3.1-16) and (3.1-13), we obtain

$$r_p = \frac{n_1\cos\theta_2 - n_2\cos\theta_1}{n_1\cos\theta_2 + n_2\cos\theta_1}, \qquad (3.1\text{-}17)$$

$$t_p = \frac{2n_1\cos\theta_1}{n_1\cos\theta_2 + n_2\cos\theta_1}, \qquad (3.1\text{-}18)$$

where again we assume $\mu_2 = \mu_1$. Equations (3.1-11), (3.1-12), (3.1-17), and (3.1-18) are known as *Fresnel formulas*, r_s, r_p are the Fresnel reflection

coefficients, and t_s, t_p are the Fresnel transmission coefficients. These formulas are general and apply to any two media. We shall show later that they
are valid even for total reflection and reflection from the surface of absorbing
media such as metals. Notice that we ignored the continuity conditions on **B**
and **D** when we derived the Fresnel formulas. This is legitimate because these
boundary conditions are redundant (see Problem 3.6).

3.1.3   Reflectance and Transmittance

The Fresnel formulas give the ratios of the amplitude of the reflected wave
and the transmitted wave to the amplitude of the incident wave. To find out
how much energy is reflected from the boundary and transmitted into the
second medium, we need to consider the ratios of the Poynting power flow
of the reflected and the transmitted waves to that of the incident wave. The
power flow parallel to the boundary surface is unaffected and is a constant
throughout the medium. Consequently, as far as the reflection and transmission are concerned, we only consider the normal component of the
time-averaged Poynting's vectors of the incident, reflected, and transmitted
waves. Thus, reflectance and transmittance are defined as

$$R_s = \left| \frac{\mathbf{x} \cdot \mathbf{S}'_{1s}}{\mathbf{x} \cdot \mathbf{S}_{1s}} \right|, \qquad R_p = \left| \frac{\mathbf{x} \cdot \mathbf{S}'_{1p}}{\mathbf{x} \cdot \mathbf{S}_{1p}} \right|, \qquad (3.1\text{-}19)$$

$$T_s = \left| \frac{\mathbf{x} \cdot \mathbf{S}_{2s}}{\mathbf{x} \cdot \mathbf{S}_{1s}} \right|, \qquad T_p = \left| \frac{\mathbf{x} \cdot \mathbf{S}_{2p}}{\mathbf{x} \cdot \mathbf{S}_{1p}} \right|, \qquad (3.1\text{-}20)$$

where **S** denotes the time-averaged Poynting's vector, and subscripts 1 and
2 denote media 1 and 2, respectively. The subscripts s and p denote s and p
waves, respectively, and **x** is the unit vector normal to the boundary surface.
The time-averaged Poynting's vector for a plane wave with a real wave vector
is given, according to (1.4-21), by

$$\mathbf{S} = \frac{\mathbf{k}}{2\omega\mu} |\mathbf{E}|^2, \qquad (3.1\text{-}21)$$

where the medium is assumed to be dielectric so that the index of refraction,
n, is a real number, **k** is the wave vector, and **E** is the amplitude of the plane
wave. Thus, according to the definition, reflectance and transmittance are
related to the Fresnel coefficients by the equations

$$R_s = |r_s|^2, \qquad (3.1\text{-}22)$$

$$R_p = |r_p|^2, \qquad (3.1\text{-}23)$$

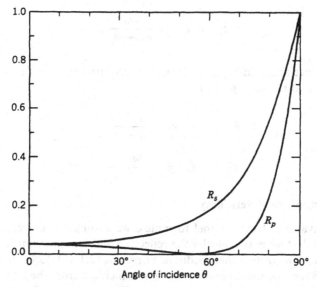

Figure 3.4 The reflectances R_s and R_p versus angle of incidence θ for $n_1 = 1.0$ and $n_2 = 1.5$.

$$T_s = \frac{n_2 \cos \theta_2 |t_s|^2}{n_1 \cos \theta_1},$$

$$T_p = \frac{n_2 \cos \theta_2 |t_p|^2}{n_1 \cos \theta_1}. \qquad (3.1\text{-}24)$$

We must remember that the transmittance formulas (3.1-24) are only valid for pure dielectric media with real n_1 and n_2 and θ_1 and θ_2. It can be shown that Eqs. (3.1-22), (3.1-23) and (3.1-24) are in agreement with the law of conservation of energy, that is $R + T = 1$. The variation of the reflectance R_s and R_p with the angle of incidence is shown in Fig. 3.4 for the case $n_1 = 1.0$, $n_2 = 1.50$. In the above discussion are assumed that the Poynting power in medium 1 can be decomposed into the incident component and the reflected component. This is only true when the medium is lossless (see Problem 3.7).

3.1.4 Normal Incidence

For normal incidence ($\theta_1 = 0$), there is no difference between the s and p waves, and the Fresnel formulas become

$$r_s = r_p = \frac{n_1 - n_2}{n_1 + n_2}, \qquad (3.1\text{-}25)$$

$$t_s = t_p = \frac{2n_1}{n_1 + n_2} . \tag{3.4-26}$$

The reflectance and transmittance for dielectrics with real n_1, n_2 are, according to (3.1-23) and (3.1-24), given by

$$R_s = R_p = \left(\frac{n_1 - n_2}{n_1 + n_2}\right)^2 , \tag{3.1-27}$$

$$T_s = T_p = \frac{4n_1 n_2}{(n_1 + n_2)^2} . \tag{3.1-28}$$

3.1.5 Principle of Reversibility

In the derivation of the Fresnel formulas, we assumed two waves in each medium so that we may consider the reflection and transmission coefficients for light incident from either medium. Formulas (3.1-11), (3.1-12), (3.1-17), and (3.1-18) are for the case when the light is incident from the first medium. These coefficients will be denoted r_{12} and t_{12}. The subscripts 12 mean that the light is incident from medium 1 onto medium 2. The corresponding coefficients for the case when the light is incident from medium 2 onto medium 1 are defined as

$$r_{21} = \left(\frac{E_2}{E_2'}\right)_{E_1 = 0} , \tag{3.1-29}$$

$$t_{21} = \left(\frac{E_1'}{E_2'}\right)_{E_1 = 0} . \tag{3.1-30}$$

It can be shown from Eqs. (3.1-7) and (3.1-13) that the following relation holds for both s wave and p waves, provided that the incident angle θ_2 in medium 2 corresponds to the angle of refraction when the light is incident from medium 1:

$$r_{21} = -r_{12}, \tag{3.1-31}$$

$$\frac{t_{21}}{n_2 \cos \theta_2} = \frac{t_{12}}{n_1 \cos \theta_1} . \tag{3.1-32}$$

Furthermore, it can be shown that

$$t_{12} t_{21} - r_{12} r_{21} = 1. \tag{3.1-33}$$

Equations (3.1-31) and (3.1-32) have been derived by Stokes for the purpose of explaining the perfect blackness of the central spot in Newton's rings. His

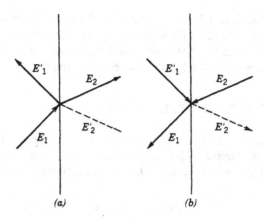

Figure 3.5 Stokes treatment of the principle of reversibility.

argument based on a law of mechanics known as the *principle of reversibility* according to which the result of a time reversal is to cause all the particles to reverse their velocity and retrace their previous motion. The results are obtained from the following argument.

Referring to Fig. 3.5, let E_1 be the incident wave. According to the definition, E_1', E_2, and E_2' are given by

$$E_1' = r_{12}E_1, \qquad E_2 = t_{12}E_1, \qquad E_2' = 0. \qquad (3.1\text{-}34)$$

Now we imagine that time is reversed so that E_2 and E_1' retrace their paths and produce the backward-propagating wave E_1. If we assume that the law of reflection and transmission holds for the time-reversed waves, we expect that the reverse-propagating E_1' and E_2 must produce their own reflected waves and transmitted waves. In other words, we expect that

$$E_1 = E_1'r_{12} + E_2t_{21}, \qquad E_2' = E_1't_{12} + E_2r_{21}. \qquad (3.1\text{-}35)$$

By combining Eqs. (3.1-34) and (3.1-35), we obtain the results (3.1-31) and (3.1-33). These results can also be obtained directly from the Fresnel formulas.

From Eqs. (3.1-31) and (3.1-32), it follows that the Fresnel reflectance and transmittance for incidence $1 \rightarrow 2$ are equal to those of incidence $2 \rightarrow 1$ provided that these two media are dielectrics with real n_1 and n_2 and the incidence angles obey Snell's refraction law ($n_1 \sin \theta_1 = n_2 \sin \theta_2$):

$$R_{12} = R_{21}, \qquad T_{12} = T_{21}. \qquad (3.1\text{-}36)$$

In fact, it will be shown later that the two Fresnel transmittances for an arbitrary layered medium embedded between two dielectrics are equal, whereas no similar relation holds for reflectances unless the layered medium itself is dielectric with real refractive indices (i.e., lossless).

3.2 TOTAL INTERNAL REFLECTION

We now consider the phenomenon of total internal reflection. If the incident medium has a refractive index larger than that of the second medium ($n_1 > n_2$) and if the incidence angle θ is sufficiently large, Snell's law,

$$\sin \theta_2 = \frac{n_1}{n_2} \sin \theta_1,$$

gives values of $\sin \theta_2$ that appear to be absurd since they are greater than unity. The critical angle of incidence, where $\sin \theta_2 = 1$ and $\theta_2 = 90°$, is given by

$$\theta_c = \sin^{-1} \frac{n_2}{n_1}, \qquad n_2 < n_1. \tag{3.2-1}$$

For waves incident from medium 1 at $\theta_1 = \theta_c$, the refracted wave is propagating parallel to the interface. There can be no energy flow across the interface. Therefore, at this angle of incidence, the energy of the light must be totally reflected regardless of the polarization state of the electric field vector **E**. For incident angle $\theta_1 > \theta_c$, $\sin \theta_2 > 1$, this means that θ_2 is a complex angle with a purely imaginary cosine**:

$$\cos \theta_2 = -i \left[\left(\frac{\sin \theta_1}{\sin \theta_c} \right)^2 - 1 \right]^{1/2}. \tag{3.2-2}$$

The negative sign ensures that the transmitted wave decays exponentially as x approaches $+\infty$. The Fresnel formulas for the reflection coefficients (3.1-11) and (3.1-17) become

$$r_s = \frac{n_1 \cos \theta_1 + i n_2 |\cos \theta_2|}{n_1 \cos \theta_1 - i n_2 |\cos \theta_2|}, \tag{3.2-3}$$

$$r_p = \frac{-i n_1 |\cos \theta_2| - n_2 \cos \theta_1}{-i n_1 |\cos \theta_2| + n_2 \cos \theta_1}. \tag{3.2-4}$$

**When $\sin \theta_2 > 1$, the angle θ_2 loses its physical interpretation as the angle of refraction for the transmitted wave. It only means that $k_{2z} = k_2 \sin \theta_2$ and $k_{2x} = k_2 \cos \theta_2$.

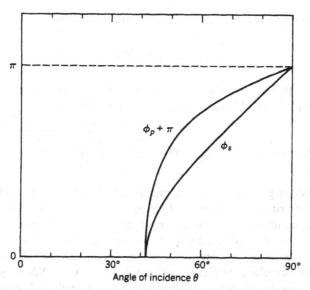

Figure 3.6 The ϕ_p and ϕ_s versus angle of incidence for $n_1 = 1.5$ and $n_2 = 1.0$.

If we examine Eq. (3.2-3) and (3.2-4) carefully, we notice that these two reflection coefficients are complex numbers of unit modulus, that is, $|r_s| = |r_p| = 1$. This means that all the light energies are totally reflected from the surface. The amplitude of the reflected wave is only different from the incident amplitude by a phase factor.

These changes of phase upon total internal reflection can be calculated from (3.2-3) and (3.2-4). We define $r_s = \exp(i\phi_s)$ and $r_p = \exp(i\phi_p)$.

These phases are given by

$$\phi_s = 2 \tan^{-1} \left(\frac{\sin^2 \theta_1 - \sin^2 \theta_c}{1 - \sin^2 \theta_1} \right)^{1/2}, \tag{3.2-5}$$

$$\phi_p = -\pi + 2 \tan^{-1} \left[\left(\frac{\sin^2 \theta_1 - \sin^2 \theta_c}{1 - \sin^2 \theta_1} \right)^{1/2} \left(\frac{n_1}{n_2} \right)^2 \right], \quad n_2 < n_1. \tag{3.2-6}$$

The variation of ϕ_s and ϕ_p versus the angle of incidence are shown in Fig. 3.6. Notice that these two waves undergo different phase changes upon total internal reflection. Linearly polarized light will thus become elliptically polarized upon total reflection. If χ and χ' denote the ellipse of polarization of the incidence wave and the reflected wave, respectively, that is,

$$\chi = \frac{E_{1p}}{E_{1s}}, \quad \chi' = \frac{E'_{1p}}{E'_{1s}} \tag{3.2-7}$$

then according to the definition, χ and χ' are related by

$$\chi' = e^{i(\phi_p - \phi_s)} \chi. \qquad (3.2\text{-}8)$$

The relative phase difference $\Delta \equiv \phi_p - \phi_s$ can be obtained from either (3.2-3) and (3.2-4) or (3.2-5) and (3.2-6) and is given by

$$\Delta = \phi_p - \phi_s = -\pi + 2\tan^{-1} \frac{\cos\theta_1 \sqrt{\sin^2\theta_1 - \sin^2\theta_c}}{\sin^2\theta_1}. \qquad (3.2\text{-}9)$$

This expression gives $\Delta = -\pi$ at grazing incidence $\theta_1 = \frac{1}{2}\pi$ and for incidence at the critical angle ($\theta_1 = \theta_c$) (see Fig. 3.6). Upon total reflection, relative phase difference, Δ may be employed to produce circularly polarized light from linearly polarized light. If the incident wave has a polarization state $\chi = 1$ (i.e., linearly polarized in a direction that has an inclination angle of 45° from the plane of incidence) and $\Delta = \pm\frac{1}{2}\pi$ or $N\Delta = \pm\frac{1}{2}\pi$ (N is number of successive total reflections), the final reflected wave will be circularly polarized (see Problem 3.5).

Total internal reflection is the basis for dielectric waveguiding in layered structures (see Chapter 11). Light waves may also be guided in a dielectric fiber with a higher refractive index than that of the cladding material, provided that the light undergoes total reflection at the boundary and that it also satisfies some mode conditions.

3.2.1 Evanescent Waves

We have shown that when the incident angle is greater than the critical angle θ_c, a wave will be totally reflected from the surface. Reflectances R_s and R_p are equal to unity and transmittances T_s and T_p vanish. Equation (3.1-24) cannot be used because θ_2 is complex. If we examine the Fresnel transmission coefficients t_s (3.1-12) and t_p (3.1-18) at total reflection, we notice that t_s and t_p are not vanishing (i.e., $t_{s,p} \neq 0$). This means that even though the light energies are totally reflected, the electromagnetic fields still penetrate into the second medium. In fact, zero transmission of light energies only means that the normal component of Poynting's vector vanishes, and the power flow is parallel to the boundary surface. It is interesting to calculate the electric field vectors in the second medium and evaluate the Poynting power flow.

The electric field of the transmitted wave is proportional to the real part of the complex quantities

$$\exp\left[i\omega t - ik_2(z\sin\theta_2 + x\cos\theta_2)\right], \qquad (3.2\text{-}10)$$

which, when θ_2 is eliminated by means of Eq. (3.2-2) and Snell's law (3.1-4), takes the form

$$\exp\left[i(\omega t - k_1 z \sin \theta_1) - qx\right], \tag{3.2-11}$$

where

$$q = k_2 \left[\left(\frac{\sin \theta_1}{\sin \theta_c} \right)^2 - 1 \right]^{1/2}, \qquad k_{1,2} = \frac{\omega}{c} n_{1,2}. \tag{3.2-12}$$

We notice that since $\theta_1 > \theta_c$, q is a positive number, and the electric field vector decreases exponentially as x increases, that is, as the distance from the surface increases. Equations (3.2-11) and (3.2-12) also show that the transmitted wave is actually propagating parallel to the boundary surface. The wavefronts are described by the surface $z = $ constant. Such a wave (3.2-11) is called an *evanescent wave*. The attenuation occurs within a distance $\sim q^{-1}$, which, except near $\theta_1 = \theta_c$, is only several wavelengths.

Let E_2 be the amplitude of the evanescent electric field vector in the second medium. The magnetic field vector can be obtained from (3.1-6) and is given by

$$\mathbf{H}_2 = \frac{1}{\omega\mu} \mathbf{k}_2 \times \mathbf{E}_2, \tag{3.2-13}$$

where $\mathbf{k}_2$ is the complex wave vector of the evanescent wave, that is,

$$\mathbf{k}_2 = \frac{\omega}{c} n_2 \left(\mathbf{z} \sin \theta_2 + \mathbf{x} \cos \theta_2 \right), \tag{3.2-14}$$

and $\mathbf{x}$, $\mathbf{z}$ are unit vectors. The x component of this complex wave vector is a pure imaginary number $-iq$. The time-averaged normal component of Poynting's vector in the second medium can be evaluated from (1.3-14) and is given by

$$\mathbf{n} \cdot \mathbf{S} = \frac{1}{2\omega\mu} \operatorname{Re}\left[(\mathbf{n} \cdot \mathbf{k}_2) |E_2|^2 \right] e^{-2qx},$$

where $\mathbf{n}$ is the unit vector normal to the boundary surface. Since $\mathbf{n} \cdot \mathbf{k}_2 = -iq$ is a pure imaginary number, $\mathbf{n} \cdot \mathbf{S} = 0$, and no power is transmitted across the boundary. This is consistent with our earlier result showing that the transmittances are zero at total reflection.

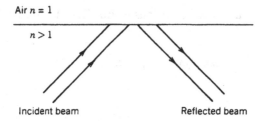

Figure 3.7   The lateral displacement of a beam at total internal reflection. (Goos–Hanchen shift).

3.2.2   Goos–Hanchen Shift

We have just shown that the normal component of Poynting's vector vanishes at total internal reflection. The parallel component S_x, however, is not zero and is given by

$$S_x = \frac{\beta}{2\omega\mu} |E_2|^2 e^{-2qx}, \tag{3.2-15}$$

where β is the z component of the wave vector $\mathbf{k}_1$. This energy is confined near the boundary surface and is created by the incident wave. Here E_2 is the amplitude of the field vector in the second medium and is related to the incident field vector E_1 by the Fresnel formulas (3.1-10), (3.1-16), (3.1-12), and (3.1-18). Thus, if an optical beam is incident at an angle greater than the critical angle θ_c, light will penetrate into the second medium with a depth of penetration on the order of $1/q$. From the energy point of view, the penetration is necessary in order to create the evanescent field. Because of this penetration, a totally reflected optical beam of finite cross section will be displaced laterally relative to the incident beam at the boundary surface (see Fig. 3.7). This shift has been observed and is known as the *Goos–Hanchen shift* [2, 3].

In what follows, we will present a simple picture enabling us to compute this lateral shift. For the sake of clarity, we will treat only two-dimensional problems and neglect the y dependence (recall that the y axis is normal to the plane of incidence). In order to compute the lateral shift, we need to compare the beam profile of both the reflected wave and the incident wave at the plane $x = 0$. The electric field (either s or p wave) of the incident beam can be represented as a Fourier integral and is of the form

$$E_i = \int A(\beta) e^{i(\omega t - \beta z - k_x x)} \, d\beta, \tag{3.2-16}$$

where

$$\beta = k_1 \sin \theta_1, \tag{3.2-17}$$

$$k_x = k_1 \cos \theta_1 = \left[\left(\frac{\omega}{c} n_1 \right)^2 - \beta^2 \right]^{1/2}. \tag{3.2-18}$$

We use β instead of θ_1 as the variable of integration for the purpose of simplifying the algebra. The Fresnel reflection coefficients r_s and r_p can be expressed in terms of β implicitly as follows:

$$r_s = \frac{k_x + iq}{k_x - iq}, \tag{3.2-19}$$

$$r_p = \frac{-in_1^2 q - n_2^2 k_x}{-in_1^2 q + n_2^2 k_x}, \tag{3.2-20}$$

where q is given by (3.2-12) and is expressed in terms of β as

$$q = \left[\beta^2 - \left(\frac{\omega}{c} n_2 \right)^2 \right]^{1/2}. \tag{3.2-21}$$

Let the reflection coefficient (either s or p wave) be of the form

$$r = \varrho(\beta) e^{i\phi(\beta)}, \tag{3.2-22}$$

where $\varrho(\beta)$ and $\phi(\beta)$ are functions of β and consequently are functions of the angle of incidence θ_1. The electric field of the reflected beam can thus be written as

$$E_r = \int A(\beta) \varrho(\beta) e^{i\phi(\beta)} e^{i(\omega t - \beta z + k_x x)} \, d\beta. \tag{3.2-23}$$

We will now assume that the divergence angle of the incident beam is small, so that $A(\beta)$ is sharply peaked around β_0 with a corresponding angle of incidence $\theta_0 > \theta_c$. We also assume that $A(\beta)$ is so sharply peaked that all the Fourier components are totally reflected. In other words, $\varrho(\beta)$ remains unity throughout the range of integration in (3.2-23). Under these conditions, the electric fields of the incidence wave and the reflected wave at the boundary plane $x = 0$ are given by, respectively,

$$E_i = e^{i(\omega t - \beta_0 z)} \int A(\beta) e^{-i(\beta - \beta_0)z} \, d\beta, \tag{3.2-24}$$

$$E_r = e^{i(\omega t - \beta_0 z)} \int A(\beta) e^{-i(\beta - \beta_0)z + i\phi(\beta)} \, d\beta. \tag{3.2-25}$$

If we examine these two Fourier representations of the fields, we find that a beam shift or profile change results from the dispersion of the phase shift $\phi(\beta)$ (i.e., variation of ϕ with β). To calculate qualitatively the effect of phase dispersion on the reflected beam shape, we need to evaluate the integral (3.2-25). Since $A(\beta)$ is only significant in a small range around β_0, one may expand $\phi(\beta)$ around the value β_0 in terms of a Taylor series:

$$\phi(\beta) = \phi(\beta_0) + \left(\frac{\partial\phi}{\partial\beta}\right)_0 (\beta - \beta_0) + \cdots, \qquad (3.2\text{-}26)$$

where $(\partial\phi/\partial\beta)_0$ is the derivative of ϕ with respect to β at β_0. Then we can substitute for $\phi(\beta)$ from (3.2-26) into (3.2-25). This leads to

$$E_r = e^{i[\omega t - \beta_0 z + \phi(\beta_0)]} \int A(\beta) e^{-i(\beta - \beta_0)[z - (\partial\phi/\partial\beta)_0]} \, d\beta, \qquad (3.2\text{-}27)$$

where we neglected the higher order terms in $\beta - \beta_0$. We now define a beam profile function for the incident wave at the boundary plane ($x = 0$) as

$$F(z) = \int A(\beta) e^{-i(\beta - \beta_0)z} \, d\beta \qquad (3.2\text{-}28)$$

so that the incidence optical beam at the boundary plane ($x = 0$) can be written as

$$E_i = F(z) e^{i(\omega t - \beta_0 z)}. \qquad (3.2\text{-}29)$$

By using definition (3.2-28) and from (3.2-27), the reflected beam can be written as

$$E_r = F\left[z - \left(\frac{\partial\phi}{\partial\beta}\right)_0\right] e^{i\phi(\beta_0)} \, e^{i(\omega t - \beta_0 z)} . \qquad (3.2\text{-}30)$$

This equation shows that, upon reflection, the beam is displaced along the boundary surface by an amount

$$d = \left(\frac{\partial\phi}{\partial\beta}\right)_0. \qquad (3.2\text{-}31)$$

The beam remains undistorted, provided approximation (3.2-26) holds.

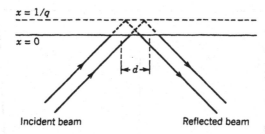

Figure 3.8 Goos–Hanchen shift and the energy penetration.

We are now ready to evaluate the Goos–Hanchen shift for both s and p waves at total internal reflection. Upon total reflection, the phase changes for plane waves can be expressed in terms of β as

$$\phi_s = 2 \tan^{-1} \left(\frac{\beta^2 - k_2^2}{k_1^2 - \beta^2} \right)^{1/2} \tag{3.2-32}$$

$$\phi_p = -\pi + 2 \tan^{-1} \left[\left(\frac{\beta^2 - k_2^2}{k_1^2 - \beta^2} \right)^{1/2} \left(\frac{n_1}{n_2} \right)^2 \right], \tag{3.2-33}$$

for s and p waves, respectively. We remember that $n_2 < n_1$ and $k_2 < \beta < k_1$. According to (3.2-31), beam displacements d_s and d_p for s and p waves, respectively, are obtained directly from (3.2-32) and (3.2-33) by simply differentiating with respect to β. Thus, we obtain the following expressions for the Goos–Hanchen shift:

$$d_s = 2 \frac{1}{q} \tan \theta_1, \tag{3.2-34}$$

$$d_p = \frac{2(1/q) \tan \theta_1}{(\sin \theta_1 / \sin \theta_c)^2 - \cos^2 \theta_1}. \tag{3.2-35}$$

In these formulas, θ_1 is the incident angle, and $1/q$ is the distance from the boundary where the amplitude of the evanescent wave decreases to $1/e$ of its value at the surface ($x = 0$).

From the expressions of lateral displacement [(3.2-34) and (3.2-35)], we see that the incidence beam actually penetrates into the second medium without deviation and is reflected at $x = 1/q$ for the s wave and at $z = (1/q)/[\sin \theta_1 / \sin \theta_c)^2 - \cos^2 \theta_1]$ for the p wave (see Fig. 3.8). The variation of the Goos–Hanchen shift with the angle of incidence is shown in Fig. 3.9. We see that the displacement of the beam is greater the closer its angle of incidence is to the critical angle θ_c and to the grazing angle $\theta_1 = \frac{1}{2}\pi$.

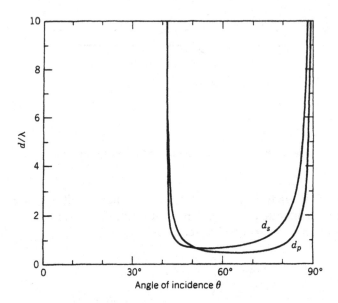

Figure 3.9   The variation of Goos–Hanchen shift with angle of incidence for $n_1 = 1.5$ and $n_2 = 1.0$.

The Goos–Hanchen shift of a light beam upon total internal reflection at a glass–air boundary has been observed [2, 3]. Although the lateral displacement at each reflection may be of several wavelengths, multiple reflections were used in the experiment in order to make the displacements observable.

3.3   POLARIZATION BY REFLECTION; BREWSTER ANGLE

The Fresnel reflection coefficients r_s and r_p vary differently as a function of the angle of incidence (see Fig. 3.3). The reflectance of the s wave (TE wave) R_s is always greater than the reflectance of the p wave (TM wave) R_p (i.e., $R_p < R_s$) except at normal incidence ($\theta_1 = 0$) and grazing incidence ($\theta_1 = \frac{1}{2}\pi$). Furthermore, the Fresnel reflection coefficient r_p vanishes when the incidence angle is such that

$$n_2 \cos \theta_1 = n_1 \cos \theta_2. \tag{3.3-1}$$

If we eliminate θ_2 in (3.3-1) by using the Snell law (3.1-4), we obtain the following expression for this particular angle of incidence:

$$\theta_B = \tan^{-1} \frac{n_2}{n_1}. \tag{3.3-2}$$

This angle of incidence is known as *Brewster's angle*, in honor of the first researcher of this phenomenon. From (3.3-1) and (3.3-2), it can be shown that $r_p = 0$ corresponds to the case when $\theta_1 + \theta_2 = \frac{1}{2}\pi$. In other words, the Fresnel reflectance for the p wave vanishes when the propagation vectors of the transmitted wave and the reflected wave are mutually orthogonal.

If a plane wave with a mixture of s and p waves is incident on the plane interface between two dielectric media at the Brewster angles, the reflected radiation is linearly polarized with the electric field vector perpendicular to the plane of incidence. This phenomenon has been used in many optical systems to produce plane-polarized beams. The transmitted wave is of course predominantly polarized parallel to the plane of incidence because the p wave is totally transmitted. Many laser windows are aligned at the Brewster angle for the purpose of discriminating against the s wave and ensuring oscillation with p polarization.

3.4   REFLECTANCE AT SURFACE OF ABSORBING MEDIUM

We have discussed in detail the reflection and refraction at the interface between two dielectrics. Let us now examine the corresponding phenomena when an electromagnetic radiation is incident onto the surface of an absorbing medium such as metal. The index of refraction of absorbing media is complex. Let $n_2 = n - ik$ be the refractive index of the absorbing medium and $n_1 = 1$ be the index of the incident medium (air). A direct application of the Snell law (3.1-4) leads to

$$\sin \theta_1 = (n - ik) \sin \theta_2, \tag{3.4-1}$$

where θ_1 is the angle of incidence and θ_2 is the angle of refraction. The angle θ_2 is complex and no longer represents the direction of propagation. It is thus convenient to rewrite Snell's law as

$$k_{1z} = \frac{2\pi}{\lambda} \sin \theta_1 = k_{2z} \tag{3.4-2}$$

and the Fresnel formulas (3.1-11) and (3.1-17) as

$$r_s = \frac{k_{1x} - k_{2x}}{k_{1x} + k_{2x}}, \tag{3.4-3}$$

$$r_p = \frac{n_1^2 k_{2x} - n_2^2 k_{1x}}{n_1^2 k_{2x} + n_2^2 k_{1x}}, \tag{3.4-4}$$

where k_{1x} and k_{2x} are the normal (x component) of the wave vectors in medium 1 and 2, respectively. Let β be the tangential (z component) of the

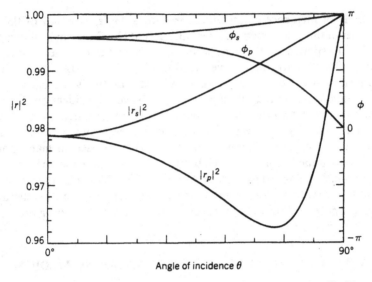

Figure 3.10   Reflectances and phase shifts of silver surface versus angle of incidence.

wave vectors (i.e., $\beta = k_{1z} = 2\pi \sin \theta_1/\lambda = \omega \sin \theta_1/c$). Then k_{1x} and k_{2x} can be written as

$$k_{1x} = \left[\left(\frac{\omega}{c} n_1\right)^2 - \beta^2\right]^{1/2} = \left[\left(\frac{\omega}{c}\right)^2 - \beta^2\right]^{1/2}, \qquad (3.4\text{-}5)$$

$$k_{2x} = \left[\left(\frac{\omega}{c} n_2\right)^2 - \beta^2\right]^{1/2} = \left[\left(\frac{\omega}{c}\right)^2 (n - ik)^2 - \beta^2\right]^{1/2}. \qquad (3.4\text{-}6)$$

At normal incidence, the reflection coefficients (3.4-3) and (3.4-4) become

$$r_s = r_p = \frac{1 - (n - ik)}{1 + (n - ik)}. \qquad (3.4\text{-}7)$$

The reflectance at normal incidence is thus

$$R = |r_s|^2 = |r_p|^2 = \frac{(1 - n)^2 + k^2}{(1 + n)^2 + k^2}, \qquad (3.4\text{-}8)$$

and the phase shift upon reflection is

$$\phi = \tan^{-1} \frac{2k}{1 - n^2 - k^2}. \qquad (3.4\text{-}9)$$

For silver at $\lambda = 0.5\,\mu m$, $n = 0.05$, and $k = 2.87$ (see Table 2.1), Eqs. (3.4-8) and (3.4-9) yield $R = 0.98$ and $\phi = 142°$. For metals with $n \to 0$ or $(n - ik) \to \infty$, the reflectance approaches unity.

Figure 3.10 shows the reflectances $|r_s|^2$ and $|r_p|^2$ as functions of angle of incidence as well as the phase shifts upon reflectance. Note that $|r_s|^2$ is a monotonically increasing function of θ, whereas $|r_p|^2$ reaches a minimum at certain angles of incident.

REFERENCES AND SUGGESTED READINGS

1. M. Born, and E. Wolf, *Principles of Optics*, 4th ed., Pergamon, New York, 1970.
2. L. M. Brekhovolskikh, *Waves in Layered Media*, Academic, New York, 1960.
3. J. A. Dobrowolski, Coatings and filters, in *Handbook of Optics*, W. G. Driscoll, ed., McGraw-Hill, New York, 1978.
4. O. S. Heavens, *Optical Properties of Thin Solid Films*, Dover, New York, 1965.
5. Z. Knittl, *Optics of Thin Films*, Wiley, New York, 1976.
6. H. A. Macleod, *Thin Film Optical Filters*, American Elsevier Pub. Co, New York, 1969.

PROBLEMS

3.1 *Reflectance for unpolarized light.* The reflectance for an unpolarized light incident on a transparent dielectric surface is the average of the reflectances for s- and p-polarized light, that is,

$$R = \tfrac{1}{2}(R_s + R_p).$$

Let the medium of incidence be air with a refractive index $n_{air} = 1$ and n be the index of refraction of the dielectric.

(a) Show that the average reflectance at the Brewster angle is

$$R_B = \frac{1}{2}\left(\frac{1 - n^2}{1 + n^2}\right)^2.$$

(b) Let R_0 be the reflectance at normal incidence. Compare R_B and R_0 by evaluating the ratio R_B/R_0 at various values of n: $n = 1.5, 2.0, 4.0$.

(c) Find the value of n when $R_B = R_0$.

(d) Show that for most optical materials R is an increasing function of the angle of incidence.

3.2 *Beam displacement by plane-parallel plate.* When a well-collimated laser beam traverses a glass plate with plane surfaces that are parallel to each other, it emerges to its original direction but with a lateral displacement *d*.

(a) Show that

$$d = t \sin \theta \left(1 - \frac{\cos \theta}{n \cos \theta'}\right) = t \sin \theta \left(1 - \frac{\cos \theta}{\sqrt{n^2 - \sin^2 \theta}}\right),$$

where *t* is the thickness of the plate, *n* is the refractive index of the plate, θ' is the angle of refraction in the plate and θ is the angle of incidence.

(b) Show that *d* is an increasing function of θ.

(c) Evaluate *d* at normal incidence, Brewster angle, and grazing incidence ($\theta \simeq 90°$).

3.3 *Refractometers.* A refractometer is used to determine the bulk refractive index. One of the most accurate refractometers is based on the method of minimum deviation by a prism.

(a) Show that minimum deviation occurs at the angle of incidence when the refracted ray inside the prism makes equal angles with the two prism surfaces.

(b) Let α be the apex angle and δ be the angle of minimum deviation. Show that

$$n = \frac{\sin \frac{1}{2}(\alpha + \delta)}{\sin \frac{1}{2}\alpha}.$$

This equation is used to determine the refractive index *n* to a Δn accurate to $\sim 10^{-4}$.

(c) If the right side of the prism is coated with a thick layer of silver, show that the light ray will retrace back when the incident angle θ satisfies the condition

$$n = \left(\frac{\sin \theta}{\sin \alpha}\right).$$

This equation can also be used to determine the index of refraction *n*.

3.4 *Phase change upon total reflection*

(a) Derive (3.2-5), (3.2-6), and (3.2-9).

(b) Show that the relative phase differences Δ is $-\pi$ for grazing incidence $\theta_1 = \frac{1}{2}\pi$ as well as for the critical angle θ_c; but for intermediate values of the angle of incidence, it is not $-\pi$: that is, the reflected light is elliptically polarized when the incident light is plane polarized.

(c) Differentiate (3.2-9) with respect to θ_1 and show that the relative difference of phase Δ is maximized when the angle of incidence θ_1 satisfies the equation

$$\sin^2 \theta_1 = \frac{2n_2^2}{n_1^2 + n_2^2} \qquad (n_2 < n_1).$$

(d) Show that the maximum value of Δ is given by

$$\Delta_m = -\pi + 2 \tan^{-1} \frac{n_1^2 - n_2^2}{2n_1 n_2}.$$

3.5 *Fresnel rhomb.* The phase change produced upon total reflection may be utilized to obtain circularly polarized light. The scheme devised by Fresnel is as follows:

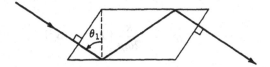

In order to transform a plane-polarized light into one that is circularly polarized, each total reflection must produce a relative phase change of $\pm\frac{1}{4}\pi$ or $\pm\frac{1}{4}\pi + N\pi$ so that the net phase change is $\pm\frac{1}{2}\pi$ upon two total reflections. Also, the incident beam and the emergent beam must be perpendicular to the rhomb surfaces.

(a) For glass with index $n_1 = 1.51$, solve, numerically, Eq. (3.2-9) for θ_1, with $\Delta = -\frac{1}{4}\pi$. *Answer:* $\theta_1 = 48°37'$ or $54°37'$.

(b) In general, a Fresnel rhomb has a very small angular aperture; that is, a slightly divergent (or convergent) beam may not be fully converted to a circularly polarized beam. To improve the acceptance angle, it is desirable to design the rhomb such that $\Delta_m = -\frac{1}{4}\pi$ so that the first-order deviation is zero [i.e., $(\partial\Delta/\partial\theta) = 0$]. Find the required index of refraction for the rhomb and the angle of the rhomb. *Answer:* $n = 1.4966$, $\theta = 51°47'$

(c) From a quantum-mechanical point of view, each circularly polarized photon carries an angular momentum $\hbar$. Therefore, according to the conservation of angular momentum, the rhomb must gain an angular momentum $\hbar$ for each circularly polarized photon emerging from the rhomb. Derive an expression for the torque experienced by the rhomb in terms of the beam intensity.

3.6 *Boundary conditions.*

(a) Show that for s-waves the continuity of B_x leads to the second equation of (3.1-7), using Snell's law.

(b) Show that for p-waves the continuity of D_x leads to the second equation of (3.1-13), using Snell's law.

(c) Show that for harmonic electromagnetic fields with a time dependence of $\exp(i\omega t)$, Eqs. (1.1-3) and (1.1-4) can be derived directly from Eqs. (1.1-1) and (1.1-2). Thus the continuity conditions on the normal components of **D** and **B** are redundant.

3.7 *Poynting's power orthogonality.* Consider an electric field

$$\mathbf{E} = \mathbf{A}e^{-ikz} + \mathbf{B}e^{ikz}$$

(a) Derive an expression for the magnetic field **H**.

(b) Show that the Poynting power along z-axis can be written

$$S = \frac{k}{2\omega\mu}[|\mathbf{A}|^2 - |\mathbf{B}|^2]$$

provided the medium is lossless (i.e., k is real).

(c) Derive the power flow along z-axis in a lossy medium with a complex k. Show that the Poynting power is *not* the algebraic sum of the Poynting power carried by the individual waves.

4

Optics of a Single Homogeneous and Isotropic Layer

In this chapter, we shall treat the optics of the simplest layered structure —a single homogeneous and isotropic layer sandwiched between two semi-infinite media. The bounding media are also isotropic and homogeneous. Specifically, we shall investigate transmittance, reflectance, and absorptance as well as phase shifts associated with the reflection and transmission of electromagnetic radiation.

4.1   ELECTROMAGNETIC TREATMENT

Referring to Fig. 4.1, we consider the reflection and transmission of electro-magnetic radiation at a thin dielectric layer between two semi-infinite media. We assume that all the media are homogeneous and isotropic, so that the whole structure can be described by

$$
n(x) = \begin{cases} n_1, & x < 0, \\ n_2, & 0 < x < d, \\ n_3 & d < x. \end{cases} \tag{4.1-1}
$$

If we further assume that the plane of incidence is the xz plane, the electric field vector of a general plane-wave solution of the wave equation can be of the form

$$
\mathbf{E}(x) \exp\left[i(\omega t - \beta z)\right] \tag{4.1-2}
$$

since the medium is homogeneous in the z direction.

The parameter β is the z component of the propagation wave vector. Since the field vectors must satisfy the boundary conditions at the interfaces, β must have the same value in all the layers so that similar to the case of a single

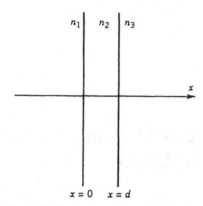

Figure 4.1 A thin homogenous layer of dielectric material.

interface, Snell's law is obeyed at each interface. If we assume that a plane wave is incident from the left, the electric field vector $\mathbf{E}(x)$ can be written as

$$
E(x) = \begin{cases}
Ae^{-ik_{1x}x} + Be^{ik_{1x}x}, & x < 0, \\
Ce^{-ik_{2x}x} + De^{ik_{2x}x}, & 0 < x < d, \\
Fe^{-ik_{3x}(x-d)}, & d < x,
\end{cases} \qquad (4.1\text{-}3)
$$

where we assume that the electric field vector is s polarized (i.e., perpendicular to the plane of incidence). The complex amplitudes A, B, C, D, and F are constants, and k_{1x}, k_{2x}, and k_{3x} are the x components of the wave vectors:

$$
k_{ix} = \left[\left(\frac{n_i \omega}{c}\right)^2 - \beta^2\right]^{1/2} = \left(\frac{\omega}{c}\right) n_i \cos\theta_i, \qquad i = 1, 2, 3, \quad (4.1\text{-}4)
$$

where θ_i is the ray angle measured from the x axis. The constant A is the amplitude of the incident wave. The constants B and F are amplitudes of the reflected and transmitted waves, respectively.

The magnetic field can be obtained from Eq. (3.1-6). The boundary conditions on the tangential components of the electric, $\mathbf{E}$, and magnetic, $\mathbf{H}$, field vectors require that E_y, E_z, H_y, and H_z must be continuous at the interfaces $x = 0$ and $x = d$. Using these boundary conditions, it is possible to obtain the amplitudes C, D, B, and F in terms of A. In applying these boundary conditions, we have to consider the s (TE) wave and the p (TM) wave separately. These two components of the electromagnetic wave are independent of each other. There is no coupling between these two components throughout the medium since every part of the medium is isotropic and homogeneous.

For the case of the s wave, the boundary conditions require that E_y and H_z be continuous at the interfaces. The electric field $\mathbf{E}(x)$ is $\mathbf{y}E(x)$, with $E(x)$ given by Eq. (4.1-3). The z component of the magnetic field is obtained by using $H_z = (i/\omega\mu)(\partial E/\partial x)$ and is given by

$$
H_z(x) = \begin{cases}
\dfrac{k_{1x}}{\omega\mu}(Ae^{-ik_{1x}x} - Be^{ik_{1x}x}), & x < 0, \\[2ex]
\dfrac{k_{2x}}{\omega\mu}(Ce^{-ik_{2x}x} - De^{ik_{2x}x}), & 0 < x < d, \quad (4.1\text{-}5) \\[2ex]
\dfrac{k_{3x}}{\omega\mu}Fe^{-ik_{3x}(x-d)}, & d < x,
\end{cases}
$$

where μ is the magnetic permeability constant.

Imposing the continuity of E_y and H_z at the interfaces $x = 0$ and $x = d$ leads to

$$A + B = C + D \qquad (4.1\text{-}6)$$

$$k_{1x}(A - B) = k_{2x}(C - D), \qquad (4.1\text{-}7)$$

$$Ce^{-ik_{2x}d} + De^{ik_{2x}d} = F, \qquad (4.1\text{-}8)$$

$$k_{2x}(Ce^{-ik_{2x}d} - De^{ik_{2x}d}) = k_{3x}F. \qquad (4.1\text{-}9)$$

These four equations can be used to solve for B, C, D, and F in terms of A. After a few steps of algebraic manipulation, we obtain

$$F = A\frac{4k_{1x}k_{2x}e^{-ik_{2x}d}}{(k_{1x} + k_{2x})(k_{2x} + k_{3x}) + (k_{1x} - k_{2x})(k_{2x} - k_{3x})e^{-2ik_{2x}d}} \qquad (4.1\text{-}10)$$

and

$$B = A\frac{(k_{1x} - k_{2x})(k_{2x} + k_{3x}) + (k_{1x} + k_{2x})(k_{2x} - k_{3x})e^{-2ik_{2x}d}}{(k_{1x} + k_{2x})(k_{2x} + k_{3x}) + (k_{1x} - k_{2x})(k_{2x} - k_{3x})e^{-2ik_{2x}d}} \qquad (4.1\text{-}11)$$

Here C and D can be written in terms of F as

$$C = \tfrac{1}{2}F\left(1 + \frac{k_{3x}}{k_{2x}}\right)e^{ik_{2x}d} \qquad (4.1\text{-}12)$$

and

$$D = \tfrac{1}{2}F\left(1 - \frac{k_{3x}}{k_{2x}}\right)e^{-ik_{2x}d}, \qquad (4.1\text{-}13)$$

respectively.

If we recall that $k_{ix} = (\omega/c)n_i\cos\theta_i$ [see Eq. (4.1-4)], we can write the Fresnel reflection and transmission coefficients of the dielectric interfaces for s waves as

$$r_{12} = \frac{k_{1x} - k_{2x}}{k_{1x} + k_{2x}}, \tag{4.1-14}$$

$$r_{23} = \frac{k_{2x} - k_{3x}}{k_{2x} + k_{3x}}, \tag{4.1-15}$$

and

$$t_{12} = \frac{2k_{1x}}{k_{1x} + k_{2x}}, \tag{4.1-16}$$

$$t_{23} = \frac{2k_{2x}}{k_{2x} + k_{3x}}, \tag{4.1-17}$$

respectively. Using these formulas, the transmission and reflection coefficients can be written as

$$t \equiv \frac{F}{A} = \frac{t_{12}t_{23}e^{-i\phi}}{1 + r_{12}r_{23}e^{-2i\phi}} \tag{4.1-18}$$

and

$$r \equiv \frac{B}{A} = \frac{r_{12} + r_{23}e^{-2i\phi}}{1 + r_{12}r_{23}e^{-2i\phi}}, \tag{4.1-19}$$

respectively. The parameter ϕ in Eqs. (4.1-18) and (4.1-19) is given by

$$\phi = k_{2x}d = \frac{2\pi d}{\lambda}n_2\cos\theta_2 \tag{4.1-20}$$

and is proportional to the thickness d and index n_2 of the layer.

A similar electromagnetic analysis for the p (or TM) wave leads to exactly the same expressions [(4.1-18) and (4.1-19)] for the transmission and reflection coefficients, respectively, except that the coefficients t_{12}, t_{23} and r_{12}, r_{23} must be those associated with the p waves.

4.2 AIRY'S FORMULAS

The expressions for transmission and reflection coefficients [(4.1-18) and (4.1-19)] can also be derived by summing the amplitudes of successive

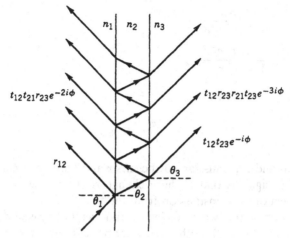

Figure 4.2 Airy summation.

reflections and refractions. Such derivation was first carried out by G. B. Airy in 1833. His method is described as follows. Consider a beam incident from the left (see Fig. 4.2). The incident beam is partially reflected and partially transmitted at the first interface. The transmitted part is subsequently reflected back and forth between the two interfaces as shown. The reflection and transmission coefficients can be obtained by adding the amplitudes of the successive reflected and transmitted rays. In doing this, it is important to include the phase factor that accounts for the geometric path difference between any two successive reflected or transmitted rays. This phase difference is given by -2ϕ, with ϕ given by Eq. (4.1-20). Let the amplitude of the incident wave be 1. Taking the phase difference into account as a factor $e^{-2i\phi}$ and adding the amplitudes of the reflected rays, we obtain

$$
\begin{aligned}
r &= r_{12} + t_{12}t_{21}r_{23}e^{-2i\phi} + t_{12}t_{21}r_{23}r_{21}r_{23}e^{-4i\phi} + \dots \\
&= r_{12} + \frac{t_{12}t_{21}r_{23}e^{-2i\phi}}{1 - r_{21}r_{23}e^{-2i\phi}} \\
&= \frac{r_{12} + r_{23}e^{-2i\phi}}{1 + r_{12}r_{23}e^{-2i\phi}},
\end{aligned}
$$

$$(4.2\text{-}1)$$

$$(4.2\text{-}2)$$

where we have used the reversibility conditions [(3.1-31) and (3.1-33)]. Similarly, the transmission coefficient t is obtained as

$$
t = t_{12}t_{23}e^{-i\phi}[1 + r_{23}r_{21}e^{-2i\phi} + (r_{23}r_{21}e^{-2i\phi})^2 + \dots]
$$

$$= \frac{t_{12}t_{23}e^{-i\phi}}{1 - r_{21}r_{23}e^{-2i\phi}} \tag{4.2-3}$$

$$= \frac{t_{12}t_{23}e^{-i\phi}}{1 + r_{12}r_{23}e^{-2i\phi}}. \tag{4.2-4}$$

We recall that ϕ is given by

$$\phi = \frac{2\pi}{\lambda} n_2 d \cos\theta_2. \tag{4.2-5}$$

The reflection and transmission coefficients r and t are, in general, complex numbers. This signifies that a phase change is experienced by the optical beam on reflection or transmission from the film.

We now examine the two coefficients r and t as functions of ϕ. According to Eqs. (4.2-2) and (4.2-4), both r and t are periodic functions of ϕ. Substituting $\phi + \pi$ for ϕ in these expressions, we obtain

$$r(\phi + \pi) = r(\phi), \tag{4.2-6}$$

$$t(\phi + \pi) = -t(\phi). \tag{4.2-7}$$

In the limit when the thickness of the layer becomes zero (i.e., $d \to 0$), the reflection and transmission coefficients should become those of the interface between media 1 and 3. In other words,

$$r(0) = r_{13}, \tag{4.2-8}$$

$$t(0) = t_{13}. \tag{4.2-9}$$

These two equations can also be derived from Eqs. (4.2-2) and (4.2-4) using $\phi = 0$ and expressions for t_{12}, t_{23} and r_{12}, r_{23}.

According to Eqs. (4.2-6)–(4.2-9), we note that

$$r(\pi) = r_{13}, \tag{4.2-10}$$

$$t(\pi) = -t_{13}. \tag{4.2-11}$$

In other words, the presence of a half-wave layer with $\phi = \pi$ does not affect the reflection and transmission of light except for a possible change of sign. We must be careful that $\phi = \pi$ only occurs at a single wavelength. Half-wave ($\frac{1}{2}\lambda$) layers are historically known as *latent layers*.

4.2.1   An Alternative Derivation

The formulas for reflection and transmission coefficients, as well as for the amplitude of waves in thin film, can be derived directly from the definition

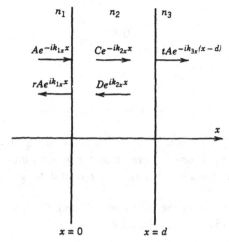

Figure 4.3 Alternative derivation of Airy's formula.

of $r_{12}, t_{12}, r_{21}, t_{21}, r_{23}$, and t_{23}. Referring to Fig. 4.3, we write the reflected wave as $rA \exp(ik_{1x}x)$ and the transmitted wave as $tA \exp[-ik_{3x}(x - d)]$, where we recall that A is the amplitude of the incident wave. At the interface $x = 0$, there are two incoming waves and two outgoing waves. The amplitudes of these waves are related by

$$C = t_{12}A + r_{21}D, \tag{4.2-12}$$

$$rA = r_{12}A + t_{21}D, \tag{4.2-13}$$

where t_{12}, t_{21}, r_{12}, and r_{21} are the transmission and reflection coefficients, respectively. At the interface $x = d$, there is only one incoming wave and two outgoing waves. The amplitudes of these waves are related by

$$tA = t_{23}Ce^{-ik_{2x}d}, \tag{4.2-14}$$

$$De^{ik_{2x}d} = r_{23}Ce^{-ik_{2x}d}, \tag{4.2-15}$$

where t_{23} and r_{23} are the transmission and reflection coefficients, respectively, associated with the interface between n_2 and n_3.

By eliminating A, C, and D from Eqs. (4.2-12)–(4.2-15), we obtain

$$r = r_{12} + \frac{t_{12}t_{21}r_{23}e^{-2i\phi}}{1 - r_{21}r_{23}e^{-2i\phi}} \tag{4.2-16}$$

and

$$t = \frac{t_{12} t_{23} e^{-i\phi}}{1 - r_{21} r_{23} e^{-2i\phi}},$$ (4.2-17)

where ϕ is given by

$$\phi = k_{2x} d = \frac{2\pi}{\lambda} n_2 d \cos\theta_2.$$ (4.2-18)

Note that Eqs. (4.2-16) and (4.2-17) are identical to Eqs. (4.2-1) and (4.2-3). In fact, these two equations are even true for general multilayer structures, provided the parameters r_{12}, t_{12}, r_{21}, t_{21}, r_{23}, and t_{23} are properly defined accordingly.

The field amplitudes in the thin films, C and D, can be obtained from Eqs. (4.2-12)–(4.2-15) and are given by

$$C = \frac{t_{12}}{1 - r_{21} r_{23} e^{-2i\phi}} A$$ (4.2-19)

and

$$D = \frac{t_{12} r_{23} e^{-2i\phi}}{1 - r_{21} r_{23} e^{-2i\phi}} A,$$ (4.2-20)

respectively.

Note that C and D are related by Eq. (4.2-15), which can also be written as

$$D = r_{23} e^{-2i\phi} C.$$ (4.2-21)

Equations (4.2-19)–(4.2-21) are useful when we want to evaluate the field distribution in thin films.

4.3   TRANSMITTANCE, REFLECTANCE, AND ABSORPTANCE

Reflectance is defined as the fraction of energy reflected from the dielectric structure and is given by

$$R = |r|^2.$$ (4.3-1)

Reflectance is meaningful only when medium 1 is nonabsorbing (i.e., n_1 is real, see also Problem 3.7). In addition, if medium 3 is also nonabsorbing, transmittance is given by

$$T = \frac{n_3 \cos\theta_3}{n_1 \cos\theta_1} |t|^2,$$ (4.3-2)

where the factor $n_3 \cos \theta_3 / n_1 \cos \theta_3$ corrects for the difference in phase velocity. Both formulas (4.3-1) and (4.3-2) can be used regardless of whether the layer (medium 2) is absorbing. Absorptance, which is defined as the fraction of energy dissipated, is given by

$$A = 1 - R - T. \qquad (4.3\text{-}3)$$

For dielectric structures with real n_1, n_2, and n_3, it can be shown using the expressions r and t that $R + T = 1$, which agrees with the conservation of energy.

4.4 EXAMPLES

We now examine the electric field distribution, reflectance, and phase shift on reflection when a plane wave is reflected from a thin dielectric film. For the sake of simplicity, we assume that the thin film is suspended in air so that we may set $n_1 = n_3 = 1$ and $n_2 = n$. In addition, we assume that the wave is normally incident, so that

$$r_{21} = r_{23} = \frac{n-1}{n+1} = -r_{12} = -r_{32} \equiv \varrho,$$

$$t_{12} = t_{32} = \frac{2}{n+1}, \qquad (4.4\text{-}1)$$

$$t_{21} = t_{23} = \frac{2n}{n+1}.$$

Using Eqs. (4.4-1), (4.2-2), and (4.2-4), we obtain

$$r = \frac{\varrho(e^{-2i\phi} - 1)}{1 - \varrho^2 e^{-2i\phi}}, \qquad (4.4\text{-}2)$$

$$t = \frac{(1 - \varrho^2)e^{-i\phi}}{1 - \varrho^2 e^{-2i\phi}}, \qquad (4.4\text{-}3)$$

where we recall that ϱ is the reflection coefficient r_{21} (or r_{23}), and we have used $t_{12}t_{23} = 1 - \varrho^2$ in arriving at Eq. (4.4-3). Note that Eqs. (4.4-2) and (4.4-3) obey the law of conservation of energy (i.e., $|r|^2 + |t|^2 = 1$).

The electric field in the region between $x = 0$ and $x = d$ is given by

$$E = Ce^{-ik_{2x}x} + De^{ik_{2x}x}. \qquad (4.4\text{-}4)$$

Using Eqs. (4.2-19), (4.2-20), and (4.4-1), this electric field can be written as

$$E = \frac{(1 - \varrho)A}{1 - \varrho^2 e^{-2i\phi}} [1 + \varrho e^{2i(k_{2x}x - \phi)}]e^{-ik_{2x}x}, \qquad (4.4\text{-}5)$$

where we recall that $\phi = k_{2x}d$ and A is the amplitude of the incident wave. Note that E is a function of x between 0 and d. The magnitude of E is maximum at $x = d$ (i.e., $\partial|E|/\partial x = 0$ at $x = d$). The electric field E at $x = d$ is given by

$$E(x = d) = tA. \qquad (4.4\text{-}6)$$

The same maximum also occurs at

$$x = d - \frac{\pi}{k_{2x}}, d - \frac{2\pi}{k_{2x}}, d - \frac{3\pi}{k_{2x}}, \ldots, \qquad (4.4\text{-}7)$$

provided that the thickness d is large enough.

The magnitude of E is minimum at

$$x = d - \frac{\pi}{2k_{2x}}, d - \frac{3\pi}{2k_{2x}}, d - \frac{5\pi}{2k_{2x}}, \ldots, \qquad (4.4\text{-}8)$$

provided that the thickness d is large enough. The electric field E at these minima is

$$E = \frac{-(1 - \varrho)^2 A}{1 - \varrho^2 e^{-2i\phi}} e^{-2i\phi}. \qquad (4.4\text{-}9)$$

The electric field at the front surface ($x = 0$) is

$$E(x = 0) = \frac{1 - \varrho}{1 - \varrho^2 e^{-2i\phi}} (1 + \varrho e^{-2i\phi})A. \qquad (4.4\text{-}10)$$

Note that the electric field at $x = 0$ is a periodic function of ϕ and varies between A and $(1 - \varrho)^2 A/(1 + \varrho^2)$. Such an electric field is minimum [i.e., $\partial|E(x = 0)|/\partial\phi = 0$] at $\phi = \frac{1}{2}\pi, \frac{3}{2}\pi, \frac{5}{2}\pi, \ldots$ and is maximum at $\phi = 0$, $\pi, 2\pi, 3\pi, \ldots$. These extrema are correlated with the extrema of the transmittance (or reflectance). According to Eqs. (4.4-2) and (4.4-3), r is zero at $\phi = 0, \pi, 2\pi, \ldots$, and $|r|$ is maximum at $\phi = \frac{1}{2}\pi, \frac{3}{2}\pi, \frac{5}{2}\pi, \ldots$.

On the other hand, t is maximum at $\phi = 0, \pi, 2\pi, \ldots$ and is minimum at $\phi = \frac{1}{2}\pi, \phi = \frac{3}{2}\pi$, and so on. These reflection and transmission coefficients, along with the electric field, are listed in Table 4.1.

Table 4.1. Reflection and Transmission Coefficients and Electric Field for Various Values of ϕ

ϕ	r	$|t|$	ψ	$E(x=0)$	$E(x=d)$
0	0	1	0	A	A
$\dfrac{\pi}{2}$	$-\dfrac{2\varrho}{1+\varrho^2}$	$\dfrac{1-\varrho^2}{1+\varrho^2}$	$\dfrac{\pi}{2}$	$\dfrac{(1-\varrho)^2}{1+\varrho^2}A$	$-i\dfrac{1-\varrho^2}{1+\varrho^2}A$
π	0	1	π	A	$-A$
$\dfrac{3\pi}{2}$	$-\dfrac{2\varrho}{1+\varrho^2}$	$\dfrac{1-\varrho^2}{1+\varrho^2}$	$-\dfrac{\pi}{2}$	$\dfrac{(1-\varrho)^2}{1+\varrho^2}A$	$i\dfrac{1-\varrho^2}{1+\varrho^2}A$
2π	0	1	0	A	A

If we define

$$t = |t|e^{-i\psi}, \tag{4.4-11}$$

the phase shift ψ on transmission can be obtained from Eq. (4.4-3) and is given by

$$\psi = \phi + \tan^{-1}\frac{\varrho^2 \sin 2\phi}{1 - \varrho^2 \cos 2\phi}. \tag{4.4-12}$$

The phase shifts ψ at various values of ϕ are also listed in Table 4.1. Figure 4.4 shows the transmittance $|t|^2$ and phase shift ψ as functions of ϕ for the case when $\varrho^2 = 0.6$. Both $|t|^2$ and ψ are periodic functions of ϕ with period 2π.

It is interesting to examine the electric field in the thin film between $x = 0$ and $x = d$. At $\phi = \pi$ (half-wave layer), the transmittance is 1, and the electric field can be written, according to Eq. (4.4-5), as

$$E = \frac{A}{1 + \varrho}[d^{-ik_{2x}x} + \varrho e^{ik_{2x}x}]. \tag{4.4-13}$$

This field has an amplitude of A and varies between $+A$ at $x = 0$ and $-A$ at $x = d$. On the other hand, at $\phi = \frac{1}{2}\pi$, when the transmittance is minimum, the electric field becomes, according to Eq. (4.4-5),

$$E = \frac{1 - \varrho}{1 + \varrho^2} A[e^{-ik_{2x}x} - \varrho e^{ik_{2x}x}]. \tag{4.4-14}$$

This field varies between $(1 - \varrho)^2 A/(1 + \varrho^2)$ at $x = 0$ and $-i(1 - \varrho^2)A/(1 + \varrho^2)$ at $x = d$. For $n = 4$ and hence $\varrho = 0.6$, these field amplitudes are

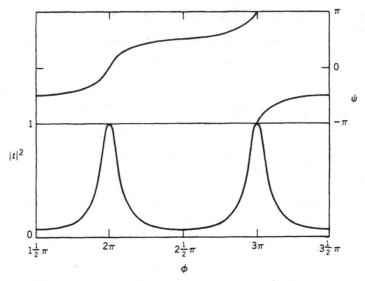

Figure 4.4 $|t|^2$ and ψ versus ϕ for $n = 4.0$ (or $\varrho^2 = 0.6$).

0.12A at $x = 0$ and $-i0.47$ at $x = d$. Note that in the case of resonant transmission ($\phi = \pi$ and $|t|^2 = 1$), the field in the thin film is much stronger compared with the case of minimum transmission ($\phi = \frac{1}{2}\pi$). Figure 4.5 shows the absolute magnitude of the electric field (i.e., $|E|$) in the thin layer.

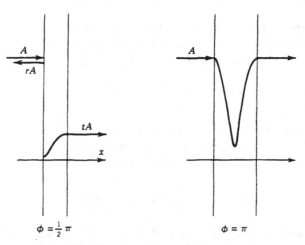

Figure 4.5 $|E|$ versus x in layer for $\phi = \frac{1}{2}\pi$ and π.

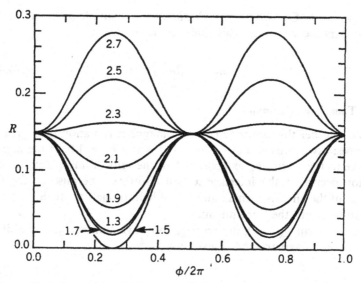

Figure 4.6 Reflectance versus ϕ for $n_1 = 1.0$, $n_2 = 2.25$, and various values of n_2.

4.4.1 A Thin Film on a Substrate (Example)

Referring to Fig. 4.6, we now consider the reflection and transmission of normally incident light onto a thin film deposited on top of a transparent substrate. The formulation of such a problem is described in the previous section. We consider a special case when $n_1 = 1.0$, $n_3 = 2.25$, and n_2 varies from 1.3 to 2.7. Figure 4.6 shows the reflectance as a function of ϕ for these values of n_2. Notice that the reflectance for each given n_2 is a periodic function of ϕ with a minimum occurring at $\phi = \frac{1}{2}\pi, \frac{3}{2}\pi, \ldots$ for $n_2 < n_3$ and at $\phi = \pi, 2\pi, \ldots$ for $n_2 > n_3$. Also notice that for $n_2 = \sqrt{n_1 n_3}$, the minimum at $\phi = \frac{1}{2}\pi, \frac{3}{2}\pi, \ldots$ becomes zero. Thus, a quarter-wave layer with $n = \sqrt{n_1 n_2}$ can be used to completely eliminate the reflection of light. The antireflection coating will be discussed again in Chapter 7.

The antireflection of light can be understood from Eq. (4.2-2). According to this formula, reflection is zero when

$$r_{12} + r_{23}e^{-2i\phi} = 0. \qquad (4.4\text{-}15)$$

For a pure dielectric structure, all the indices are real numbers. The reflection coefficients r_{12}, r_{23} for normal incidence are also real. For the case when $n_1 < n_2 < n_3$, both r_{12} and r_{23} are negative numbers. Thus, ϕ must be $\frac{1}{2}\pi$,

$\frac{3}{2}\pi$, ... to satisfy Eq. (4.4-15). In addition, the magnitude of r_{12} and r_{23} must equal to cancel each order. This leads to the condition

$$n_2 = \sqrt{n_1 n_3}. \qquad (4.4\text{-}16)$$

4.4.2   Tunneling (Example)

We now consider the tunneling of electromagnetic radiation through an air gap between two prisms. Let $n_1 = n_3 = 1.5$ and $n_2 = 1.0$. If the angle of incidence in the prism, θ_1, is greater than the critical angle, total internal reflection occurs at the interface $x = 0$. However, because of the finite thickness of the air gap, a small amount of light tunnels through the air gap and transmits into the third medium.

When $\theta_1 > \sin^{-1}(n_2/n_1)$, the propagation wave number k_{2x} in the air gap becomes pure imaginary and is given by

$$k_{2x} = -i\frac{2\pi}{\lambda}\sqrt{n_1^2 \sin^2 \theta_1 - n_2^2} \equiv iq, \qquad (4.4\text{-}17)$$

where q is the same parameter defined in Eq. (3.2-12).

The term $\exp(-i\phi)$ in Eq. (4.2-4) thus becomes $\exp(-qd)$, which decays exponentially for large d. Thus, the transmittance can be written approximately as

$$T \sim |t_{12}t_{23}|^2 e^{-2qd}, \qquad (4.4\text{-}18)$$

provided $qd \gg 1$. Figure 4.7 shows the transmittance for both s and p waves as a function of d/λ at $\theta_1 = 45°$. Note that the transmittance is insignificant for $d > 2\lambda$. Resonant tunneling of electromagnetic radiation through air gaps will be discussed later in Section 6.5.

4.4.3   A Thin Silver Film (Example)

As another example of using the Airy formulas, we consider the reflection and transmission of electromagnetic radiation through a thin metal film. Using an index of refraction of $n = 0.05 - i2.87$ at $\lambda = 0.5\,\mu m$, we calculate the transmittance, reflectance, and absorptance for normal incidence with a substrate index $n_3 = 1.5$. Figure 4.8 shows the transmittance T, reflectance R, and absorptance $A = 1 - R - T$ as functions of the film thickness.

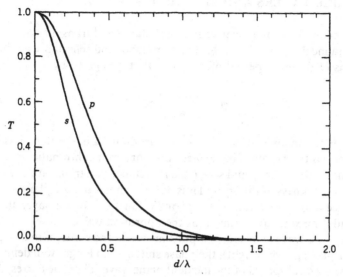

Figure 4.7 T_s and T_p versus d/λ for $n_1 = n_3 = 1.50$, $n_2 = 1.0$, and $\theta_1 = 45°$.

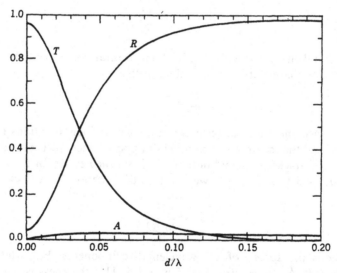

Figure 4.8 Transmittance, reflectance, and absorptance of a thin silver film as functions of thickness.

4.5 THICK LAYERS AND SPECTRAL AVERAGING

In the previous section, we derived the reflectance and transmittance of a thin film for monochromatic light. Both transmittance and reflectance are periodic functions of ϕ with a period of π. Recall that ϕ is given by

$$\phi = \frac{2\pi}{\lambda} nd \cos \theta, \qquad (4.5\text{-}1)$$

where n is the index of refraction of the medium, d is the thickness of the layer, and θ is the ray angle in the medium. For a monochromatic plane wave, ϕ is uniquely determined and so are the reflectance and transmittance regardless of the thickness of the layer. Thus, for a monochromatic plane wave, the reflectance and transmittance are periodic functions of the layer thickness and exhibit maxima and minima at the appropriate thicknesses.

If the plane-wave light source is polychromatic and contains many spectral components (e.g., white light), the phase shift ϕ is no longer well defined. Let $\Delta\lambda$ be the spectral spread of the incident plane wave. Then the corresponding spread of phase shift, $\Delta\phi$, for a given layer thickness is given approximately by

$$\Delta\phi = -\frac{\Delta\lambda}{\lambda} \phi, \qquad (4.5\text{-}2)$$

where λ is taken as the wavelength of the central component. When the spread of the phase shift is greater than π, that is,

$$|\Delta\phi| > \pi, \qquad (4.5\text{-}3)$$

the reflectance and transmittance are no longer periodic functions of layer thickness, and the maxima and minima disappear. Since $\Delta\phi$ is proportional to ϕ (or layer thickness), condition (4.5-3) is satisfied easier for thick layers. Using Eqs. (4.5-1) and (4.5-2), we find that the criterion for a thick layer is

$$d > \frac{\lambda^2}{2n\,\Delta\lambda}, \qquad (4.5\text{-}4)$$

where $\Delta\lambda$ is the spread of the wavelength components. For white light with $\lambda = 6000\,\text{Å}$, $\Delta\lambda = 4000\,\text{Å}$ and $n = 1.5$. This thickness is $0.3\,\mu\text{m}$. In other words, for white light, a glass plate $0.3\,\mu\text{m}$ thick is a thick layer. A polychromatic light with a wavelength spread of $\Delta\lambda$ has a coherence length of approximately $\lambda^2/\Delta\lambda$. Thus, according to Eq. (4.5-4), a layer is thick when the thickness is greater than the coherence length of the source.

4.5.1 Spectral Averaging

To calculate the transmittance (or reflectance) of a thick plate for a poly-chromatic light source, we need to do spectral averaging. Suppose that the spectral distribution is nearly uniform over a range (λ_1, λ_2). Then the average transmittance is given by

$$\langle T \rangle = \frac{1}{\lambda_2 - \lambda_1} \int_{\lambda_1}^{\lambda_2} T(\lambda) \, d\lambda, \tag{4.5-5}$$

where $T(\lambda)$ is the transmittance at λ. Since λ is directly related to ϕ by Eq. (4.5-1), Eq. (4.5-5) is equivalent to

$$\langle T \rangle = \frac{1}{\pi} \int_0^{\pi} T \, d\phi. \tag{4.5-6}$$

Using Eqs. (4.3-2) and (4.2-4), it can be shown that (see Problem 4.3)

$$\langle T \rangle = \frac{T_{12} T_{23}}{1 - R_{12} R_{23}}, \tag{4.5-7}$$

where T_{12}, T_{23} and R_{12}, R_{23} are the transmittance and reflectance of the interfaces, respectively.

PROBLEMS

4.1 *Locus of r on complex plane*

(a) Show that the locus of r Eq. (4.2-2) on the complex plane for various values of ϕ is a circle even for complex r_{12} and r_{23}. Derive the expressions for the location of the center and the radius. *Hint*: Use Eq. (4.2-2) and find a complex constant c such that the absolute value of $r - c$ remains constant. The absolute values gives the radius, and c is the center of the circle.
Answer:

$$c = \frac{r_{12} - r_{12}^* |r_{23}|^2}{1 - |r_{12} r_{23}|^2},$$

$$\text{radius} = |(1 - c r_{12}) r_{23}|.$$

(b) Show that for real r_{12} and r_{23}, the maxima (or minima) of $|r^2|$ occur at $\phi = \frac{1}{2}\pi, \frac{3}{2}\pi, \ldots$ or at $\phi = \pi, 2\pi, 3\pi, \ldots$.

4.2 *Frustrated total reflection.* Referring to the following figure, we consider a low-index thin film of thickness t sandwiched between two
high-index prisms such that total reflections take place at the film–prism
interfaces.

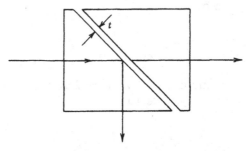

Because of the finite thickness of the low-index film, the evanescent
wave may reach the second surface and cause light to be transmitted
(optical tunneling). Assume that the incident condition is such that
$n_1 \sin \theta_1 > n_2$ and $n_3 = n_1$.

(a) Let $r_{12} = e^{i\phi_{12}}$ and $r_{23} = e^{i\phi_{23}}$, where ϕ_{12} and ϕ_{23} are phase shifts.
Show that the reflectance is

$$R = \frac{e^{2qt} + e^{-2qt} + 2 \cos (\phi_{12} - \phi_{23})}{e^{2qt} + e^{-qt} + 2 \cos (\phi_{12} + \phi_{23})},$$

where q is given by Eq. (4.4-17).

(b) Plot reflectance versus layer thickness (t/λ) for the case of a dense
flint glass prism $(n_1 = n_3 = 1.72)$ and MgF_2 layer $(n_2 = 1.37)$ at an
angle of incidence of 60°. Show that the structure acts as a low-pass
filter, that is, one that transmits long wavelengths only.

4.3 *Spectral averaging and thick layers*

(a) *Spectral averaging.* Show that

$$\langle t^*t \rangle = \frac{1}{\pi} \int_0^\pi t^*t \, d\phi = \frac{|t_{12}t_{23}|^2}{1 - |r_{12}r_{23}|^2}$$

by carrying out the integral. *Hint:* Let $z = e^{-2i\phi}$ and carry out a
contour integration around the unit circle on the complex plane and
use the Cauchy integral formula

$$\oint \frac{f(z)}{z - z_0} \, dz = 2\pi i f(z_0).$$

(b) *Incoherent summation of multiple reflections.* The transmittance may also be obtained by adding, instead of the field amplitudes, the intensities of the multiple reflections:

$$\langle t^*t \rangle = |t_{12}t_{23}|^2 + |t_{12}t_{23}|^2 |r_{12}r_{23}|^2 + |t_{12}t_{23}| |r_{12}r_{23}|^4 + \cdots.$$

Show that $\langle t^*t \rangle$ is identical to that of the spectral averaging (a).

(c) Find the expression of spectral-averaged transmissivity $\langle T \rangle$ for a glass plate with index n at normal incidence. *Answer:* $\langle T \rangle = 2n/(1 + n^2)$.

5

Matrix Formulation for Isotropic Layered Media

The method described in Section 4.1 can be used to solve for the amplitudes of electric and magnetic fields of any isotropic layered media. However, when the number of layers becomes too large, the analysis becomes very complicated because of the large number of equations involved. In this chapter, we will introduce a matrix method that is a systematic approach to solving such a problem. The matrix method is especially useful when a computer is available that can handle the matrix algebra. It is also very useful when a large portion of the structure is periodic. The optics of periodic layered media will be discussed in Chapter 6.

5.1 2 × 2 MATRIX FORMULATION

5.1.1 2 × 2 Matrix Formulation for a Thin Film

We will now revisit the problem of the reflection and transmission of electromagnetic radiation through a thin film using the 2 × 2 matrix method. The dielectric structure is described by (see Fig. 5.1)

$$n(x) = \begin{cases} n_1, & x < 0, \\ n_2, & 0 < x < d, \\ n_3, & d < x, \end{cases}$$

where n_1, n_2, and n_3 are the refractive indices and d is the thickness of the film. Since the whole medium is homogeneous in the z direction (i.e., $\partial n/\partial z = 0$), the electric field that satisfies Maxwell's equations has the form

$$E = E(x)e^{i(\omega t - \beta z)}, \qquad (5.1\text{-}1)$$

where β is the z component of the wave vector and ω is the angular frequency. In Eq. (5.1-1), we assume that the electromagnetic wave is propagating in the

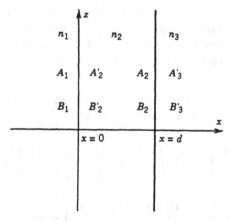

Figure 5.1 A thin layer of dielectric medium.

xz plane, and we further assume that the electric field is either an s wave (with $\mathbf{E} \parallel \mathbf{y}$) or a p wave (with $\mathbf{H} \parallel \mathbf{y}$).

The electric field $E(x)$ consists of a right-traveling wave and a left-traveling wave and can be written as

$$E(x) = Re^{-ik_x x} + Le^{ik_x x} \equiv A(x) + B(x), \qquad (5.1\text{-}2)$$

where $\pm k_x$ are the x components of the wave vector and R and L are constants in each homogeneous layer. Let $A(x)$ represent the amplitude of the right-traveling wave and $B(x)$ be that of the left-traveling one. To illustrate the matrix method, we define

$$
\begin{aligned}
A_1 &= A(0^-),\\
B_1 &= B(0^-),\\
A_2' &= A(0^+),\\
B_2' &= B(0^+),\\
A_2 &= A(d^-),\\
B_2 &= B(d^-),\\
A_3' &= A(d^+),\\
B_3' &= B(d^+),
\end{aligned}
\qquad (5.1\text{-}3)
$$

where 0^- represents the left side of the interface, $x = 0$, and 0^+ represents the right side of the same interface. Similarly, d^- and d^+ are defined for the interface at $x = d$. Note that $E(x)$ for the s wave is a continuous function of x. However, as a result of the decomposition of Eq. (5.1-2), $A(x)$ and $B(x)$ are no longer continuous at the interfaces. If we represent the two amplitudes of $E(x)$ as column vectors, the column vectors shown in Fig. 5.1 are related by

$$
\begin{pmatrix} A_1 \\ B_1 \end{pmatrix} = D_1^{-1} D_2 \begin{pmatrix} A_2' \\ B_2' \end{pmatrix} \equiv D_{12} \begin{pmatrix} A_2' \\ B_2' \end{pmatrix}, \tag{5.1-4}
$$

$$
\begin{pmatrix} A_2' \\ B_2' \end{pmatrix} = P_2 \begin{pmatrix} A_2 \\ B_2 \end{pmatrix} = \begin{pmatrix} e^{i\phi_2} & 0 \\ 0 & e^{-i\phi_2} \end{pmatrix} \begin{pmatrix} A_2 \\ B_2 \end{pmatrix}, \tag{5.1-5}
$$

$$
\begin{pmatrix} A_2 \\ B_2 \end{pmatrix} = D_2^{-1} D_3 \begin{pmatrix} A_3' \\ B_3' \end{pmatrix} \equiv D_{23} \begin{pmatrix} A_3' \\ B_3' \end{pmatrix}, \tag{5.1-6}
$$

where D_1, D_2, and D_3 are the dynamical matrices introduced in Section 3.1 and given by

$$
D_\alpha = \begin{cases} \begin{pmatrix} 1 & 1 \\ n_\alpha \cos \theta_\alpha & -n_\alpha \cos \theta_\alpha \end{pmatrix} & \text{for } s \text{ wave,} \\[2em] \begin{pmatrix} \cos \theta_\alpha & \cos \theta_\alpha \\ n_\alpha & -n_\alpha \end{pmatrix} & \text{for } p \text{ wave,} \end{cases} \tag{5.1-7}
$$

where $\alpha = 1, 2, 3$ and θ_α is the ray angle in each layer and is related to β and $k_{\alpha x}$ by

$$
\beta = n_\alpha \frac{\omega}{c} \sin \theta_\alpha, \qquad k_{\alpha x} = n_\alpha \frac{\omega}{c} \cos \theta_\alpha. \tag{5.1-8}
$$

P_2 is the so-called propagation matrix, which accounts for propagation through the bulk of the layer, and ϕ_2 is given by

$$
\phi_2 = k_{2x} d. \tag{5.1-9}
$$

The matrices D_{12} and D_{23} may be regarded as *transmission matrices* that link the amplitudes of the waves on the two sides of the interfaces and are given by

$$
D_{12} = \begin{pmatrix} \frac{1}{2}\left(1 + \frac{k_{2x}}{k_{1x}}\right) & \frac{1}{2}\left(1 - \frac{k_{2x}}{k_{1x}}\right) \\[1em] \frac{1}{2}\left(1 - \frac{k_{2x}}{k_{1x}}\right) & \frac{1}{2}\left(1 + \frac{k_{2x}}{k_{1x}}\right) \end{pmatrix} \quad \text{for } s \text{ wave} \tag{5.1-10}
$$

and

$$D_{12} = \begin{pmatrix} \frac{1}{2}\left(1 + \frac{n_2^2 k_{1x}}{n_1^2 k_{2x}}\right) & \frac{1}{2}\left(1 - \frac{n_2^2 k_{1x}}{n_1^2 k_{2x}}\right) \\ \frac{1}{2}\left(1 - \frac{n_2^2 k_{1x}}{n_1^2 k_{2x}}\right) & \frac{1}{2}\left(1 + \frac{n_2^2 k_{1x}}{n_1^2 k_{2x}}\right) \end{pmatrix} \quad \text{for } p \text{ wave.} \quad (5.1\text{-}11)$$

The expressions for D_{23} are similar to those of D_{12}, except that the subscript indices need to be replaced with 2 and 3. Equations (5.1-10) and (5.1-11) can be written formally as (see Problem 5.1)

$$D_{12} = \frac{1}{t_{12}}\begin{pmatrix} 1 & r_{12} \\ r_{12} & 1 \end{pmatrix} \quad (5.1\text{-}12)$$

where t_{12} and r_{12} are the Fresnel transmission and reflection coefficients, respectively, and are given by

$$r_{12} = \begin{cases} \dfrac{k_{1x} - k_{2x}}{k_{1x} + k_{2x}} & \text{for } s \text{ wave} \\[3mm] \dfrac{n_1^2 k_{2x} - n_2^2 k_{1x}}{n_1^2 k_{2x} + n_2^2 k_{1x}} & \text{for } p \text{ wave} \end{cases} \quad (5.1\text{-}13)$$

and

$$t_{12} = \begin{cases} \dfrac{2 k_{1x}}{k_{1x} + k_{2x}} & \text{for } s \text{ wave} \\[3mm] \dfrac{2 n_1^2 k_{2x}}{n_1^2 k_{2x} + n_2^2 k_{1x}} & \text{for } p \text{ wave,} \end{cases} \quad (5.1\text{-}14)$$

respectively.

From Eqs. (5.1-4)–(5.1-6), the amplitudes A_1, B_1 and A_3', B_3' are related by

$$\begin{pmatrix} A_1 \\ B_1 \end{pmatrix} = D_1^{-1} D_2 P_2 D_2^{-1} D_3 \begin{pmatrix} A_3' \\ B_3' \end{pmatrix}. \quad (5.1\text{-}15)$$

Note that the column vectors representing the plane-wave amplitudes in each layer are related by a product of 2 × 2 matrices in sequence. Each side of an interface is represented by a dynamical matrix, and the bulk of each layer is

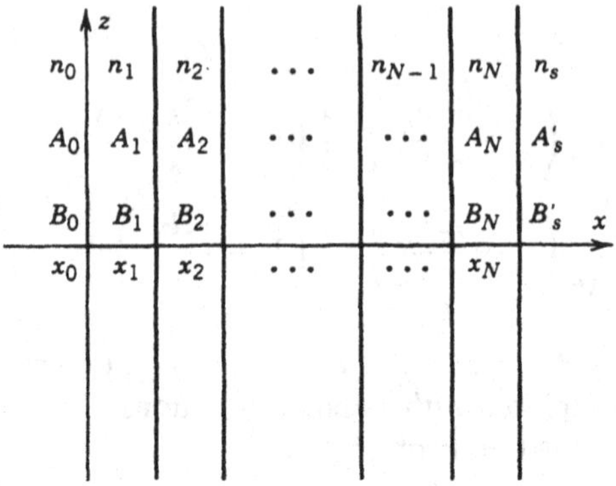

Figure 5.2   A multilayer dielectric medium.

represented by a propagation matrix. Such a recipe can be extended to the case of multilayer structures.

5.1.2   2 × 2 Matrix Formulation for Multilayer System

Referring to Fig. 5.2, we now consider the case of multilayer structures. The dielectric structure is described by

$$
n(x) = \begin{cases}
n_0, & x < x_0, \\
n_1, & x_0 < x < x_1, \\
n_2, & x_1 < x < x_2, \\
\vdots & \vdots \\
n_N, & x_{N-1} < x < x_N, \\
n_s, & x_N < x,
\end{cases}
\tag{5.1-16}
$$

where n_l is the refractive index of the lth layer, x_l is the position of the interface between the lth layer and the $(l + 1)$th layer, n_s is the substrate index of refraction, and n_0 is that of the incident medium.

The layer thicknesses d_l are related to the x_l's by

$$
\begin{aligned}
d_1 &= x_1 - x_0, \\
d_2 &= x_2 - x_1, \\
&\vdots \\
d_N &= x_N - x_{N-1}.
\end{aligned}
\tag{5.1-17}
$$

The electric field of a general plane-wave solution of the wave equation can still be written as

$$E = E(x)e^{i(\omega t - \beta z)},\qquad(5.1\text{-}18)$$

where the electric field distribution $E(x)$ can be written as

$$E(x) = \begin{cases} A_0 e^{-ik_{0x}(x-x_0)} + B_0 e^{ik_{0x}(x-x_0)}, & x < x_0, \\ A_l e^{-ik_{lx}(x-x_l)} + B_l e^{ik_{lx}(x-x_l)}, & x_{l-1} < x < x_l, \\ A_s' e^{-ik_{sx}(x-x_N)} + B_s' e^{ik_{sx}(x-x_N)}, & x_N < x, \end{cases}\qquad(5.1\text{-}19)$$

where k_{lx} is the x component of the wave vectors

$$k_{lx} = \left[\left(n_l \frac{\omega}{c}\right)^2 - \beta^2\right]^{1/2}, \qquad l = 0, 1, 2, \ldots, N, s,\qquad(5.1\text{-}20)$$

and is related to the ray angle θ_l by

$$k_{lx} = n_l \frac{\omega}{c} \cos \theta_l.\qquad(5.1\text{-}21)$$

According to Eqs. (5.1-19) and (5.1-2), A_l and B_l represent the amplitude of plane waves at interface $x = x_l$. Thus, using the same argument as in Section 5.1.1, we can write

$$\begin{pmatrix} A_0 \\ B_0 \end{pmatrix} = D_0^{-1} D_1 \begin{pmatrix} A_1 \\ B_1 \end{pmatrix},\qquad(5.1\text{-}22)$$

$$\begin{pmatrix} A_l \\ B_l \end{pmatrix} = P_l D_l^{-1} D_{l+1} \begin{pmatrix} A_{l+1} \\ B_{l+1} \end{pmatrix}, \qquad l = 1, 2, \ldots, N,$$

where $N + 1$ represents s, $A_{N+1} = A_s'$, $B_{N+1} = B_s'$ and the matrices can be written as

$$D_l = \begin{pmatrix} 1 & 1 \\ n_l \cos \theta_l & -n_l \cos \theta_l \end{pmatrix} \qquad \text{for } s \text{ wave}\qquad(5.1\text{-}23a)$$

and

$$D_l = \begin{pmatrix} \cos \theta_l & \cos \theta_{l'} \\ n_l & -n_l \end{pmatrix} \qquad \text{for } p \text{ wave},\qquad(5.1\text{-}23b)$$

respectively, and

$$P_l = \begin{pmatrix} e^{i\phi_l} & 0 \\ 0 & e^{-i\phi_l} \end{pmatrix}, \qquad (5.1\text{-}24)$$

with

$$\phi_l = k_{lx}d_l. \qquad (5.1\text{-}25)$$

The relation between A_0, B_0 and A'_s, B'_s can thus be written as

$$\begin{pmatrix} A_0 \\ B_0 \end{pmatrix} = \begin{pmatrix} M_{11} & M_{12} \\ M_{21} & M_{22} \end{pmatrix} \begin{pmatrix} A'_s \\ B'_s \end{pmatrix}, \qquad (5.1\text{-}26)$$

with the matrix given by

$$\begin{pmatrix} M_{11} & M_{12} \\ M_{21} & M_{22} \end{pmatrix} = D_0^{-1} \left[\prod_{l=1}^{N} D_l P_l D_l^{-1} \right] D_s. \qquad (5.1\text{-}27)$$

Here, we recall that N is the number of layers, A_0 and B_0 are the amplitudes of the plane waves in medium 0 at $x = x_0$, and A'_s, B'_s are the amplitudes of the plane waves in medium s at $x = x_N$.

5.1.3   Example

To illustrate the use of the matrix method, let us consider the case of a single homogeneous thin film with index n_2 and thickness d. Let n_1 and n_3 be the indices of the bounding media (see Fig. 5.1). According to Eq. (5.1-15), the matrix is given by

$$\begin{pmatrix} M_{11} & M_{12} \\ M_{21} & M_{22} \end{pmatrix} = D_1^{-1} D_2 P_2 D_2^{-1} D_3, \qquad (5.1\text{-}28)$$

where $D_l, l = 1, 2, 3$, is given by Eq. (5.1-23) and P_2 is the propagation matrix of the thin film and is given by Eq. (5.1-24) with $l = 2$. By carrying out matrix inversion and multiplications in Eq. (5.1-28) for both s waves (TE) and p waves (TM), we obtain the following expressions for the matrix elements. For s waves, the M_{ij} are given by

$$M_{11} = \frac{1}{2}\left(1 + \frac{n_3 \cos\theta_3}{n_1 \cos\theta_1}\right)\cos\phi + \frac{1}{2}i\sin\phi\left(\frac{n_2\cos\theta_2}{n_1\cos\theta_1} + \frac{n_3\cos\theta_3}{n_2\cos\theta_2}\right),$$

$$M_{12} = \frac{1}{2}\left(1 - \frac{n_3 \cos\theta_3}{n_1 \cos\theta_1}\right)\cos\phi + \frac{1}{2}i\sin\phi\left(\frac{n_2\cos\theta_2}{n_1\cos\theta_1} - \frac{n_3\cos\theta_3}{n_2\cos\theta_2}\right),$$

$$M_{21} = \frac{1}{2}\left(1 - \frac{n_3 \cos \theta_3}{n_1 \cos \theta_1}\right) \cos \phi - \frac{1}{2} i \sin \phi \left(\frac{n_2 \cos \theta_2}{n_1 \cos \theta_1} - \frac{n_3 \cos \theta_3}{n_2 \cos \theta_2}\right),$$

$$M_{22} = \frac{1}{2}\left(1 + \frac{n_3 \cos \theta_3}{n_1 \cos \theta_1}\right) \cos \phi - \frac{1}{2} i \sin \phi \left(\frac{n_2 \cos \theta_2}{n_1 \cos \theta_1} + \frac{n_3 \cos \theta_3}{n_2 \cos \theta_2}\right),$$

$$(5.1\text{-}29)$$

where ϕ is the phase change of light as it propagates through the thin film and is given by

$$\phi = \frac{2\pi n_2 d}{\lambda} \cos \theta_2. \qquad (5.1\text{-}30)$$

For p waves, the M_{ij} are given by

$$M_{11} = \frac{1}{2}\left(\frac{\cos \theta_3}{\cos \theta_1} + \frac{n_3}{n_1}\right) \cos \phi + \frac{1}{2} i \sin \phi \left(\frac{n_2 \cos \theta_3}{n_1 \cos \theta_2} + \frac{n_3 \cos \theta_2}{n_2 \cos \theta_1}\right),$$

$$M_{12} = \frac{1}{2}\left(\frac{\cos \theta_3}{\cos \theta_1} - \frac{n_3}{n_1}\right) \cos \phi + \frac{1}{2} i \sin \phi \left(\frac{n_2 \cos \theta_3}{n_1 \cos \theta_2} - \frac{n_3 \cos \theta_2}{n_2 \cos \theta_1}\right),$$

$$M_{21} = \frac{1}{2}\left(\frac{\cos \theta_3}{\cos \theta_1} - \frac{n_3}{n_1}\right) \cos \phi - \frac{1}{2} i \sin \phi \left(\frac{n_2 \cos \theta_3}{n_1 \cos \theta_2} - \frac{n_3 \cos \theta_2}{n_2 \cos \theta_1}\right),$$

$$M_{22} = \frac{1}{2}\left(\frac{\cos \theta_3}{\cos \theta_1} + \frac{n_3}{n_1}\right) \cos \phi - \frac{1}{2} i \sin \phi \left(\frac{n_2 \cos \theta_3}{n_1 \cos \theta_2} + \frac{n_2 \cos \theta_2}{n_1 \cos \theta_1}\right).$$

$$(5.1\text{-}31)$$

These expressions also apply to media with complex refractive indices and for incident conditions when θ's are complex.

For multilayer structures, the expressions for M_{ij} become complicated and are best handled by a computer program.

5.2   TRANSMITTANCE AND REFLECTANCE

Using the 2 × 2 matrix method, we now discuss the reflectance and transmittance of monochromatic plane waves through a multilayer dielectric structure. If the light is incident from medium 0, the reflection and transmission coefficients are defined as

$$r = \left(\frac{B_0}{A_0}\right)_{B_s = 0} \qquad (5.2\text{-}1)$$

and

$$t = \left(\frac{A_s}{A_0}\right)_{B_s=0}, \tag{5.2-2}$$

respectively. Here we drop the prime (') and define $A_s = A'_s$, $B_s = B'_s$.

Using the matrix equation (5.1-26) and following definitions (5.2-1) and (5.2-2), we obtain

$$r = \frac{M_{21}}{M_{11}} \tag{5.2-3}$$

and

$$t = \frac{1}{M_{11}}. \tag{5.2-4}$$

Reflectance is given by

$$R = |r|^2 = \left|\frac{M_{21}}{M_{11}}\right|^2, \tag{5.2-5}$$

provided medium 0 is lossless.

If the bounding media $(0, s)$ are both pure dielectric with real n_s and n_0, transmittance T is given by

$$T = \frac{n_s \cos \theta_s}{n_0 \cos \theta_0} |t|^2 = \frac{n_s \cos \theta_s}{n_0 \cos \theta_0} \left|\frac{1}{M_{11}}\right|^2, \tag{5.2-6}$$

provided both the incident wave and the transmitted wave have real propagation vectors (i.e., real θ_0 and θ_s).**

5.2.1   Example: Quarter-Wave Stack

To illustrate the use of the matrix method in the calculation of the reflection and transmission of a multilayer structure, we consider a layered medium consisting of N pairs of alternating quarter-waves ($n_1 d_1 = n_2 d_2 = \frac{1}{4}\lambda$) with refractive indices n_1 and n_2, respectively. Let n_0 by the index of refraction of the incident medium and n_s be the index of refraction of the substrate. The

**When $\beta > n_l(\omega/c)$, k_{lx} becomes imaginary according to Eq. (5.1-20).

reflectance R at normal incidence can be obtained as follows: According to Eq. (5.1-27), the matrix is given by

$$\begin{pmatrix} M_{11} & M_{12} \\ M_{21} & M_{22} \end{pmatrix} = D_0^{-1}[D_1 P_1 D_1^{-1} D_2 P_2 D_2^{-1}]^N D_s. \qquad (5.2\text{-}7)$$

The propagation matrix for quarter-wave layers (with $\phi_l = \frac{1}{2}\pi$) is given by

$$P_{1,2} = \begin{pmatrix} i & 0 \\ 0 & -i \end{pmatrix}. \qquad (5.2\text{-}8)$$

By using Eq. (5.1-23) for the dynamical matrices and assuming normal incidence, we obtain, after some matrix manipulation,

$$D_1 P_1 D_1^{-1} D_2 P_2 D_2^{-1} = \begin{pmatrix} \dfrac{-n_2}{n_1} & 0 \\ 0 & \dfrac{-n_1}{n_2} \end{pmatrix}. \qquad (5.2\text{-}9)$$

Carrying out the matrix multiplication in Eq. (5.2-7) and using Eq. (5.2-5), the reflectance is

$$R = \left(\frac{1 - (n_s/n_0)(n_1/n_2)^{2N}}{1 + (n_s/n_0)(n_1/n_2)^{2N}} \right)^2. \qquad (5.2\text{-}10)$$

This equation can be written as

$$R = \tanh^2 N\chi \qquad (5.2\text{-}11)$$

with

$$\chi = \ln \frac{n_1}{n_2} + \frac{1}{2N} \ln \frac{n_s}{n_0}. \qquad (5.2\text{-}12)$$

For large N, the reflectance R approaches unity exponentially as a function of N. Table 5.1 lists the reflectance of a quarter-wave stack formed by zinc sulfide and magnesium fluoride layers on a zinc sulfide substrate at normal incidence (i.e., $n_0 = 1.0$, $n_1 = 2.32$, $n_2 = 1.36$, and $n_s = 2.32$ at $\lambda = 1\ \mu m$ and $\theta_0 = 0$).

5.2.2 Example: Latent ($\frac{1}{2}\lambda$) Layers

If the wave propagation in a layered medium satisfies the condition that $k_{lx}d_l = m\pi$ ($m = 1, 2, 3, \ldots$), the propagation matrix (5.1-24) associated

Table 5.1. Reflectance of a Quarter-Wave Stack

N	R
0	0.16
1	0.55
2	0.82
3	0.93
4	0.98
5	0.99

with that layer (lth layer) becomes a unit matrix or its negative. The product $D_l P_l D_l^{-1}$ also becomes a unit matrix or its negative. From Eq. (5.1-27), we know that such a layer does not affect the matrix M, except for a possible change of sign. Such a layer will consequently not affect wave propagation and has no effect on the reflectance and transmittance of the multilayer medium. At normal incidence, this condition becomes $n_l d_l = m(\frac{1}{2}\lambda)$, i.e., the optical thickness of the layer is an integral multiple of half-wavelengths for the particular λ. If all the layers become half-wave layers at a particular wavelength (i.e., $k_{lx} d_l = m\pi$ for all l), matrix (5.1-27) becomes simply $D_0^{-1} D_s$. Reflectance and transmittance at this condition must be identical to that of the simple boundary between n_0 and n_s. Note that this normally occurs only at one wavelength.

5.3 GENERAL THEOREMS OF LAYERED MEDIA

In this section, we derive some general theorems on layered media. Since most of these theorems can be derived from the properties of the matrix element M_{ij}, it is useful to examine first the symmetry relation between these elements.

From Eqs. (5.1-29) and (5.1-31), we note that the matrix elements M_{ij} satisfy the relations

$$M_{21} = M_{12}^*, \qquad M_{22} = M_{11}^*, \qquad (5.3\text{-}1)$$

provided n_1, n_2, n_3 and θ_1, θ_3 are real. In fact, it can be shown that Eq. (5.3-1) holds for a general dielectric structure, provided all the layers are pure dielectric with real n_l and θ_0 and θ_s are real (see Problem 5.2).

The propagation matrix P_l [Eq. (5.1-24)] is a unimodular matrix,

$$\det(P_l) = |P_l| = 1. \qquad (5.3\text{-}2)$$

The matrix product $D_i P_i D_i^{-1}$ is merely a transformation of the propagation matrix and is also unimodular. Thus, the determinant of the matrix M is very simple and is given by

$$|M| = \frac{n_s \cos \theta_s}{n_0 \cos \theta_s}, \qquad (5.3\text{-}3)$$

which is independent of all the inner layers and depends only on the bounding media. The matrix M becomes unimodular when $n_s = n_0$, that is,

$$|M| = 1 \quad \text{when} \quad n_s = n_0. \qquad (5.3\text{-}4)$$

We now examine the principle of reversibility.

5.3.1 Left and Right Incidence Theorem

For a given dielectric structure defined by Eq. (5.1-16), the reflection and transmission coefficients defined by Eqs. (5.2-1) and (5.2-2), respectively, may be considered as functions of β:

$$r = r(\beta), \qquad t = t(\beta). \qquad (5.3\text{-}5)$$

Let r' and t' be the reflection and transmission coefficients, respectively, when light is incident from the right side (medium s) with the same β. There exist some interesting relationships between these four coefficients. According to Eq. (5.1-26), these four coefficients are written as

$$r = \left(\frac{B_0}{A_0}\right)_{B_s=0} = \frac{M_{21}}{M_{11}}, \qquad t = \left(\frac{A_s}{A_0}\right)_{B_s=0} = \frac{1}{M_{11}}, \qquad (5.3\text{-}6)$$

$$r' = \left(\frac{A_s}{B_s}\right)_{A_0=0} = -\frac{M_{12}}{M_{11}}, \qquad t' = \left(\frac{B_0}{B_s}\right)_{A_0=0} = \frac{|M|}{M_{11}}, \qquad (5.3\text{-}7)$$

where $|M|$ is the determinant of the matrix. Here, we recall that the primes (') on A_s and B_s have been dropped. Let T and T' be the transmittances of the layered structure when light is incident from the left medium and right medium, respectively. These two transmittances are given by

$$T' = \frac{n_0 \cos \theta_0}{n_s \cos \theta_s} |t'|^2, \qquad T = \frac{n_s \cos \theta_s}{n_0 \cos \theta_0} |t|^2, \qquad (5.3\text{-}8)$$

respectively. Using Eq. (5.3-8), and the expression for $|M|$ in Eq. (5.3-3), we obtain

$$T' = T \qquad (5.3\text{-}9)$$

and

$$t' = |M| t = \frac{n_s \cos \theta_s}{n_0 \cos \theta_0} t. \tag{5.3-10}$$

Equations (5.3-9) and (5.3-10) are known as the left-and-right incidence theorem, which states that the transmittance and the phase shift when light is incident from the left medium is identical to those when light is incident from the right medium with the same β. This theorem is true, provided n_0, n_s and θ_0, θ_s are real. When $n_s = n_0$, we have, according to Eq. (5.3-10)

$$t' = t. \tag{5.3-10a}$$

5.3.2 Principles of Reversibility

We have derived the Stokes equations (3.1-31) and (3.1-33) for the Fresnel coefficients at a dielectric interface. For the case of a dielectric multilayer structure with real index of refraction (i.e., no absorption), the functional relations between these four coefficients can be obtained in a similar way. Since these four coefficients are, in general, complex, time reversal is accompanied by a phase conjugation. The argument used in Fig. 3.5 leads to

$$tt'^* + rr^* = 1, \qquad tr'^* + rt^* = 0, \tag{5.3-11}$$

which must be satisfied by these four coefficients. It can be shown that the explicit expressions obtained by the matrix method Eqs. (5.3-6) and (5.3-7) agree with condition (5.3-11) (see Problem 5.3), provided the symmetry relation (5.3-1) holds.

5.3.3 Conservation of Energy

In the case when all the layers and the bounding media are pure dielectrics with real n_l, the conservation of energy requires that

$$R + T = 1. \tag{5.3-12}$$

The expressions for r and t [Eqs. (5.3-6) and (5.3-7)] obtained by using the matrix method agree with the conservation of energy.

PROBLEMS

5.1 *Dynamical matrices*
    (a) Derive Eqs. (5.1-10) and (5.1-11).

(b) Derive Eq. (5.1-12) using Eqs. (5.1-10), (5.1-11), (5.1-13), and (5.1-14).

(c) Derive Eq. (5.1-12) directly from the definition of t_{12} and r_{12}. *Hint:* Use the set of equations

$$B_1 = r_{12}A_1 + t_{21}B_2', \qquad A_2' = t_{12}A_1 + r_{21}B_2',$$

and Eqs. (3.1-31) and (3.1-33).

5.2 *Symmetry property of matrix elements*

(a) Show that

$$Q_l \equiv D_l P_l D_l^{-1} = \begin{pmatrix} \cos\phi & \dfrac{i}{\alpha}\sin\phi \\ i\alpha\sin\phi & \cos\phi \end{pmatrix},$$

where $\alpha = n_l\cos\theta_l$ and $\phi = (\omega/c)\alpha d_l = (\omega/c)d_l n_l\cos\theta_l$. Note that Q_l is unimodular (i.e., $\det Q_l = 1$).

(b) Show that the matrix Q_l has the form

$$Q_l = \begin{pmatrix} c & ia \\ ib & c \end{pmatrix},$$

where a, b, and c are real, provided n_l is real and $\cos\theta_l$ is real or imaginary.

(c) Show that all the matrices of the form shown in (b) form a group. In other words, if Q_1 and Q_2 are in the group, $Q_1 Q_2$ must also be in the group.

(d) Using (a), (b), and (c), show that the matrix elements of M in Eq. (5.1-27) satisfy $M_{21} = M_{12}^*$ and $M_{22} = M_{11}^*$, provided all n_0, n_l, n_s are real and θ_0 and θ_s are real.

5.3 *Stokes equations*

(a) Show that the explicit expressions for r, t, r', and t' [Eqs. (5.3-6) and (5.3-7)] obey the principle of reversibility [Eq. (5.3-12)], provided the whole structure is lossless.

(b) Show that in the case when absorption is present, the Stokes equations can be generalized to

$$tt'^\# + rr^\# = 1, \qquad tr'^\# + rt^\# = 0,$$

where the superscript $\#$ indicates complex conjugation and the reversal of all the signs of the imaginary part of n_l.

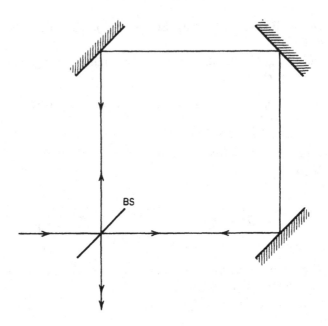

Figure 5.3 A Sagnac interferometer.

5.4 *Sagnac interferometer.*   The principle of reversibility [Eq. (5.3-11)] can be employed to explain the perfect blackness of the central spots in Newton's ring. Here, we use the same equation to explain the nullness at the output of a Sagnac interferometer. Referring to Fig. 5.3, a Sagnac interferometer consists of a beamsplitter and three mirrors arranged in a ring geometry. An input beam is divided into two at the beamsplitter. One beam travels along the clockwise path, while the other travels along the counterclockwise path. These two beams recombined at the beamsplitter.

(a) Show that the field amplitude of the output is proportional to

$$(rr' + tt)\,\exp\,(ikL),$$

where L is the loop length and k is the wave number. The parameters r, r', and t are the same as those defined in Section 5.3.

(b) Use Eq. (5.3-11) to show that rr' and tt are always opposite in sign. In other words, they are different in phase by 180°.

(c) Show that if the beamsplitter is 50%–50%, the output is always zero.

(d) Show that such an interferometer can be used for rotation sensing.

5.5 Show that total reflection (i.e., $R = 1$) is impossible for a finite dielectric layered medium made of materials with a finite index of refraction and bounded by two semi-infinite media of the same refractive index.

6

Optics of Periodic Layered Media

Periodic layered media are a special class of layered media in which layers of dielectric material are stacked in a periodic fashion. The simplest example of a periodic medium consists of alternating layers of two different materials with equal thicknesses. Wave propagation in these media exhibits many interesting and potentially useful phenomena. These include Bragg reflection, holography, and optical stop bands. The diffraction of x rays in crystals is a good example of wave propagation in periodic layered media. In the animal kingdom, the remarkable colors of some beetles and butterflies and the silvery skin and scales of some deep-sea fish are believed to be due to reflection of light from periodic layered media [1]. In addition to the existence of these naturally occurring periodic materials, periodic layered media can also be grown artificially by using various techniques, including molecular beam epitaxy (MBE), metal–organic chemical vapor phase deposition (MOCVD), and atomic layer epitaxy (ALE). In this chapter, we discuss the propagation of optical waves in periodic layered media that consist of isotropic materials of thicknesses comparable to the wavelength of the electromagnetic radiation. For media with layer thicknesses comparable to atomic dimensions, many optical anomalies occur. This latter class of structures, known as *superlattices*, will be discussed in Chapter 12.

6.1 PERIODIC LAYERED MEDIA

The propagation of electromagnetic radiation in periodic media has been a subject of interest because of its several useful applications. It is known that light experiences Fresnel reflection from dielectric interfaces where the refractive index is discontinuous. A periodic layered medium may be considered as a periodic array of dielectric interfaces that exhibit discontinuity. As a result of the large number of reflections from these interfaces, it is possible to achieve extremely high reflectance even though a single reflection from an individual interface is very feeble. In this section, we treat the propagation of electromagnetic radiation in a simple periodic layered

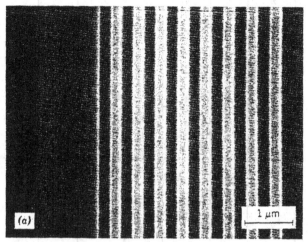

Figure 6.1 A typical periodic layered medium consisting of alternating layers of GaAs and $Al_xGa_{1-x}As$ grown on a GaAs substrate by MBE [1]. (From A. Yariv and P. Yeh, *Optical Waves in Crystals*, Wiley, New York, 1984, p. 481. Copyright © 1984. Reprinted by permission of John Wiley & Sons, Inc.)

medium that consists of alternating layers of transparent materials with different refractive indices. A typical example of such a medium is shown in Fig. 6.1. We take the x axis along the direction normal to the layers and assume that the materials are nonmagnetic. The index-of-refraction profile is given by

$$n(x) = \begin{cases} n_2, & 0 < x < b; \\ n_1, & b < x < \Lambda; \end{cases} \tag{6.1-1}$$

with

$$n(x) = n(x + \Lambda), \tag{6.1-2}$$

where b and $a = \Lambda - b$ are the thicknesses of the layers and Λ is the period. The geometry of the structure is sketched in Fig. 6.2.

To solve for the propagation of electromagnetic radiation in these media, we use the 2×2 matrix formulation derived in Chapter 5. Following Eqs. (5.1-18) and (5.1-19), we write the electric field as

$$E = E(x)e^{i(\omega t - \beta z)}, \tag{6.1-3}$$

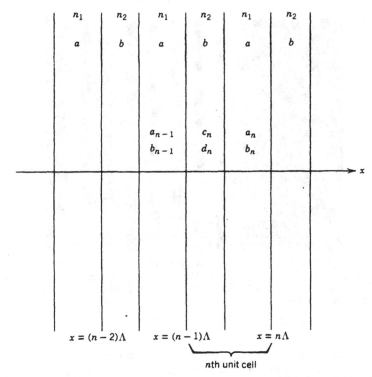

Figure 6.2 A schematic drawing of a periodic layered medium and the plane-wave amplitudes associated with nth unit cell and its neighboring layers. (Adapted from A. Yariv and P. Yeh, *Optical Waves in Crystals*, Wiley, New York, 1984, p. 166. Copyright © 1984. By permission of John Wiley & Sons, Inc.)

where β is the z component of the wave vector. The electric field distribution $E(x)$ within each homogeneous layer can be expressed as the sum of an incident plane wave and a reflected plane wave. The complex amplitudes of these two waves constitute the components of a column vector. The electric field in the nth unit cell can thus be written as

$$
E(x) = \begin{cases} a_n e^{-ik_{1x}(x-n\Lambda)} + b_n e^{ik_{1x}(x-n\Lambda)}, & n\Lambda - a < x < n\Lambda; \\ c_n e^{-ik_{2x}(x-n\Lambda+a)} + d_n e^{ik_{2x}(x-n\Lambda+a)}, & (n-1)\Lambda < x < n\Lambda - a; \end{cases}
$$

$$(6.1\text{-}4)$$

with

$$k_{1x} = \left[\left(\frac{n_1 \omega}{c}\right)^2 - \beta^2\right]^{1/2} = \frac{n_1 \omega}{c} \cos \theta_1,$$

$$k_{2x} = \left[\left(\frac{n_2 \omega}{c}\right)^2 - \beta^2\right]^{1/2} = \frac{n_2 \omega}{c} \cos \theta_2, \qquad (6.1\text{-}5)$$

where θ_1 and θ_2 are the ray angles in the layers.

Using the matrix method derived in Section 5.1, the constant a_n, b_n, c_n, and d_n are related by

$$\begin{pmatrix} a_{n-1} \\ b_{n-1} \end{pmatrix} = D_1^{-1} D_2 P_2 \begin{pmatrix} c_n \\ d_n \end{pmatrix} \qquad (6.1\text{-}6)$$

and

$$\begin{pmatrix} c_n \\ d_n \end{pmatrix} = D_2^{-1} D_1 P_1 \begin{pmatrix} a_n \\ b_n \end{pmatrix}, \qquad (6.1\text{-}7)$$

where

$$P_1 = \begin{pmatrix} e^{ik_{1x}a} & 0 \\ 0 & e^{-ik_{1x}a} \end{pmatrix} \qquad (6.1\text{-}8)$$

and

$$P_2 = \begin{pmatrix} e^{ik_{2x}b} & 0 \\ 0 & e^{-ik_{2x}b} \end{pmatrix}. \qquad (6.1\text{-}9)$$

The dynamical matrices D_1 and D_2 are independent of the layer thickness and are given by Eqs. (5.1-23). Substituting Eq. (5.1-23) for D's and Eq. (6.1-8) for P_1 and carrying out the matrix multiplication, we obtain, for the s waves (or TE waves, with the E vector perpendicular to the xz plane),

$$\begin{pmatrix} a_{n-1} \\ b_{n-1} \end{pmatrix} = \frac{1}{2} \begin{pmatrix} e^{ik_{2x}b}\left(1 + \dfrac{k_{2x}}{k_{1x}}\right) & e^{-ik_{2x}b}\left(1 - \dfrac{k_{2x}}{k_{1x}}\right) \\ e^{ik_{2x}b}\left(1 - \dfrac{k_{2x}}{k_{1x}}\right) & e^{-ik_{2x}b}\left(1 + \dfrac{k_{2x}}{k_{1x}}\right) \end{pmatrix} \begin{pmatrix} c_n \\ d_n \end{pmatrix}, \qquad (6.1\text{-}10)$$

and similarly,

$$
\begin{pmatrix} c_n \\ d_n \end{pmatrix} = \frac{1}{2} \begin{pmatrix} e^{ik_{1x}a}\left(1 + \dfrac{k_{1x}}{k_{2x}}\right) & e^{-ik_{1x}a}\left(1 - \dfrac{k_{1x}}{k_{2x}}\right) \\ e^{ik_{1x}a}\left(1 - \dfrac{k_{1x}}{k_{2x}}\right) & e^{-ik_{1x}a}\left(1 + \dfrac{k_{1x}}{k_{2x}}\right) \end{pmatrix} \begin{pmatrix} a_n \\ b_n \end{pmatrix}. \qquad (6.1\text{-}11)
$$

By eliminating

$$
\begin{pmatrix} c_n \\ d_n \end{pmatrix},
$$

the matrix equation

$$
\begin{pmatrix} a_{n-1} \\ b_{n-1} \end{pmatrix} = \begin{pmatrix} A & B \\ C & D \end{pmatrix} \begin{pmatrix} a_n \\ b_n \end{pmatrix} \qquad (6.1\text{-}12)
$$

is obtained. The matrix elements are

$$
A = e^{ik_{1x}a}\left[\cos k_{2x}b + \frac{1}{2} i \left(\frac{k_{2x}}{k_{1x}} + \frac{k_{1x}}{k_{2x}} \right) \sin k_{2x}b \right],
$$

$$
B = e^{-ik_{1x}a}\left[\frac{1}{2} i \left(\frac{k_{2x}}{k_{1x}} - \frac{k_{1x}}{k_{2x}} \right) \sin k_{2x}b \right],
$$

$$
C = e^{ik_{1x}a}\left[-\frac{1}{2} i \left(\frac{k_{2x}}{k_{1x}} - \frac{k_{1x}}{k_{2x}} \right) \sin k_{2x}b \right],
$$

$$
D = e^{-ik_{1x}a}\left[\cos k_{2x}b - \frac{1}{2} i \left(\frac{k_{2x}}{k_{1x}} + \frac{k_{1x}}{k_{2x}} \right) \sin k_{2x}b \right], \qquad (6.1\text{-}13)
$$

and, according to Eq. (6.1-5), can be viewed as functions of β.

The matrix in Eq. (6.1-12) is the unit cell translation matrix that relates the complex amplitudes of the incident plane wave a_{n-1} and the reflected plane wave b_{n-1} in one layer of a unit cell to those of the equivalent layer in the next unit cell. Because this matrix relates the fields of two equivalent layers with the same index of refraction, it is unimodular, that is,

$$
AD - BC = 1. \qquad (6.1\text{-}14)
$$

It is important to note that the matrix that relates

$$
\begin{pmatrix} c_{n-1} \\ d_{n-1} \end{pmatrix} \quad \text{to} \quad \begin{pmatrix} c_n \\ d_n \end{pmatrix}
$$

is different from the matrix in Eq. (6.1-12). These matrices, however, possess the same trace. As will be shown later, the trace of the translation matrix is directly related to the band structure of the stratified periodic medium.

The matrix elements (A, B, C, D) for TM waves (the H vector perpendicular to xz plane) are slightly different from those of the TE waves. They are given by

$$A_{TM} = e^{ik_{1x}a}\left[\cos k_{2x}b + \frac{1}{2}i\left(\frac{n_2^2 k_{1x}}{n_1^2 k_{2x}} + \frac{n_1^2 k_{2x}}{n_2^2 k_{1x}}\right)\sin k_{2x}b\right],$$

$$B_{TM} = e^{-ik_{1x}a}\left[\frac{1}{2}i\left(\frac{n_2^2 k_{1x}}{n_1^2 k_{2x}} - \frac{n_1^2 k_{2x}}{n_2^2 k_{1x}}\right)\sin k_{2x}b\right],$$

$$C_{TM} = e^{ik_{1x}a}\left[-\frac{1}{2}i\left(\frac{n_2^2 k_{1x}}{n_1^2 k_{2x}} - \frac{n_1^2 k_{2x}}{n_2^2 k_{1x}}\right)\sin k_{2x}b\right],$$

$$D_{TM} = e^{-ik_{1x}a}\left[\cos k_{2x}b - \frac{1}{2}i\left(\frac{n_2^2 k_{1x}}{n_1^2 k_{2x}} + \frac{n_1^2 k_{2x}}{n_2^2 k_{1x}}\right)\sin k_{2x}b\right]. \qquad (6.1\text{-}15)$$

As noted in the preceding, only one column vector is independent. We can choose it, as an example, as the column vector of the n_1 layer in the zeroth unit cell. The remaining column vectors of the equivalent layers are given as

$$\begin{pmatrix} a_n \\ b_n \end{pmatrix} = \begin{pmatrix} A & B \\ C & D \end{pmatrix}^{-n} \begin{pmatrix} a_0 \\ b_0 \end{pmatrix}. \qquad (6.1\text{-}16)$$

By using Eq. (6.1-14), the preceding equation can be simplified to

$$\begin{pmatrix} a_n \\ b_n \end{pmatrix} = \begin{pmatrix} D & -B \\ -C & A \end{pmatrix}^n \begin{pmatrix} a_0 \\ b_0 \end{pmatrix}. \qquad (6.1\text{-}17)$$

The column vector for the n_2 layer can always be obtained by using Eq. (6.1-7); more generally, we can specify the field uniquely by specifying any a_i and b_j.

6.2 BLOCH WAVES AND BAND STRUCTURES

Wave propagation in periodic media is very similar to the motion of electrons in crystalline solids. In fact, formulation of the Kronig–Penney model [3] used in the elementary band theory of solids is mathematically identical to that of the electromagnetic radiation in periodic layered media. Thus, some of the physical concepts used in solid-state physics such as Bloch waves, Brillouin zones, and forbidden bands can also be used here. A periodic layered medium is equivalent to a one-dimensional lattice that is invariant under lattice translation. In other words

$$n^2(x + \Lambda) = n^2(x) \tag{6.2-1}$$

where Λ is the period.

According to the Floquet theorem, solutions of wave equations for a periodic medium are of the form

$$E_K(x, z) = E_K(x)e^{-i\beta z}e^{-iKx}, \tag{6.2-2}$$

where $E_K(x)$ is periodic with a period Λ, that is,

$$E_K(x + \Lambda) = E_K(x). \tag{6.2-3}$$

The subscript K indicates that the function $E_K(x)$ depends on K. The constant K is known as the *Bloch wave number*. The problem at hand is thus that of determining K and $E_K(x)$.

In terms of our column vector representation and from Eq. (6.1-4), the periodic condition (6.2-3) for the Bloch wave is simply

$$\begin{pmatrix} a_n \\ b_n \end{pmatrix} = e^{-iK\Lambda} \begin{pmatrix} a_{n-1} \\ b_{n-1} \end{pmatrix}. \tag{6.2-4}$$

It follows from Eqs. (6.1-12) and (6.2-4) that the column vector of the Bloch wave satisfies the following eigenvalue problem:

$$\begin{pmatrix} A & B \\ C & D \end{pmatrix} \begin{pmatrix} a_n \\ b_n \end{pmatrix} = e^{iK\Lambda} \begin{pmatrix} a_n \\ b_n \end{pmatrix}. \tag{6.2-5}$$

The phase factor $e^{iK\Lambda}$ is thus the eigenvalue of the translation matrix (A, B, C, D) and is given by

$$e^{iK\Lambda} = \tfrac{1}{2}(A + D) \pm \{[\tfrac{1}{2}(A + D)]^2 - 1\}^{1/2}. \tag{6.2-6}$$

The eigenvectors corresponding to the eigenvalues of Eq. (6.2-6) are obtained from Eq. (6.2-5) and are

$$\begin{pmatrix} a_0 \\ b_0 \end{pmatrix} = \begin{pmatrix} B \\ e^{iK\Lambda} - A \end{pmatrix} \tag{6.2-7}$$

times any arbitrary constant. The Bloch waves that result from Eq. (6.2-7) can be considered as the eigenvectors of the translation matrix with eigenvalues $e^{\pm iK\Lambda}$ given by Eq. (6.2-6). The two eigenvalues in Eq. (6.2-6) are the inverse of each other since the translation matrix is unimodular. Equation (6.2-6) gives the dispersion relation between ω, β, and K for the Bloch wavefunction

$$K(\beta, \omega) = \frac{1}{\Lambda} \cos^{-1} \left[\tfrac{1}{2}(A + D) \right]. \tag{6.2-8}$$

Regimes where $|\tfrac{1}{2}(A + D)| < 1$ correspond to real K and thus to propagating Bloch waves; when $|\tfrac{1}{2}(A + D)| > 1$, however, $K = m\pi/\Lambda + iK_i$ which has an imaginary part K_i so that the Bloch wave is evanescent. These are the so-called forbidden bands of the periodic medium. The band edges are the regimes where $|\tfrac{1}{2}(A + D)| = 1$.

According to Eqs. (6.1-4) and (6.2-4), the final results for the Bloch wave in the n_1 layer of the nth unit cell is

$$E_K(x)e^{-iKx} = \left[(a_0 e^{-ik_{1x}(x-n\Lambda)} + b_0 e^{ik_{1x}(x-n\Lambda)})e^{iK(x-n\Lambda)} \right]e^{-iKx}, \tag{6.2-9}$$

where a_0 and b_0 are given by Eq. (6.2-7). This completes the solution of the Bloch waves. Note that the expression inside the square brackets in Eq. (6.2-9) is periodic.

The band structure for a typical periodic layered medium as obtained from Eq. (6.2-8) is shown in Figs. 6.3 and 6.4 for TE and TM waves, respectively. It is interesting to note that the TM forbidden bands shrink to zero when $\beta = (\omega/c)n_2 \sin \theta_B$ with θ_B the Brewster angle, since at this angle the incident and reflected waves are uncoupled. The dispersion relation ω versus K for the special case $\beta = 0$ (i.e., normal incidence) is shown in Fig. 6.5. The dispersion relation between ω and K in this case can be written as

$$\cos K\Lambda = \cos k_1 a \cos k_2 b - \frac{1}{2}\left(\frac{n_2}{n_1} + \frac{n_1}{n_2}\right) \sin k_1 a \sin k_2 b, \tag{6.2-10}$$

with $k_1 = (\omega/c)n_1$ and $k_2 = (\omega/c)n_2$.

Equation (6.2-10) can be solved approximately for K in the forbidden band, since in the first forbidden gap $\operatorname{Re} K = \pi/\Lambda$, we put

$$K\Lambda = \pi \pm ix. \tag{6.2-11}$$

Let ω_0 be the center of the forbidden band such that

$$k_1 a = k_2 b = \tfrac{1}{2}\pi. \tag{6.2-12}$$

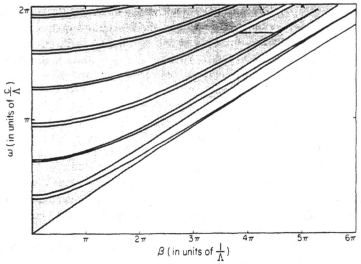

Figure 6.3 Band structure of a typical periodic layered medium in the $\beta\omega$ plane for TE waves. The dark zones are allowed bands in which $|\cos K\Lambda| < 1$. (From A. Yariv and P. Yeh, *Optical Waves in Crystals*, Wiley, New York, 1984, p. 171. Copyright © 1984. Reprinted by permission of John Wiley & Sons., Inc.)

A structure with this condition is known as a quarter-wave stack. At frequency ω_0, Eq. (6.2-10) becomes

$$\cos K\Lambda = -\frac{1}{2}\left(\frac{n_1}{n_2} + \frac{n_2}{n_1}\right). \tag{6.2-13}$$

By substituting Eq. (6.2-11) in Eq. (6.2-13) and solving for x, we obtain

$$x = \log\left|\frac{n_2}{n_1}\right| \simeq \frac{2(n_2 - n_1)}{n_2 + n_1}, \tag{6.2-14}$$

where the last equality holds when $|n_2 - n_1| \ll n_{1,2}$. This is the imaginary part of $K\Lambda$ at the center of the forbidden band. Within the forbidden band, x varies from zero at the band edges to this maximum value (6.2-14) at ω_0.

Let y be the normalized frequency deviation from the center of the forbidden band, ω_0;

$$y = \frac{\omega - \omega_0}{c} n_1 a = \frac{\omega - \omega_0}{c} n_2 b. \tag{6.2-15}$$

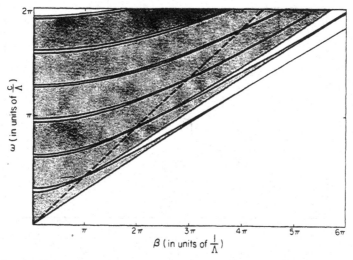

Figure 6.4 Band structure of a typical periodic layered medium in the $\beta\omega$ plane for TM waves. Dark zones are allowed bands in which $|\cos k\Lambda| < 1$. (From A. Yariv and P. Yeh, *Optical Waves in Crystals*, Wiley, New York, 1984, p. 172. Copyright © 1984. Reprinted by permission of John Wiley & Sons, Inc.)

Substitution of Eqs. (6.2-11) and (6.2-15) in Eq. (6.2-10) leads to

$$\cosh x = \frac{1}{2}\left(\frac{n_2}{n_1} + \frac{n_1}{n_2}\right)\cos^2 y - \sin^2 y. \qquad (6.2\text{-}16)$$

This is an expression for the imaginary part of $K\Lambda$ in the forbidden band as a function of frequency. The band edges are obtained by setting $x = 0$. The result is

$$y_{\text{edge}} = \pm\sin^{-1}\frac{n_2 - n_1}{n_2 + n_1}. \qquad (6.2\text{-}17)$$

In terms of frequency, the bandgap $\Delta\omega_{\text{gap}}$ is thus given by

$$\Delta\omega_{\text{gap}} = \omega_0\frac{4}{\pi}\sin^{-1}\frac{|(n_2 - n_1)|}{n_2 + n_1} \simeq \omega_0\frac{2}{\pi}\frac{\Delta n}{n}, \qquad (6.2\text{-}18)$$

whereas the imaginary part of $K\Lambda$ at the center of the forbidden band is

$$(K_i\Lambda)_{\text{max}} = 2\frac{|n_2 - n_1|}{n_2 + n_1} \simeq \frac{\Delta n}{n}, \qquad (6.2\text{-}19)$$

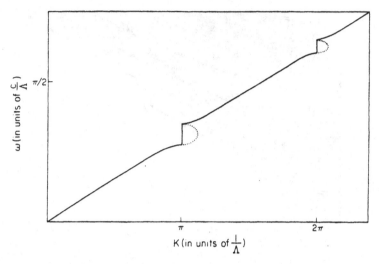

Figure 6.5 The dispersion relation between K and ω when $\beta = 0$ (normal incidence). Dashed curves give the imaginary part of K in arbitrary units. (From A. Yariv and P. Yeh, *Optical Waves in Crystals*, Wiley, New York, 1984, p. 173. Copyright © 1984. Reprinted by permission of John Wiley & Sons, Inc.)

where $\Delta n = |n_2 - n_1|$ and $n = \frac{1}{2}(n_1 + n_2)$. As an exercise, the student may want to find the Fourier expansion coefficients of the dielectric constant for the periodic layered medium. The result is (see Problem 6.5)

$$\frac{|\varepsilon_1|}{\varepsilon_0} = \frac{4}{\pi} \frac{|n_2 - n_1|}{n_2 + n_1}. \qquad (6.2\text{-}20)$$

We notice that the fundamental bandgap of a periodic layered medium is directly proportional to the amplitude of the fundamental spatial component of the Fourier expansion of the dielectric constant. In a similar fashion, the higher order bandgaps are proportional to the amplitudes of the corresponding spatial components of the Fourier expansion.

6.3 BRAGG REFLECTORS

When a monochromatic plane wave is incident onto a periodic layered medium, a Bloch wave is generated in the medium according to the discussion in the last section. If the Bloch wave falls in the so-called forbidden bands, such a wave is evanescent and cannot propagate in the medium. Thus, the light energy is expected to be totally reflected, and the medium acts as a

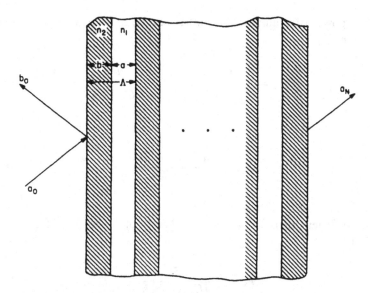

Figure 6.6 Schematic drawing of a Bragg reflector. (From A. Yariv and P. Yeh, *Optical Waves in Crystals*, Wiley, New York, 1984, p. 175. Copyright © 1984. Reprinted by permission of John Wiley & Sons, Inc.)

high-reflectance reflector for the incident wave. By properly designing the periodic layered medium, it is possible to achieve extremely high reflectance for some selected spectral region.

We now consider the reflection and transmission of electromagnetic radiation through a periodic layered medium. Such a structure exhibits resonance reflection very much like the diffraction of x rays by crystal lattice planes. Therefore, it is called a Bragg reflector. For the sake of illustrating the basic properties of Bragg reflectors, we consider a periodic layered medium that consists of N unit cells (i.e., N pairs of layers) and is bounded by homogeneous media of index n_1. The geometry of such a structure is sketched in Fig. 6.6. The coefficient of reflection is given by

$$r_N = \left(\frac{b_0}{a_0}\right)_{b_N = 0}.$$

(6.3-1)

From Eq. (6.1-16), we have the following relation:

$$\begin{pmatrix} a_0 \\ b_0 \end{pmatrix} = \begin{pmatrix} A & B \\ C & D \end{pmatrix}^N \begin{pmatrix} a_N \\ b_N \end{pmatrix}.$$

(6.3-2)

The Nth power of a unimodular matrix can be simplified by the following matrix identity:

$$\begin{pmatrix} A & B \\ C & D \end{pmatrix}^N = \begin{pmatrix} AU_{N-1} - U_{N-2} & BU_{N-1} \\ CU_{N-1} & DU_{N-1} - U_{N-2} \end{pmatrix}, \qquad (6.3\text{-}3)$$

where

$$U_N = \frac{\sin (N+1)K\Lambda}{\sin K\Lambda}, \qquad (6.3\text{-}4)$$

with K given by Eq. (6.2-8).

The coefficient of reflection is immediately obtained from Eqs. (6.3-1)–(6.3-3) as

$$r_N = \frac{CU_{N-1}}{AU_{N-1} - U_{N-2}}. \qquad (6.3\text{-}5)$$

The reflectance is obtained by taking the absolute square of r_N,

$$|r_N|^2 = \frac{|C|^2}{|C|^2 + (\sin K\Lambda / \sin NK\Lambda)^2}. \qquad (6.3\text{-}6)$$

In Eq. (6.3-6) we have an analytic expression of the reflectance of a multilayer reflector. The derivation of Eq. (6.3-6) is left as an exercise for the student (see Problem 6.1). The term $|C|^2$ is directly related to the reflectance of a single unit cell by

$$|r_1|^2 = \frac{|C|^2}{|C|^2 + 1} \qquad (6.3\text{-}7)$$

or

$$|C|^2 = \frac{|r_1|^2}{1 - |r_1|^2}. \qquad (6.3\text{-}8)$$

The $|r_1|^2$ for a typical Bragg reflector is usually much less than 1. As a result, $|C|^2$ is roughly equal to $|r_1|^2$. The second term in the denominator of Eq. (6.3-6) is a fast varying function of K or, alternatively, of β and ω. Therefore, it dominates the structure of the reflectance spectrum. Between any two forbidden bands there are exactly $N - 1$ nodes where the reflectance vanishes. The peaks of the reflectance occur at the centers of the forbidden bands. There are exactly $N - 2$ sidelobes that are all under the envelope $|C|^2/[|C|^2 + (\sin K\Lambda)^2]$. At the band edges, $K\Lambda = m\pi$, the reflectance is given by

$$|r_N|^2 = \frac{|C|^2}{|C|^2 + (1/N)^2}.$$

(6.3-9)

In the forbidden gap, $K\Lambda$ is a complex number

$$K\Lambda = m\pi + iK_i\Lambda.$$

(6.3-10)

The reflectance formula of Eq. (6.3-5) becomes

$$|r_N|^2 = \frac{|C|^2}{|C|^2 + (\sinh K_i\Lambda/\sinh NK_i\Lambda)^2}.$$

(6.3-11)

For large N, the second term in the denominator approaches zero exponentially as $e^{-2(N-1)K_i\Lambda}$. It follows that the reflectance in the forbidden gap is near unity for a Bragg reflector with a substantial number of periods.

The TE and TM waves have different band structures and reflectances. For TM waves incident at the Brewster angle, there is no reflected wave. This is due to the vanishing of the dynamical factor $|C|^2$ at that angle.

The reflectance for some typical Bragg reflectors as a function of frequency and angle of incidence are shown in Figs. 6.7 and 6.8.

6.3.1 Incidence from Air

If the light is incident from a medium (e.g., air) with an index of refraction different from n_1, the reflection coefficient is given by

$$r = \frac{r_{a1} + r_N}{1 + r_{a1}r_N},$$

(6.3-12)

where r_N is the coefficient given by Eq. (6.3-1) and r_{a1} is the Fresnel reflection coefficient between the air and medium 1 (see Problem 6.2).

At the center of each stop band, the period of the layered medium satisfies the condition

$$K\Lambda = m\pi,$$

(6.3-13)

where m is an integer and K is approximately k_x when $n_1 \simeq n_2$. The light waves will be highly reflected since successive reflections from the neighboring interfaces will be in phase with one another and therefore will be constructively superimposed. This situation is analogous to the Bragg reflection

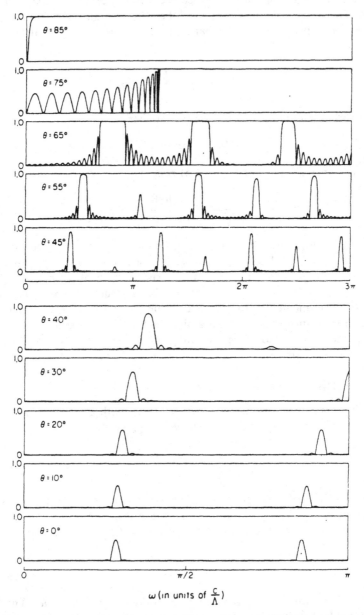

Figure 6.7 Reflectance spectrum of a 15-period Bragg reflector for TE waves at various angles of incidence. Indices of refraction of layers are taken as $n_1 = 3.4$ and $n_2 = 3.6$, and layer thicknesses are such that $a = b = \frac{1}{2}\Lambda$. (From A. Yariv and P. Yeh, *Optical Waves in Crystals*, Wiley, New York, p. 178. Copyright © 1984. Reprinted by permission of John Wiley & Sons, Inc.)

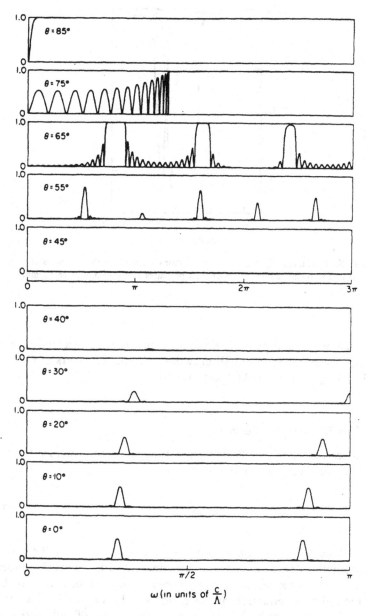

ω (in units of $\frac{c}{\Lambda}$)

Figure 6.8 Reflectance spectrum of a 15-period Bragg reflector for TM waves of various angles of incidence. The indices of refraction of layers are $n_1 = 3.4$ and $n_2 = 3.6$, and the layer thicknesses are such that $a = b = \frac{1}{2}\Lambda$. The Brewster angle θ_B is 46.6°. (From A. Yariv and P. Yeh, *Optical Waves in Crystals*, Wiley, New York, 1984, p. 179. Copyright © 1984. Reprinted by permission of John Wiley & Sons, Inc.)

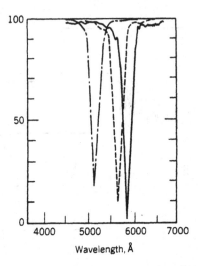

Figure 6.9   Spectral transmittance of Bragg reflector consisting of 85 layer pairs of alternating polystyrene and polyvinyl alcohol. The applied pressures are 0, 15.5, and 18.5 g/mm² [5].

of x rays from crystal planes. High-reflectance Bragg reflectors have been demonstrated by using the MBE growth of alternating layers of GaAs and $Al_{0.3}Ga_{0.7}As$ grown on GaAs substrates [4].

6.3.2   Tunable Bragg Reflectors

As discussed earlier, peak reflectance of a Bragg reflector occurs at the center of a forbidden band and can be written as

$$\lambda = 2n\Lambda \qquad (6.3-14)$$

for normal incidence, where n is the average index of refraction. We note that the wavelength of peak reflectance scales linearly with the average index of refraction and the period Λ. Thus, a Bragg reflector can be made tunable if either the index of refraction or the period can be varied. Such a tunable Bragg reflector has been fabricated using thin films of polystyrene and polyvinyl alcohol, and the spectral tuning is achieved by varying the layer thickness via compression [5]. Figure 6.9 shows the transmittance of such a Bragg filter.

6.4 FORM BIREFRINGENCE

Birefringence occurs naturally in crystals or materials that consist of periodic arrays of long or nonspherical molecules. The birefringent properties can be explained in terms of the electrical properties of molecules. Birefringence may also occur as a result of stratification, which destroys optical isotropy. Since the layers involved are normally much larger than molecules, such phenomenon is called *form birefringence*.

In the preceding sections, we have derived some of the most important characteristics of Bloch waves propagating in the periodic layered medium. An exact expression for the dispersion relation (6.2-8) between K, β, and ω was derived. This dispersion relation can be represented by contours of constant frequency in the βK plane, as in Fig. 6.10. It can be seen that these contours are more or less circular with only a slight ellipticity. The origin corresponds to the contour of zero frequency. In the long-wavelength regime ($\lambda \gg \Lambda$), these are similar to the dispersion curves of electromagnetic waves in a negative uniaxial crystal. The birefringence property of a periodic layered medium will now be discussed. These contours become distorted and modified at shorter wavelengths and near the boundaries of the Brillouin zone ($K\Lambda = m\pi$), where the wavelength is comparable to the dimension of a unit cell and the electromagnetic waves interact strongly with the periodic medium (Fig. 6.10).

Let us now consider the propagation of electromagnetic waves in an infinite medium consisting of alternating layers of two different homogeneous and isotropic substances. Although each individual layer is isotropic, the whole structure behaves as an anisotropic medium. The TE and TM waves are found to propagate with different effective phase velocities, and the periodic medium is birefringent.

If the period Λ is sufficiently small compared to the wavelength, the whole structure behaves as if it were homogeneous and uniaxially anisotropic. The wave given by Eq. (6.2-9) thus behaves as if it were a plane wave.

In Fig. 6.10, the contours of constant ω are plotted in the βK plane. These are sections of the normal surfaces with the βK plane for various frequencies. It is evident from inspection that at the long-wavelength limit ($\lambda \gg \Lambda$), the dispersions of a layered medium is qualitatively similar to that of a negative uniaxial crystal.

To demonstrate this analogy, we substitute Eqs. (6.1-13) and (6.1-15) for A and D in Eq. (6.2-8) and take the limit of $k_{1x}a \ll 1$, $k_{2x}b \ll 1$, and $K\Lambda \ll 1$ and expand all the transcendental functions. After neglecting higher order terms, we obtain

$$\frac{K^2}{n_o^2} + \frac{\beta^2}{n_o^2} = \frac{\omega^2}{c^2} \quad \text{(TE)}, \qquad (6.4\text{-}1)$$

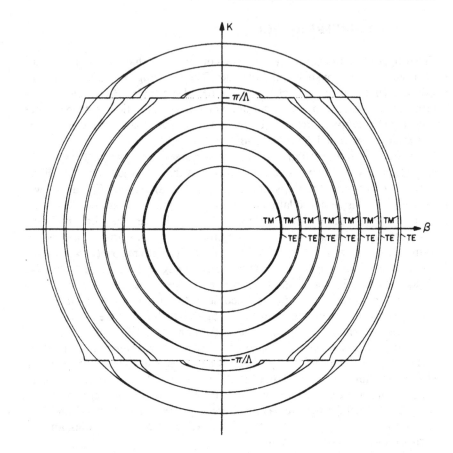

Figure 6.10 Contours of constant frequency in βK plane. (From A. Yariv and P. Yeh, *Optical Waves in Crystals*, Wiley, New York, 1984, p. 206. Copyright © 1984. Reprinted by permission of John Wiley & Sons, Inc.)

$$\frac{K^2}{n_o^2} + \frac{\beta^2}{n_c^2} = \frac{\omega^2}{c^2} \quad \text{(TM)}, \tag{6.4-2}$$

with

$$n_o^2 = \frac{a}{\Lambda} n_1^2 + \frac{b}{\Lambda} n_2^2, \tag{6.4-3}$$

$$\frac{1}{n_c^2} = \frac{a}{\Lambda} \frac{1}{n_1^2} + \frac{b}{\Lambda} \frac{1}{n_2^2}. \tag{6.4-4}$$

Equations (6.4-1) and (6.4-2) represent the two shells of the normal surface in the βK plane. One surface (6.4-1) applies to a TE wave and is a sphere, while the TM normal surface (6.4-2) is an ellipsoid of revolution. The TE waves thus are formally analogous to the so-called ordinary waves in a uniaxial crystal, while TM waves are the extraordinary waves. The normal surface becomes more complicated at higher frequencies. It consists of two oval surfaces osculating each other at the intersections with the K axis as long as the frequency is below the first forbidden gap. For frequencies higher than the forbidden gap, the oval surfaces break into several sections. The break points occur at

$$K = m\frac{\pi}{\Lambda}, \qquad m = \text{integer}, \qquad (6.4\text{-}5)$$

which is the Bragg condition.

Equations (6.4-3) and (6.4-4) are very important in the optical theory of dichroic polarizers as well as the synthesis of negative uniaxial crystals with prescribed properties. To illustrate this, let us consider a periodic layered medium that consists of alternating metal and dielectric materials. Let n_1 be the complex refractive index of the metal layer and n_2 be that of the dielectric layer. If the metal is a good conductor, $|n_1|^2 \gg n_2^2$. Thus, according to Eqs. (6.4-3) and (6.4-4), the ordinary and extraordinary refractive indices for long-wavelength light ($\lambda \gg \Lambda$) become

$$n_o \simeq n_1 \left(\frac{a}{\Lambda}\right)^{1/2}, \qquad (6.4\text{-}6)$$

$$n_e \simeq n_2 \left(\frac{b}{\Lambda}\right)^{-1/2}. \qquad (6.4\text{-}7)$$

We see that the ordinary refractive index n_o is like that of a metal and the extraordinary refractive index n_e is like that of a dielectric. Therefore, the ordinary wave (TE) will be reflected as if the medium is metallic, and the extraordinary wave (TM) will be transmitted. Physically, the TE waves have their electric field vectors parallel to the layers and induce currents in the metal layers; whereas the TM waves have their electric field vectors perpendicular to the layers. Since the metal layers are separated by dielectric insulating layers, no current flow will be induced.

The form birefringence has been observed in a periodic layered medium consisting of 0.1235-μm-thick AlAs and 0.1062-μm-thick GaAs grown by MBE [6]. The measured birefringence ($n_{\text{TE}} - n_{\text{TM}}$) as a function of wavelength is shown in Fig. 6.11.

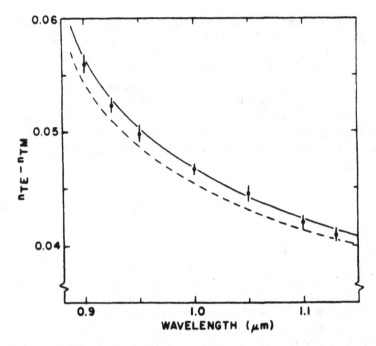

Figure 6.11 Measured form birefringence of a GaAs–AlAs layered medium. The Points are experimental results. The dashed curve is a plot of the theoretical prediction of the birefringence obtained from Eqs. (6.4-3) and (6.4-4). The solid curve is calculated from an expansion of Eq. (6.2-8). (From A. Yariv and P. Yeh, *Optical Waves in Crystals*, Wiley, New York, 1984, p. 208. Copyright © 1984. Reprinted by permission of John Wiley & Sons, Inc.)

6.5 RESONANT TUNNELING

Equation (6.3-6), derived in Section 6.3, is very useful for studying the tunneling of electromagnetic radiation through layered media. The subject of tunneling was studied in Section 4.4. The transmittance was estimated by Eq. (4.4-18). Referring to Fig. 6.12, we revisit the same problem by using the reflectance formula Eq. (6.3-6). If we assume that all the layers are lossless so that both n_1 and n_2 are real, the transmittance can be written as

$$|t_N|^2 = 1 - |r_N|^2 = \frac{1}{1 + |C|^2 (\sin N K\Lambda / \sin K\Lambda)^2}, \qquad (6.5\text{-}1)$$

where we recall that N is the number of periods and C and $K\Lambda$ are given by Eqs. (6.1-13) and (6.2-8).

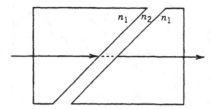

Figure 6.12 Schematic drawing of a low-index layered sandwiched between media of higher index of refraction.

Let the indices of refraction be such that $n_2 < n_1$ and the incidence condition be such that

$$n_2 \frac{\omega}{c} < \beta < n_1 \frac{\omega}{c} \tag{6.5-2}$$

so that the electromagnetic wave is evanescent in layers with index of refraction n_2. The structure depicted in Fig. 6.12 can be considered as a special case of the Bragg reflector with a period $N = 1$. The transmittance of such a structure can be written according to Eq. (6.5-1),

$$|t_1|^2 = \frac{1}{1 + |C|^2}, \tag{6.5-3}$$

where, according to Eq. (6.1-13), $|C|$ is given by

$$|C| = \frac{1}{2}\left(\frac{p}{q} + \frac{q}{p}\right) \sinh qb, \tag{6.5-4}$$

where b is the thickness of the barrier layer.

In the preceding equation, we have used $k_{1x} = p$ and $k_{2x} = -iq$, so that

$$p = \left[\left(\frac{n_1 \omega}{c}\right)^2 - \beta^2\right]^{1/2},$$

$$q = \left[\beta^2 - \left(\frac{n_2 \omega}{c}\right)^2\right]^{1/2}. \tag{6.5-5}$$

If the barrier layer n_2 is thick enough so that $qb \gg 1$, the transmittance $|t_1|^2$ can be written, according to Eq. (6.5-3), as

$$|t_1|^2 = \frac{16}{\left(\dfrac{p}{q} + \dfrac{q}{p}\right)^2} e^{-2qb} \tag{6.5-6}$$

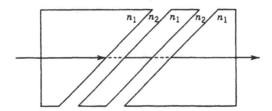

Figure 6.13 Schematic drawing of a layered structure consisting of two barrier layers separated by a layer of high index of refraction.

where we have used $\sinh qb \simeq \frac{1}{2} \exp(qb)$. Notice that transmittance is insignificant when the barrier layer is thick enough ($qb \gg 1$). Eq. (6.5-6) agrees with the earlier result (4.4-18).

6.5.1 Resonant Tunneling

According to Eqs. (6.5-3) and (6.5-6), transmittance under the incidence condition [Eq. (6.5-2)] is always less than unity and becomes small when the barrier layer is thick. We now consider the case when there are two barrier layers separated by a high index layer, as shown in Fig. 6.13. The transmittance of such a structure can still be described by Eq. (6.5-1) by taking $N = 2$.

Substituting $N = 2$ into Eq. (6.5-1), we obtain the transmittance of a double layer,

$$|t_2|^2 = \frac{1}{1 + 4|C|^2 \cos^2 K\Lambda}, \qquad (6.5\text{-}7)$$

with $\cos K\Lambda$ given by

$$\cos K\Lambda = \cos pa \cosh qb - \frac{1}{2}\left(\frac{p}{q} - \frac{q}{p}\right)\sin pa \sinh qb, \qquad (6.5\text{-}8)$$

where we recall that b is the thickness of the barrier layer and a is the thickness of the layer with the high index of refraction.

According to Eq. (6.5-7), we notice that resonant tunneling of electromagnetic radiation ($|t_2|^2 = 100\%$) occurs when $\cos K\Lambda = 0$. Since $\cos K\Lambda$ is a sinusoidal function of pa, $\cos K\Lambda = 0$ can always be achieved by varying the thickness a. Figure 6.14 plots the transmittance $|t_2|^2$ as a function of the layer thickness a. Since $\cos K\Lambda$ is a periodic function a, resonant tunneling can occur at several layer thicknesses a.

The transmission of electromagnetic radiation through a double barrier is very similar to that of a Fabry–Perot etalon, which will be discussed in the

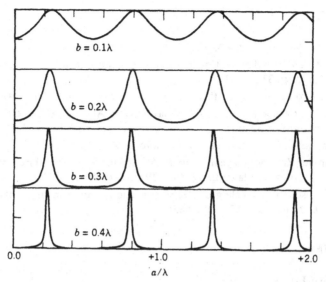

Figure 6.14 Resonant tunneling of electromagnetic radiation through a double barrier. Parameters: $n_1 = 1.0$, $n_2 = 1.50$, and $\beta = 1.20(\omega/c)$. Curves (from top to bottom) are for barriers with thicknesses b of 0.1λ, 0.2λ, 0.3λ, and 0.4λ.

next chapter. Here, each barrier acts like a high-reflectance mirror. If the separation between the mirrors is a multiple of half-wavelengths, high transmission occurs. In the case of the double barrier, resonant tunneling does not occur at exactly integral multiples of half-wavelengths. This is a result of the finite phase shift upon reflection from each individual barrier. According to Eqs. (6.5-7) and (6.5-8), resonant tunneling occurs at

$$\cot pa = \frac{1}{2}\left(\frac{p}{q} - \frac{q}{p}\right)\tanh qb. \tag{6.5-9}$$

Resonant tunneling can also occur in layered structures that contain multiple barrier layers. According to Eq. (6.5-1), resonant tunneling occurs when

$$\frac{\sin NK\Lambda}{\sin K\Lambda} = 0. \tag{6.5-10}$$

It is believed that the resonant tunneling of electromagnetic radiation is related to the excitation of guided modes supported by the high-index layers [7].

REFERENCES

1. E. Denton, *Sci. Am.* **224**, (1), 65 (1971).
2. P. Yeh, A. Yariv, and A. Y. Cho, Optical surface waves in periodic layered media, *Appl. Phys. Lett.* **32**, 104 (1978).
3. R. de L. Kronig and W. G. Penney, *Proc. Roy. Soc. (London)* **139**, 499 (1931).
4. J. P. van der Ziel and M. Illegems, Multilayer GaAs–Al$_{0.3}$Ga$_{0.7}$As dielectric quarter-wave stacks grown by molecular beam epitaxy, *Appl. Opt.* **14**, 2627 (1975).
5. M. Kimura, K. Okahara, and T. Miyamoto, *J. Appl. Phys.* **50**, 1222 (1979).
6. J. P. van der Ziel, M. Illegems, and R. M. Mikulyak, Optical birefringence of thin GaAs–AlAs multilayer films, *Appl. Phys. Lett.* **28**, 735 (1976).
7. P. Yeh, Resonant Tunneling of electromagnetic radiation in superlattice structures, *J. Opt. Soc. Am.* **A2**, 568 (1985).

PROBLEMS

6.1   Derive Eq. (6.3-6)

6.2   The reflection coefficient (6.3-5) is derived for the case when the incident medium has an index of refraction n_1. Consider the incidence from air by assuming the existence of an infinitesimally thin layer with index n_1. Derive Eq. (6.3-12).

6.3   Derive an asymptotic expression of the reflectance at the center of the forbidden band for large N.

6.4   Design a Bragg reflector using polyvinyl alcohol ($n_1 = 1.527$) and polystyrene ($n_2 = 1.582$) for a center peak reflectance of 99% at $\lambda = 500$ nm. Find the number of layers and layer thicknesses.

6.5   Using Eq. (6.1-1) and the periodic condition (6.1-2), we may carry out Fourier expansion of the index of refraction.

   (a) Let

$$n(x) = \sum_m C_m \exp(-imGx)$$

   where $G = 2\pi/\Lambda$. Find the expansion coefficients C_m.

   (b) Let

$$n^2(x) = \sum_m D_m \exp(-imGx)$$

   where $G = 2\pi/\Lambda$. Find the expansion coefficients D_m.

   (c) Derive Eq. (6.2-20).

6.6 Show that the forbidden band disappears for a half-wave stack when $k_1 a = k_2 b = \pi$.

6.7 *Bragg condition*

(a) Show that for TE waves

$$\cos K\Lambda = \cos k_{1x}a \cos k_{2x}b - \frac{1}{2}\left(\frac{k_{1x}}{k_{2x}} + \frac{k_{2x}}{k_{1x}}\right)\sin k_{1x}a \sin k_{2x}b.$$

(b) Show that $\cos K\Lambda$ for TE waves can also be written

$$\cos K\Lambda = \cos (k_{1x}a + k_{2x}b) - \frac{(k_{1x} - k_{2x})^2}{2k_{1x}k_{2x}} \sin k_{1x}a \sin k_{2x}b.$$

(c) For $|n_1 - n_2| \ll n_1, n_2$, $K\Lambda$ can be written approximately

$$K\Lambda \simeq k_{1x}a + k_{2x}b.$$

(d) Show that an approximate form of Bragg condition can be written

$$k_{1x}a + k_{2x}b = m\pi.$$

7

Some Applications of Isotropic Layered Media

In this chapter, we will address some of the important applications that use isotropic layered media. These include the Fabry–Perot interferometer, the Gires–Tournois etalon, high-reflectance coating, antireflection coating, spectral filters, edge filters, Christiansen–Bragg filters, and ellipsometry.

7.1 FABRY–PEROT INTERFEROMETER

The Fabry–Perot interferometer, or etalon, named after its inventors, can be considered as the simplest type of optical resonator. Normally, such an instrument consists of two parallel dielectric mirrors separated at a distance l. Here, we consider a simple structure that consists of a plane-parallel plate of thickness l and refractive index n immersed in a medium of index n'. The reflection and transmission coefficients r and t are, according to Eqs. (4.2-2) and (4.2-4), given by

$$r = \frac{(1 - e^{-2i\phi})\sqrt{R}}{1 - Re^{-2i\phi}}, \tag{7.1-1}$$

$$t = \frac{Te^{-i\phi}}{1 - Re^{-2i\phi}}, \tag{7.1-2}$$

where we used the fact that $r_{23} = -r_{12}$ as well as the definitions

$$R = r_{12}^2 = r_{23}^2, \qquad T = t_{12}t_{23}, \tag{7.1-3}$$

where R and T are, respectively, the fraction of the intensity reflected and transmitted at each interface and will be referred to in the following discussion as the mirror's reflectance and transmittance. If the incident intensity (watts per square meter) is taken as unity, we obtain, from Eq. (7.1-1), the following expression for the fraction of the incident intensity that is reflected:

$$|r|^2 = \frac{4R \sin^2 \phi}{(1 - R)^2 + 4R \sin^2 \phi}. \tag{7.1-4}$$

Moreover, from Eq. (7.1-2),

$$|t|^2 = \frac{(1 - R)^2}{(1 - R)^2 + 4R \sin^2 \phi}, \tag{7.1-5}$$

for the transmitted fraction. Our basic model contains no loss mechanism, so conservation of energy requires that $|r|^2 + |t|^2$ be equal to 1, as is indeed the case.

Let us consider the transmission characteristics of a Fabry–Perot etalon. According to Eq. (7.1-5), transmittance is unity whenever

$$\phi = \frac{2\pi n}{\lambda} l \cos \theta = m\pi, \qquad m = \text{integer.} \tag{7.1-6}$$

By using $\lambda = c/v$, condition (7.1-6) for maximum transmission can be written as

$$v_m = m \frac{c}{2nl \cos \theta}, \qquad m = \text{integer,} \tag{7.1-7}$$

where c is the velocity of light in a vacuum and v is the optical frequency. For a fixed l and v, Eq. (7.1-7) defines the unity transmission (resonance) frequencies of the etalon. Two neighboring resonance frequencies are separated by the so-called *free spectral range*

$$\Delta v \equiv v_{m+1} - v_m = \frac{c}{2nl \cos \theta}. \tag{7.1-8}$$

Theoretical transmission plots of a Fabry–Perot etalon are shown in Fig. 7.1. The maximum transmission is unity, as stated previously. The minimum transmission, on the other hand, approaches zero as R approaches unity.

If we allow for the existence of losses in the etalon medium, we find that the peak transmission is less than unity. Let α be the bulk absorption coefficient. Then Eq. (7.1-2) becomes

$$t = \frac{Te^{-i\phi}e^{-\alpha l}}{1 - Re^{-2i\phi}e^{-2\alpha l}}. \tag{7.1-9}$$

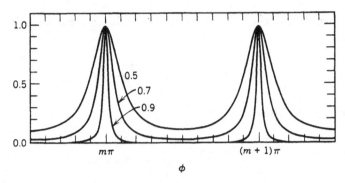

Figure 7.1 Transmission characteristics of a Fabry–Perot etalon for mirror reflectance $R =$ 0.5, 0.7, 0.9.

The maximum transmission drops from unity to

$$|t|^2_{max} = \frac{(1 - R)^2 e^{-2\alpha l}}{(1 - Re^{-2\alpha l})^2}. \qquad (7.1\text{-}10)$$

Figure 7.2 shows the transmission characteristics of a Fabry–Perot etalon by taking into account the effect of cavity losses.

If we write the amplitude transmission coefficient (7.1-2) as

$$t = |t|e^{-i\psi} \qquad (7.1\text{-}11)$$

and examine the phase shift ψ as a function of the cavity length l (or equivalently ϕ), we note that a strong dispersion of ψ versus ϕ exists at resonance. This is shown in Fig. 7.3.

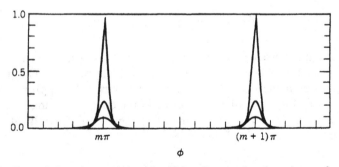

Figure 7.2 Transmission characteristics of a Fabry–Perot etalon for mirror reflectance of $R = 0.9$ and cavity loss 0, 0.05, and 0.1 (i.e., $\exp(-\alpha l) = 1.0, 0.95, 0.9$).

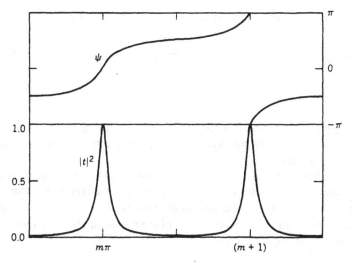

Figure 7.3 Transmittance and phase shift of a Fabry–Perot etalon as functions of cavity length (or equivalently ϕ).

In the preceding discussion, we used a plane-parallel plate of dielectric medium as a simplest type of Fabry–Perot interferometer. In practice, the Fabry–Perot interferometers, or etalons, are made of two high-reflectance mirrors that are parallel and are separated by a distance l. The transmission characteristics are identical to those of the plane-parallel plate. The intra-cavity electric fields, however, are very different in these two cases. According to our earlier discussion in Section 4.4, field amplitudes are equal to that of the incident wave at maximum transmission ($|t|^2 = 1$, see Table 4.1). This is no longer true in the case of the practical Fabry–Perot etalon, which consists of two mirrors separated at a distance in air.

Let I_i be the incident intensity and I_o be the transmitted intensity. Inside the Fabry–Perot etalon, we may split the electric field as the sum of a right-traveling wave and a left-traveling wave. Let I_1 and I_2 be the intensities of these waves, respectively. These intensities are related by

$$I_o = |t|^2 I_i, \tag{7.1-12}$$

$$I_o = TI_1 = (1 - R)I_1, \tag{7.1-13}$$

$$I_2 = RI_1, \tag{7.1-14}$$

where we recall that R is the mirror reflectance and $T = 1 - R$. According to these equations, the intracavity intensities are given by

$$I_1 = \frac{|t|^2}{1 - R} I_i,$$ (7.1-15)

$$I_2 = \frac{R|t|^2}{1 - R} I_i.$$ (7.1-16)

For mirror reflectance $R = 0.99$, the intracavity intensity I_1 in such a Fabry–Perot etalon is 100 times that of the incident intensity when maximum transmission occurs.

A Fabry–Perot etalon can be used as an optical spectrum analyzer. We now consider the characteristics of such a spectrum analyzer in terms of its resolving power and useful spectral range. Taking, for practical purposes, the case of normal incidence ($\theta = 0°$), we find the maximum transmission occurs when $\phi = m\pi$, or equivalently,

$$l = m \frac{\lambda}{2},$$ (7.1-17)

where m is an integer and n is taken as 1 for air. In other words, maximum transmission occurs when the cavity length is an exact integral multiple of a half-wavelength. Equation (7.1-17) can also be written as

$$v = m \frac{c}{2l},$$ (7.1-18)

where c is the velocity of light and v is the frequency of light. According to Eq. (7.1-18), a Fabry–Perot etalon at a fixed cavity length l can allow maximum transmission for a number of frequencies. These frequencies are separated by the so-called free spectral range (FSR),

$$\text{FSR} = \frac{c}{2l}.$$ (7.1-19)

For a Fabry–Perot etalon with $l = 15\,\text{cm}$, the free spectral range is 1 GHz.

If the cavity length is changed by dl, the corresponding change in the peak transmission frequency is given by

$$dv = \text{FSR} \frac{dl}{\lambda/2},$$ (7.1-20)

According to this equation, we can tune the peak transmission frequency of a Fabry–Perot etalon by one free spectral range by changing the cavity length by half a wavelength.

This property is employed in operating the Fabry–Perot etalon as a spectrum analyzer. The optical signal to be analyzed passes through the etalon as its length is being swept. If the width of the transmission peaks of the etalon is small compared to that of the variation of the spectral characteristics of the incident optical beam, the output of the etalon will constitute a replica of the spectral profile of the signal. In this application, it is important that the spectral width of the signal beam be smaller than the free spectral range of the etalon ($c/2l$) so that the ambiguity due to simultaneous transmission through more than one transmission peak can be avoided. For the same reason, the total cavity length scan is limited to $dl < \frac{1}{2}\lambda$.

The resolving power of a Fabry–Perot spectrum analyzer is limited by the finite width of its transmission peaks. We now calculate the full width of these peaks at their half-maximum values (FWHM). The FWHM is often taken as the resolution of the etalon. According to Eq. (7.1-5), half-maximum transmission (i.e., $|t|^2 = 0.5$) occurs at

$$\sin^2\phi = \frac{(1 - R)^2}{4R}.$$
(7.1-21)

For most high-resolution etalons, R is very close to 1. Thus, the solutions of Eq. (7.1-21) are given approximately as

$$\phi = m\pi \pm \frac{1 - R}{2\sqrt{R}}.$$
(7.1-22)

The FWHM in terms of frequency $\Delta v_{1/2}$ is thus given by

$$\Delta v_{1/2} = \frac{c}{2l}\frac{1 - R}{\pi\sqrt{R}}.$$
(7.1-23)

Or, defining finesse as

$$F \equiv \frac{\pi\sqrt{R}}{1 - R},$$
(7.1-24)

we obtain

$$\Delta v_{1/2} = \frac{c}{2lF}.$$
(7.1-25)

Finesse F, often used as a measure of the resolution of a Fabry–Perot etalon, is, according to Eq. (7.1-25), the ratio of the separation between peaks to the width of a transmission peak.

7.2 GIRES–TOURNOIS INTERFEROMETER

We now know that the intensity transmission of a given Fabry–Perot inter-ferometer is a function of wavelength. Thus, when a white light is incident onto a Fabry–Perot interferometer, the spectral distribution of the trans-mitted light is different from that of the incident one. In addition to the spectral modulation, the phase of the transmitted light is also being spectrally modulated. The phase dispersion is especially strong at peak transmission (see Fig. 7.3).

Most Fabry–Perot etalons are made of the two identical mirrors. In fact, the symmetry of the mirror reflectance is important to obtain high finesse (see Problem 7.1). Such symmetrical structures give high transmission only when the cavity length is an integral number of half-wavelengths. Any asymmetry in the mirror reflectance will lead to a decrease in either transmission or finesse.

We now consider an extreme case of an asymmetrical Fabry–Perot etalon, which consists of a rear mirror of 100% reflectance and a front mirror of finite reflectance. The front mirror is a partially reflecting dielectric coating with $R < 1$. This is the so-called Gires–Tournois etalon. The reflectance of the etalon is obviously 100% because light cannot pass through the second mirror and the whole stack is lossless. All the electromagnetic energy will be reflected provided the mirror reflectance remains 100% in the spectral regimes of interest. In fact, the reflection coefficient can be written, according to Eq. (4.2-2), as

$$r = \exp(-i\Phi) = \frac{-\sqrt{R} + e^{-2i\phi}}{1 - \sqrt{R}e^{-2i\phi}}, \qquad (7.2\text{-}1)$$

where we have taken $r_{12} = -\sqrt{R}$ and $r_{23} = 1$, and ϕ is given by

$$\phi = \frac{2\pi}{\lambda} nl. \qquad (7.2\text{-}2)$$

The phase shift on reflection Φ defined can be expressed in terms of ϕ as

$$\Phi = 2\tan^{-1}\left(\frac{1 + \sqrt{R}}{1 - \sqrt{R}} \tan\phi\right). \qquad (7.2\text{-}3)$$

In the limit when the reflectance of the front mirror vanishes ($R = 0$), this phase Φ is reduced to 2ϕ, which is simply a round-trip optical phase gained by the light beam. When the reflectance R is greater than zero ($R > 0$), the phase Φ is substantially increased because of the multiple reflections in the asymmetric Fabry–Perot cavity.

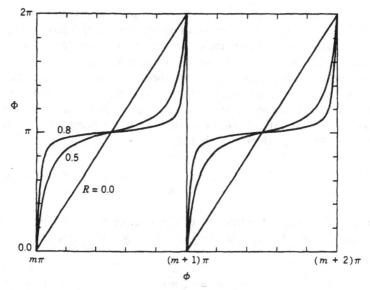

Figure 7.4 Phase shift of Gires–Tournois etalon with front mirror reflectance of $R = 0.0$. 0.5, 0.8.

Figure 7.4 plots the phase shift Φ versus the cavity length for various mirror reflectances R. Note that phase dispersion is very strong at resonance when $\phi = m\pi$.

7.3 ANTIREFLECTION COATING

It is known that when an electromagnetic radiation reaches a dielectric boundary between two media of different refractive indices, n_1 and n_2, respectively, Fresnel reflection occurs. The reflection is a result of the dielectric discontinuity and can be reduced or even eliminated by the proper coating of thin dielectric layers. We will discuss the case of simple dielectric coatings of a few layers of homogeneous material to eliminate reflections. In Chapter 9, we will discuss inhomogeneous layers that can also provide antireflection.

The intensity of reflected electromagnetic radiation at a boundary between two dielectric media is given by the reflectance

$$R = \frac{(n_2 - n_1)^2}{(n_2 + n_1)^2}. \tag{7.3-1}$$

For the case of a glass–air interface, this reflectance is approximately 4% in the visible spectral regime. In the case of optical systems that have many such dielectric interfaces, intensity transmission can be very low. Thus, it is important to eliminate or reduce such surface reflection.

The single homogeneous layer antireflection coating is the first and simplest dielectric structure to achieve this goal and is probably still the most widely used. The basic principle of such antireflection is briefly described in Section 4.4.2. Such antireflection consists of a single homogeneous layer a quarter-wavelength in thickness, that is,

$$t = \frac{\lambda}{4n},\qquad (7.3\text{-}2)$$

where n is the index of refraction of the coating layer. To achieve zero reflectance at the desired wavelength, the index of refraction must be given by

$$n = \sqrt{n_1 n_2}.\qquad (7.3\text{-}3)$$

In practice, materials with such an index of refraction may not exist. Nevertheless, using available materials with an index of refraction close to that given in Eq. (7.3-3), a great reduction in reflectance is obtained. This has been shown in Fig. 4.6 for $n_1 = 1.0$ and $n_2 = 2.25$. According to Eq. (4.2-2), the residual reflectance at the desired wavelength is given by

$$R = \frac{(n_1 n_2 - n^2)^2}{(n_1 n_2 + n^2)^2}.\qquad (7.3\text{-}4)$$

Note that residual reflectance is of the second order of the deviation in the index. We also notice that by using materials with an index of refraction between n_1 and n_2, a single-layer coating will reduce reflectance regardless of the layer thickness. Thus, the reflectance of a coated surface is always lower than that of an uncoated one, provided that the index of refraction of the coating is between that of the two media. In some sense, the coating layer smooths dielectric discontinuity.

It is possible to achieve antireflection by using more than one layer. There has been much previous work on multilayer antireflections. Here, we will limit ourselves to some basic properties of two-layer coatings. Interested readers are referred to Ref. 2 for more details. Generally speaking, the addition of layers offers the possibility of achieving zero reflectance at the desired wavelength using available materials, and the possibility of achieving low reflectance over a broader spectral region.

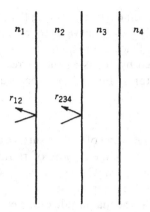

Figure 7.5 Schematic drawing of a two-layer coating.

Referring to Fig. 7.5, we consider the reflectance property of a two-layer coating. By generalizing Eq. (4.2-2), the reflection coefficient r of such a structure can be written as

$$r = \frac{r_{12} + r_{234}e^{-2i\phi_2}}{1 + r_{12}r_{234}e^{-2i\phi_2}}, \tag{7.3-5}$$

with r_{234} given by

$$r_{234} = \frac{r_{23} + r_{34}e^{-2i\phi_3}}{1 + r_{23}r_{34}e^{-2i\phi_3}}, \tag{7.3-6}$$

and

$$\phi_2 = \frac{2\pi}{\lambda}\,n_2 t_2,$$

$$\phi_3 = \frac{2\pi}{\lambda}\,n_3 t_3, \tag{7.3-7}$$

where n_2, n_3 and t_2, t_3 are the refractive indices and thicknesses of layers 2 and 3, respectively, and r_{12}, r_{23}, and r_{34} are the Fresnel reflection coefficients of the three dielectric interfaces, respectively. Here r_{234} is the reflection coefficient of the composite structure that consists of a layer of medium 3 sandwiched between medium 2 and medium 4. Such a reflection coefficient r_{234} can be obtained from Eq. (4.2-2) by using the appropriate subscripts. According to the discussion in the earlier part of this section, the magnitude of r_{234} can be changed continuously by varying the thickness of layer 3. This offers a degree of freedom to achieve zero reflectance by matching the magnitude of r_{12} and r_{234}. Once the magnitudes of r_{12} and r_{234} are the same, zero reflectance is achieved by choosing the correct thickness such that the two terms in the numerator of Eq. (7.3-5) cancel each other.

7.3.1 Example

Consider the antireflection coating of a glass surface at $\lambda = 0.5 \mu m$ using two layers of MgF_2 and MgO. Let the indices of refraction be $n_1 = 1.0$ (air), $n_2 = 1.37$ (MgF_2), $n_3 = 1.76$ (MgO), and $n_4 = 1.46$ (fused silica) (see also Table 1.1). According to the recipe described, we must choose the layer thickness of MgO such that the magnitude of r_{234} equals to that of r_{12}, which is -0.156 in this case. The magnitude of r_{234} for MgF_2–MgO–glass varies between $r_{234} = 0.032$ at zero thickness ($t_3 = 0$) and $r_{234} = 0.215$ at quarter-wave thickness (710 Å), according to Eq. (7.3-6).

By choosing a thickness of 356 Å for the MgO layer, r_{234} is a complex number given by [according to Eq. (7.3-6)]

$$r_{234} = 0.156 \exp[-i\,2.51], \qquad (7.3\text{-}8)$$

where 0.156 is the magnitude of r_{234}. To achieve antireflection, we must now select the thickness of layer 2 such that

$$r_{12} + r_{234} \exp[-2i\phi_2] = 0. \qquad (7.3\text{-}9)$$

Using $r_{12} = -0.156$ and Eqs. (7.3-7) and (7.3-8), this leads to a thickness of 1096 Å for the MgF_2 layer.

7.3.2 Example

By reversing the order of the two coating layers, we consider the possibility of antireflecting the same glass surface at $\lambda = 0.5 \mu m$, as in the example using two layers of MgF_2 and MgO. The indices of refraction are now $n_1 = 1.0$ (air), $n_2 = 1.76$ (MgO), $n_3 = 1.37$ (MgF_2), and $n_4 = 1.46$ (fused silica) (see also Table 1.1). Here, we must choose the layer thickness of MgF_2 such that the magnitude of r_{234} equals to that of r_{12}, which is -0.275 in this case. The magnitude of r_{234} for MgF_2–MgO–glass varies between $r_{234} = 0.093$ at zero thickness and $r_{234} = -0.156$ at the quarter-wave thickness (910 Å), according to Eq. (7.3-6). We see that it is impossible to match the magnitude of r_{234} and r_{12} in this case.

In most optical systems, antireflection is designed for normal incidence. It is important to keep in mind that the reflectance of any layered structure is a function of the angle of incidence. Therefore, antireflection at normal incidence does not mean zero reflectance at other angles of incidence. In addition, the reflectance at oblique incidence depends on the polarization state. Such dependence on polarization and angle of incidence can also be used for applications. These include thin-film polarizers, wave retardation plates, and ellipsometry.

7.4 ELLIPSOMETRY

As mentioned earlier, the complex reflection coefficients for the s- and p-polarized light are, in general, different for oblique incidence. Thus, a polarized light that is linearly polarized and contains both components of s and p waves may have a different state of polarization after reflecting from a layered structure. Ellipsometry is an experimental technique that studies such transformation of the polarization state upon reflection from or transmission through a layered structure. Such a technique is very useful for accurate measurement of the index of refraction and the thickness of thin layers (i.e., monolayer or atomic layer). In addition, because of the non-perturbing nature, such a technique is very suitable for in situ monitoring during the process of layer deposition [3].

Ellipsometry can be used to measure both the index of refraction and the thickness of either the transparent or absorbing layers. In most cases, a circularly polarized light or a linearly polarized light with equal s and p components is incident obliquely onto a layered structure. The reflected light is elliptically polarized since the two reflection coefficients differ in both phase and magnitude. By measuring the polarization state of the reflected light, the refractive index and the thickness may be found.

We now examine the polarization state of the reflected beam. Let r_s and r_p be the complex reflection coefficients for the perpendicular (s) and the parallel (p) component of the incident light, respectively. The polarization state of the reflected light can be represented by the following complex number:

$$\chi = \frac{r_p}{r_s} = e^{i\Delta} \tan\psi, \qquad (7.4\text{-}1)$$

where Δ and ψ are the so-called *ellipsometric angles*. A complete description of the polarization state that includes the orientation, sense of revolution, and ellipticity can be expressed in terms of Δ and ψ [4]. The angle ψ is limited between 0 and $\frac{1}{2}\pi$, whereas Δ is limited between π and $-\pi$. These two ellipsometric angles can be obtained experimentally by using a phase

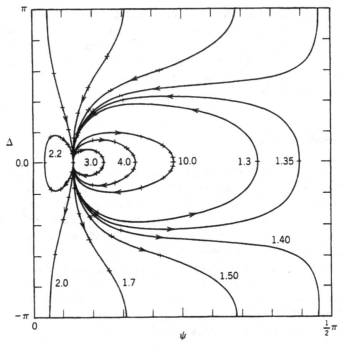

Figure 7.6 Iso-index contours in the plane of ellipsometric angles. Each contour corresponds to a film of fixed index of refraction with thickness increasing in the direction of the arrow. The substrate is a dielectric of refractive index $n = 2.5$, and the angle of incidence is $60°$.

compensator and a polarizer. The index of refraction and the film thickness can then be deduced from the data. Figure 7.6 plots contours of equal refractive index in the $\psi\Delta$ plane. Each curve corresponds to a film of fixed index of refraction with thickness l increasing in the direction of the arrow. Note that the iso-index contours have a common origin at $l = 0$, which corresponds to the film-free surface. The ellipsometric angles ψ and Δ are cyclic functions of thickness, and the curves repeat periodically with every $180°$ change in ϕ. For practical purposes, each point in the plane corresponds to a unique value for the index of refraction of the film. The film thickness, however, may need two independent measurements to uniquely determine its value. We also note that the uncertainty in determining the index of refraction is relatively large when the data point is near the origin.

   Although measurements of the ellipsometric angles ψ and Δ can yield information about the index of refraction and film thickness, there are other independent measurements that can also be used to obtain the same parameters of the film. For example, some ellipsometers use a rotating analyzer and detector combination to obtain information about the polarization state of the reflected light. Let θ be the angle between the analyzer and the plane of incidence. Then the intensity detected is given by

$$I = I_0[1 + A_2\cos(2\theta) + B_2\sin(2\theta)],$$

where I_0, A_2, and B_2 are constants that depend on the two complex reflection coefficients and the polarization state of the incident light. For a circularly polarized incident light, these parameters are given by

$$I_o = \tfrac{1}{2}(|r_p|^2 + |r_s|^2)I_i,$$

$$A_2 = \frac{|r_p|^2 - |r_s|^2}{|r_p|^2 + |r_s|^2},$$

$$B_2 = i\frac{r_p^* r_s - r_s^* r_p}{|r_p|^2 + |r_s|^2},$$

(7.4-2)

where we recall that r_p and r_s are complex reflection coefficients and I_i is incident intensity. Note that the sign of B_2 depends on the sense of revolution (or the handedness) of the circular polarization. Figure 7.7 plots the iso-index contours in the $A_2 B_2$ plane.

The A_2 and B_2 parameters may be easier to obtain experimentally. However, they are not necessarily more accurate in terms of deducing the index of refraction and the film thickness. By comparing Figs. 7.6 and 7.7, we note that all the iso-index contours are closed loops in the $A_2 B_2$ plane. This leads to the intersection of several iso-index contours and thus to ambiguity in determining the index of refraction.

7.5 HIGH-REFLECTANCE COATING

High-reflectance mirrors are desirable in many applications. These include high-finesse Fabry–Perot interferometers and low-loss laser resonators. Mirrors made of metallic films such as silver, aluminum, or gold are generally of high reflectance. For example, a silver mirror can achieve reflectance approaching 99% in the visible spectrum. Approximately 1% of light energy penetrates the surface of the metal and gets absorbed in the bulk of the metal. These metallic mirrors cannot be used with high-power lasers because even a small fraction of absorption can cause severe heating problems. Nor can these mirrors be used for high-finesse Fabry–Perot interferometry. Thus, there is a need to design high-reflectance mirrors by using materials that do not absorb light.

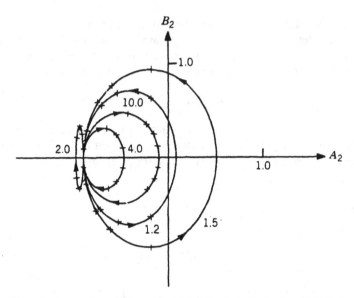

Figure 7.7 Iso-index contours of ellipsometry in $A_2 B_2$ plane. Each contour corresponds to a film of fixed index of refraction, with thickness increasing in the direction of the arrow. The substrate is a dielectric of refractive index $n = 2.5$, and the angle of incidence is 60°.

The dielectric layered structure that consists of alternating quarter-wave layers of two different materials is the simplest way of achieving high reflectance. This is the so-called Bragg reflector described in Chapter 6 and is a special type of periodic layered media. The reflection coefficient of such a Bragg reflector is given by Eq. (6.3-12), assuming that the substrate is the same material as one of the layers. Figure 7.8 and 7.9 show the reflectance spectrum for two Bragg reflectors, respectively. The peak reflectance of such a structure corresponds to the wavelength when all the layers are quarter-wave in thickness. Let t_1, t_2 and n_1, n_2 be the thicknesses and indices of refraction of the layers, respectively. Then this wavelength corresponds to

$$n_1 t_1 = \tfrac{1}{4}\lambda, \qquad n_2 t_2 = \tfrac{1}{4}\lambda. \qquad (7.5\text{-}1)$$

The peak reflectance is given by

$$R = \left(\frac{1 - (n_s/n_a)(n_2/n_1)^{2N}}{1 + (n_s/n_a)(n_2/n_1)^{2N}} \right)^2, \qquad (7.5\text{-}2)$$

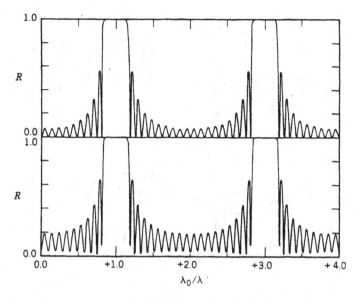

Figure 7.8 Reflectance spectrum of 10-period Bragg reflector consisting of alternating layers of dielectric media with indices of refraction $n_1 = 1.5$, $n_2 = 2.5$, and $n_s = n_1 = 1.5$. The layers of quarter-wave thickness correspond to wavelength λ_0. The top curve is for case when incident medium is n_1. The lower curve is for case of incidence from air.

where n_s represents the index of refraction of the substrate, n_a is that of the air, and N is the number of periods. For large N, this equation can be written as

$$R = \tanh^2\left(N\ln\frac{n_2}{n_1} + \frac{1}{2}\ln\frac{n_s}{n_a}\right). \qquad (7.5\text{-}3)$$

Note that peak reflectance approaches unity as N increases. The bandwidth of the Bragg reflectors is determined by the forbidden band of the periodic layered structure and is given by [according to Eq. (6.2-18)]

$$\frac{\Delta\lambda}{\lambda} = \frac{4}{\pi}\sin^{-1}\frac{|n_2 - n_1|}{n_2 + n_1}. \qquad (7.5\text{-}4)$$

This width is reduced by a factor of $2p + 1$ if the layer thicknesses are $n_1 t_1 = (2p + 1)\lambda/4$. Equation (7.5-4) can be written approximately as

$$\frac{\Delta\lambda}{\lambda} \simeq \frac{2}{\pi}\frac{\Delta n}{n}, \qquad (7.5\text{-}5)$$

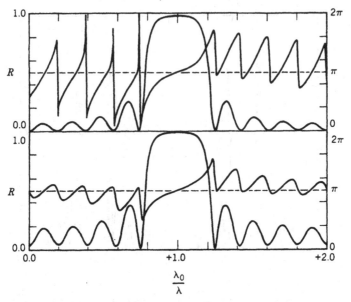

Figure 7.9 Reflectance spectrum of a five-period Bragg reflector similar to that described in Fig. 7.8. Here, the phase shift upon reflection is also shown. The lower curve is for case of incidence from air.

provided that $n_1 \simeq n_2$. Here, Δn is $|n_1 - n_2|$. Note that the bandwidth of the Bragg reflectors is determined by the fractional difference of the index of refraction of the neighboring layers or, equivalently, the fractional index modulation $\Delta n/n$ of the periodic layered medium.

Bragg reflectors can provide high reflectance over any desired spectral regime of interest by properly tailoring the layer thickness. Unlike metallic mirrors, there is a strong phase dispersion associated with Bragg reflection. The phase shift changes rapidly with wavelength outside the forbidden band. This is illustrated in Fig. 7.9 for the case of a five-period Bragg reflector. Note that the phase shift is 180° at the center of the forbidden band. For $n_1 > n_2$, this phase shift becomes 0 (see Problem 7.2).

Bragg reflectors can be made to reflect broad bands of light by stacking up several periodic layered media with different periods. In this case, each periodic medium acts as a band rejection filter for each wavelength. If the rejection bandwidths are wide enough to have substantial overlap, the whole structure can reject a broad band of light. Broad-band reflectors can also be made by using aperiodic layered medium in which the local period is an increasing (or decreasing) function of position (see Fig. 7.10). Such structures are called *chirped periodic layered medium*.

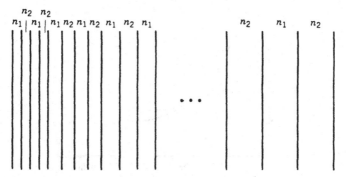

n_2 n_2
n_1 | n_1 | n_1 n_2 n_1 n_2 n_1 n_2 n_1 n_2 n_1 n_2

. . .

Figure 7.10 Schematic drawing of a chirped periodic layered medium.

7.6 FIELD OF VIEW OF SPECTRAL FILTERS

The Fabry–Perot etalon, as well as the Bragg reflector, can be used as a spectral filter that only allow certain wavelengths to transmit or reflect. We now examine the angular field of view (FOV) of these filters. Consider the case of a Fabry–Perot etalon that is designed to transmit a passband at wavelength λ_0. According to Eq. (7.1-6), this wavelength is related to the cavity length l by

$$nl = \tfrac{1}{2}m\lambda_0, \qquad (7.6\text{-}1)$$

where we recall that n is the index of refraction of the cavity medium. For light incident at an oblique angle θ, the passband is shifted to a new wavelength governed by the equation

$$nl\cos\theta' = \tfrac{1}{2}m\lambda, \qquad (7.6\text{-}2)$$

where θ' is the angle in the medium.

According to Eqs. (7.6-1) and (7.6-2), this new passband at an oblique angle of incidence is related to the passband at normal incidence by

$$\lambda = \lambda_0\cos\theta'. \qquad (7.6\text{-}3)$$

Note that the passband is shifted toward the shorter wavelength at an oblique angle of incidence. Even though it is derived for a Fabry–Perot etalon, Eq. (7.6-3) is generally true for all spectral filters made of layered structures. In these cases, the angle θ' is the effective (or averaged) ray angle in the layered medium. Let n be the effective index of refraction of the layered

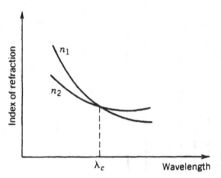

Figure 7.11 Crossing of dispersion curves.

medium. In most cases, the effective index of refraction is simply the averaged index of refraction. Equation (7.6-3) can be written as

$$\lambda = \lambda_0 \sqrt{1 - \left(\frac{\sin\theta}{n}\right)^2},$$ (7.6-4)

where θ is the angle of incidence in air.

As the passband shifts with the angle of incidence, a spectral filter is only useful for certain angular FOVs. For this purpose, the FOV of a spectral filter is defined as the full angular width at which the shift in wavelength equals half of the spectral bandwidth. According to this definition, a filter's FOV is related to its bandwidth and is given by

$$\theta = 2n\left(\frac{\Delta\lambda_{1/2}}{\lambda}\right)^{1/2},$$ (7.6-5)

where $\Delta\lambda_{1/2}$ is the bandwidth of the filter. According to this equation, the FOV of a narrow-band filter is very small. For example, a spectral filter with a bandwidth of 1 Å at $\lambda = 5000$ Å has a FOV of only 2°.

The limited FOV of narrow-band spectral filters is an inherent property of layered media in which the optical interference is based on the optical path length difference between the layers. Recently, a new class of filters was invented in which spectral discrimination is based on the dispersion of the dielectric materials. There is the so-called *Christiansen-Bragg filter* [5]. It is similar to a conventional Bragg reflector except that the two materials have different dispersion in such a way that the dispersion curve intersects at a desired wavelength (see Fig. 7.11). This layered structure is optically homogeneous only to the radiation of wavelength λ_c. Radiation of other wavelengths will see the index discontinuity at the interfaces and will be reflected, provided the layer thicknesses are properly chosen. In particular, if the layered structure is made of a chirped quarter-wave stack, this forms a broad-band reflector for all wavelengths except those near λ_c. A broad-band Bragg reflector generally has a wide FOV according to Eq. (7.6-5). In addition, the peak transmission at λ_c is a material property of the layered structure that is independent of the angle of incidence. In other words, the passband of a Christiansen–Bragg filter is independent of the angle of incidence. The details of such filters are beyond the scope of this book. Interested readers are referred to Ref. 5.

In addition to the application described in the previous sections, layered media can also be used as spectral edge filters (e.g., heat reflectors and cold mirrors), conical edge filters, beamsplitters, polarizers, wave plates, and so on.

REFERENCES

1. F. Gires and P. Tournois, An interferometer useful for pulse compression of a frequency-modulated light pulse, *C. R. Acad. Sci. (Paris)*, **258**, 6112 (1964).

2. J. A. Dobrowolski, Coatings and filters, in *Handbook of Optics*, W. G. Driscoll. eds., McGraw-Hill, New York, 1978.

3. R. M. A. Azzam and N. M. Bashara, *Ellipsometry and Polarized Light*, North-Holland, Amsterdam, 1977, Chapter 3.

4. A. Yariv and P. Yeh, *Optical Waves in Crystals*, Wiley, New York, 1984, Chapter 3.

5. P. Yeh, Christiansen–Bragg filters, *Opt. Comm.*, **35**, 9 (1980).

PROBLEMS

7.1 *Asymmetric Fabry–Perot etalon.*

(a) Derive an expression for the transmittance of an asymmetric Fabry–Perot etalon with mirror reflectance $R_1 \neq R_2$ and transmittance $T_1 \neq T_2$. *Answer:*

$$\tau = \frac{T_1 T_2}{(1 - \sqrt{R_1}\sqrt{R_2})^2 + 4\sqrt{R_1 R_2}\sin^2\phi}.$$

(b) Find the peak transmittance τ_{max} and show that the symmetric Fabry–Perot etalon always has the highest peak transmittance.

(c) Find the expression for the minimum transmittance τ_{min} and derive the expression for the rejection ratio τ_{max}/τ_{min}. Show that the symmetric Fabry–Perot etalon always has the highest rejection ratio.

7.2 *Phase shift of Bragg reflectors.* Consider a Bragg reflector made of alternating material of quarter-wave thicknesses. Show that the phase shift at the center of the stop band is 180° or 0°, depending on the magnitude of n_1 relative to n_2.

7.3 *Wave plates.* A thin film can be used as a wave plate that provides phase retardation between the s and p components. Use the iso-index contours in Fig. 7.6 to design a quarter-wave plate.

7.4 *Fabry–Perot etalon with gain.* The simplest model for a laser resonator is a Fabry–Perot etalon filled with gain medium. Let g be the gain coefficient. The transmission coefficient can be obtained by simply replacing the absorption coefficient α with $-g$ in Eq. (7.1-9):

$$t = \frac{Te^{-i\phi}e^{gl}}{1 - Re^{-2i\phi}e^{2gl}}.$$

(a) Show that transmittance can be greater than unity.
(b) Show that transmittance can reach infinity under appropriate conditions. Physically, infinite transmittance means that finite transmittance is obtained even when there is no incident light. This is exactly what happens when the laser resonator oscillates. Derive the oscillation conditions.
(c) Show that finesse F reaches infinity when such a resonator oscillates. Physically, an infinite finesse means that the classical bandwidth of a laser is zero. In practice, spontaneous emission leads to finite bandwidth of laser output.

7.5 *Two-layer antireflection coating.* The condition of zero reflectance is given by

$$r_{12} + r_{234}\exp[-2i\phi_2] = 0$$

according to Eq. (7.3-5).
(a) Show that for lossless medium, zero reflectance occurs only when

$$|r_{12}| = |r_{234}|.$$

(b) Show that the magnitude of r_{234} varies between

$$\left|\frac{n_2 - n_4}{n_2 + n_4}\right| \quad \text{and} \quad \left|\frac{n_2 n_4 - n_3^2}{n_2 n_4 + n_3^2}\right|.$$

(c) Shows that if $n_1 = 1.0$ and $|r_{24}| < |r_{12}|$, zero reflectance occurs only when

$$n_3^2 > \frac{n_2 n_4 (1 - r_{12})}{1 + r_{12}}.$$

(d) Show that, using $n_1 = 1$ and $r_{12} = (1 - n_2)/(1 + n_2)$, the necessary condition for zero reflectance becomes $n_3^2 > n_4 n_2^2$.

8

Inhomogeneous Layers

In this chapter, we consider "layered" structures that consist of layers whose index of refraction varies as a function of the depth of the layer. Layers with continuous varying index of refraction play an important role in spectral filters and broad-band antireflection coatings. Here, we will discuss in some detail the wave propagation in these inhomogeneous layers as well as the reflection and transmission properties of such structures. We will also discuss several structures which have index profiles that allow exact solutions. We will start with the WKB approximation, which is a general approach and can be applied to any index profile.

8.1 THE WKB APPROXIMATION

We consider the propagation of light in an inhomogeneous layer described by an index profile $n(x)$. The wave equation for light propagating along the x axis is given by

$$\frac{d^2}{dx^2} E(x) + \left(\frac{\omega}{c} n\right)^2 E(x) = 0, \tag{8.1-1}$$

where ω is the angular frequency and c is the velocity of light in a vacuum.

When the index profile $n(x)$ is uniform [i.e., $n(x) =$ constant], the solution is the well-known plane wave that, aside from a constant factor, can be written as

$$E(x) = \exp[-i\phi(x)], \tag{8.1-2}$$

where $\phi(x)$ is the phase and can be written as

$$\phi(x) = kx, \tag{8.1-3}$$

with k the wave number. The wave number is related to the index of refraction by

$$k = n\frac{\omega}{c} = n\frac{2\pi}{\lambda}, \qquad (8.1\text{-}4)$$

where we recall that λ is the wavelength of light in a vacuum.

This solution [Eq. (8.1-2)] is no longer valid when the index of refraction varies in space. However, in the case when the index profile $n(x)$ is a "slowly varying function" of x, the solution [Eq. (8.1-2)] may be considered as a first approximation. Substitution of Eq. (8.1-2) into the wave equation (8.1-1) yields

$$-i\left(\frac{d^2}{dx^2}\phi\right) - \left(\frac{d}{dx}\phi\right)^2 + \left(\frac{\omega}{c}n\right)^2 = 0. \qquad (8.1\text{-}5)$$

If we assume that $(d^2/dx^2)\phi$ is small, in other words,

$$\left|\frac{d^2}{dx^2}\phi\right| \ll \left(\frac{\omega}{c}n\right)^2, \qquad (8.1\text{-}6)$$

the first approximation becomes

$$\phi(x) = \pm\frac{\omega}{c}\int n(x)\, dx = \pm\frac{2\pi}{\lambda}\int n(x)\, dx. \qquad (8.1\text{-}7)$$

The condition of the validity of Eq. (8.1-6) [that $(d^2/dx^2)\phi$ be small relative to $(n\omega/c)^2$] is then

$$\left|\frac{1}{n}\frac{d}{dx}n\right| \ll \left|\frac{\omega}{c}n\right| = \left|\frac{2\pi}{\lambda}n\right|. \qquad (8.1\text{-}8)$$

Physically, this slowly varying condition means that the fractional change in the index of refraction in one wavelength should be small compared to the index of refraction.

A second approximation can now be obtained by iteration. From Eq. (8.1-7), we obtain

$$\frac{d^2}{dx^2}\phi = \pm\frac{\omega}{c}\frac{d}{dx}n. \qquad (8.1\text{-}9)$$

Substituting this for the small term in Eq. (8.1-5), we obtain

$$\left(\frac{d}{dx}\phi\right)^2 = \left(\frac{\omega}{c}n\right)^2 - i\left(\pm\frac{\omega}{c}\frac{d}{dx}n\right). \qquad (8.1\text{-}10)$$

$$\frac{d}{dx}\phi = \pm \frac{\omega}{c} n - \frac{i}{2}\frac{1}{n}\frac{d}{dx} n, \qquad (8.1\text{-}11)$$

$$\phi(x) = \pm \frac{\omega}{c} \int n(x)\, dx - \frac{i}{2} \ln n(x). \qquad (8.1\text{-}12)$$

The two choices of sign in Eq. (8.1-12) give two approximate solutions that may be combined to give the general solution

$$E(x) = \frac{1}{\sqrt{n(x)}}\left\{ A \exp\left[-i\frac{\omega}{c}\int n(x)\, dx\right] + B \exp\left[i\frac{\omega}{c}\int n(x)\, dx\right]\right\}, \qquad (8.1\text{-}13)$$

where A and B are arbitrary constants. The minus and plus sign obviously correspond to right-going and left-going waves, respectively.

There is a good physical reason for the amplitude denominator $\sqrt{n}$ in Eq. (8.1-13) in connection with the transport of energy. In a lossless medium, the flow of energy along the x axis must be independent of x. According to Eq. (1.4-21), the time-averaged Poynting's vector is given by

$$S = \tfrac{1}{2}cn\varepsilon_0 E^2. \qquad (8.1\text{-}14)$$

Since the index of refraction n is varying in an inhomogeneous medium, the amplitude must also vary correspondingly to keep the energy flow constant.

8.2   SOME EXACT SOLUTIONS

We now discuss solutions of the wave propagation [Eq. (8.1-1)] for various index profiles that support closed-form solutions. We shall adopt the description of inhomogeneity by using the index profile $n(x)$ instead of using the dielectric constant $\varepsilon(x)$.

8.2.1   Linear Layer 1

We first consider a dielectric layer whose index profile is linear and is given by

$$n(x) = n_0 + \frac{n_s - n_0}{L} x, \qquad (8.2\text{-}1)$$

where n_0 is the index of refraction at $x = 0$, n_s is the index of refraction at $x = L$, and L is the layer thickness. Figure 8.1 shows such a linear profile.

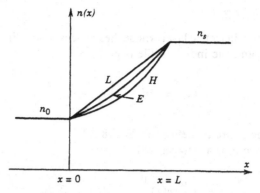

Figure 8.1 Index profiles of some inhomogeneous layers: L = linear, E = exponential, H = hyperbolic.

By introducing a new variable,

$$\xi = \frac{\omega}{c} n(x), \qquad d\xi = \alpha \, dx, \tag{8.2-2}$$

with

$$\alpha = \frac{\omega}{c} \frac{n_s - n_0}{L}. \tag{8.2-3}$$

the wave equation (8.1-1) becomes

$$\frac{d^2}{d\xi^2} E + \frac{1}{\alpha^2} \xi^2 E = 0. \tag{8.2-4}$$

The solution of Eq. (8.2-4) consists of Bessel functions of order $\frac{1}{4}$ and can be written as

$$E = \sqrt{\xi} \left[A J_{1/4} \left(\frac{1}{2\alpha} \xi^2 \right) + B Y_{1/4} \left(\frac{1}{2\alpha} \xi^2 \right) \right], \tag{8.2-5}$$

where A and B are arbitrary constants and $J_{1/4}$ and $Y_{1/4}$ are Bessel functions of the first and second kind, respectively.

8.2.2 Linear Layer 2

We now consider a dielectric layer whose dielectric profile is linear [recall that $\varepsilon(x) = \varepsilon_0 n^2(x)$], and the index profile is given by

$$n^2(x) = n_0^2 + \frac{n_s^2 - n_0^2}{L} x, \qquad (8.2\text{-}6)$$

where n_0, n_s, and L are as defined in Eq. (8.2-1).

Again, we introduce a new variable,

$$\xi = \left[\frac{\omega}{c} n(x)\right]^2, \qquad d\xi = \beta \, dx, \qquad (8.2\text{-}7)$$

with

$$\beta = \left(\frac{\omega}{c}\right)^2 \frac{n_s^2 - n_0^2}{L}. \qquad (8.2\text{-}8)$$

Substitution of Eqs. (8.2-7) and (8.2-8) into the wave equation (8.1-1) leads to

$$\frac{d^2}{d\xi^2} E + \frac{1}{\beta^2} \xi E = 0. \qquad (8.2\text{-}9)$$

The solution of Eq. (8.2-9) also consists of Bessel functions, but of order $\frac{1}{3}$, and can be written as

$$E = \sqrt{\xi}\left[AJ_{1/3}\left(\frac{2}{3\beta}\xi^{3/2}\right) + BY_{1/3}\left(\frac{2}{3\beta}\xi^{3/2}\right)\right], \qquad (8.2\text{-}10)$$

where A and B are arbitrary constants, and $J_{1/3}$ and $Y_{1/3}$ are Bessel functions of the first and second order, respectively.

8.2.3 Exponential Layer

Consider a dielectric layer whose index profile is given by

$$n(x) = n_0 \exp\left[\frac{x}{L} \ln\left(\frac{n_s}{n_0}\right)\right], \qquad (8.2\text{-}11)$$

where, again, n_0, n_s, and L are as defined in Eq. (8.2-1).

Let the new variable be

$$\xi \cdot = \frac{\omega}{c} n(x) = \frac{\omega}{c} n_0 e^{\gamma x}, \tag{8.2-12}$$

with γ given by

$$\gamma = \frac{1}{L} \ln \frac{n_s}{n_0}. \tag{8.2-13}$$

Substituting Eq. (8.2-12) into the wave equation and using

$$d\xi = \gamma\xi \, dx, \tag{8.2-14}$$

we obtain

$$\xi^2 \frac{d^2}{d\xi^2} E + \xi \frac{d}{d\xi} E + \frac{1}{\gamma^2} \xi^2 E = 0. \tag{8.2-15}$$

The solution of Eq. (8.2-15) consists of zero-order Bessel functions and can be written as

$$E = AJ_0\left(\frac{\xi}{\gamma}\right) + BY_0\left(\frac{\xi}{\gamma}\right), \tag{8.2-16}$$

where A and B are arbitrary constants.

8.2.4 Hyperbolic Layer

Consider a dielectric layer whose index profile is given by

$$n(x) = \frac{n_0}{1 - [(n_s - n_0)/n_s] \frac{x}{L}}. \tag{8.2-17}$$

To simplify the wave equation, we introduce the new variable:

$$\xi = 1 - \frac{n_s - n_0}{n_s} \frac{x}{L} \equiv 1 - \alpha x \tag{8.2-18}$$

with α given by

$$\alpha \equiv \frac{n_s - n_0}{n_s L}. \tag{8.2-19}$$

Substituting Eqs. (8.2-18) and (8.2-17) into the wave equation (8.1-1) and using $d\xi = -\alpha\, dx$, we obtain

$$\frac{d^2}{d\xi^2} E + \left(\frac{\omega n_0}{c\alpha}\right)^2 \frac{1}{\xi^2} E = 0. \qquad (8.2\text{-}20)$$

The solution of this equation can be written as

$$E = A\sqrt{\xi}\cos(m\ln\xi) + B\sqrt{\xi}\sin(m\ln\xi), \qquad (8.2\text{-}21)$$

where A and B are arbitrary constants and m is given by

$$m^2 = \left(\frac{\omega n_0}{c\alpha}\right)^2 - \frac{1}{4} = \left(\frac{\omega}{c}\right)^2 \left(\frac{n_0 n_s L}{n_s - n_0}\right)^2 - \frac{1}{4}. \qquad (8.2\text{-}22)$$

For reference purposes, the asymptotic forms of the various kinds of Bessel functions will be given for small and large values of their argument. Only the two leading terms will be given for simplicity:
For $z \ll 1$,

$$J_0(z) = 1 - \tfrac{1}{4}z^2,$$

$$J_1(z) = \tfrac{1}{2}z - \tfrac{1}{16}z^3,$$

$$Y_0(z) = \frac{2}{\pi}\left\{\ln\frac{z}{2} + 0.5772 \ldots\right\},$$

$$Y_1(z) = \frac{2}{\pi}\left\{\frac{1}{z} - \tfrac{1}{2}z\left(\ln\frac{z}{2} - \tfrac{1}{2} + 0.5772 \ldots\right)\right\}, \qquad (8.2\text{-}23)$$

$$J_\nu(z) = \frac{1}{\Gamma(\nu+1)}\left(\frac{z}{2}\right)^\nu, \qquad \nu \neq -1, -2, -3, \ldots,$$

$$Y_\nu(z) = -\frac{\Gamma(\nu)}{\pi}\left(\frac{2}{z}\right)^\nu, \qquad \nu > 0,$$

where ν is assumed to be real.
For $z \gg 1$,

$$J_\nu(z) = \left(\frac{2}{\pi z}\right)^{1/2}\cos\left(z - \frac{\nu}{2}\pi - \tfrac{1}{4}\pi\right),$$

$$Y_\nu(z) = \left(\frac{2}{\pi z}\right)^{1/2}\sin\left(z - \frac{\nu}{2}\pi - \tfrac{1}{4}\pi\right). \qquad (8.2\text{-}24)$$

The transition from the small-z behavior to the large-z asymptotic form occurs in the region $z \sim v$.

8.3 REFLECTANCE AND TRANSMITTANCE OF INHOMOGENEOUS LAYERS

The reflection and transmission of electromagnetic radiation from an inhomogeneous layer can be treated by matching the boundary conditions at $x = 0$ and $x = L$. To illustrate this, let us consider the exponential layer with an index profile given by Eq. (8.2-11). According to Eq. (8.2-16), the electric field can be written as

$$
E = \begin{cases}
Ce^{-ik_0 x} + De^{ik_0 x}, & x < 0, \\[2mm]
AJ_0\left(\dfrac{\xi}{\gamma}\right) + BY_0\left(\dfrac{\xi}{\gamma}\right), & 0 < x < L, \\[2mm]
Fe^{-ik_s x}, & L < x,
\end{cases}
\tag{8.3-1}
$$

where $k_0 = \omega n_0/c$, $k_s = n_s\omega/c$ are wave numbers, ξ is given by Eq. (8.2-12), C is the amplitude of the incident wave, D is the amplitude of the reflected wave, and F is the amplitude of the transmitted wave. Here, we recall that we are treating the case of normal incidence only.

The boundary condition requires that both the electric field and its derivative must be continuous at any point. Imposing these conditions at $x = 0$ and $x = L$ leads to

$$
\begin{aligned}
C + D &= AJ_0(y_0) + BY_0(y_0), \\
-ik_0(C - D) &= -k_0 AJ_1(y_0) - k_0 BY_1(y_0), \\
Fe^{-ik_s L} &= AJ_0(y_s) + BY_0(y_s), \\
-ik_s Fe^{-ik_s L} &= -k_s AJ_1(y_s) - k_s BY_1(y_s),
\end{aligned}
\tag{8.3-2}
$$

where

$$
y_0 = \frac{k_0}{\gamma} = \frac{2\pi n_0 L}{\lambda \ln(n_s/n_0)}, \qquad
y_s = \frac{k_s}{\gamma} = \frac{2\pi n_s L}{\lambda \ln(n_s/n_0)}.
\tag{8.3-3}
$$

In arriving at Eq. (8.3-2), we have used the following relationship of Bessel functions:

$$
J_0'(y) = -J_1(y), \qquad Y_0'(y) = -Y_1(y),
\tag{8.3-4}
$$

where the prime indicates differentiation.

The four equations in (8.3-2) can be solved for the four unknowns A, B, D, and F. Once these four amplitudes are found, the reflection and transmission coefficients can be obtained from the formulas

$$r = \frac{D}{C} \tag{8.3-5}$$

and

$$t = \frac{F}{C}, \tag{8.3-6}$$

respectively.

After a few steps of algebraic operation, we arrive at

$$r = \frac{[J_0(y_0) - iJ_1(y_0)][Y_0(y_s) - iY_1(y_s)] - [Y_0(y_0) - iY_1(y_0)][J_0(y_s) - iJ_1(y_s)]}{[J_0(y_0) + iJ_1(y_0)][Y_0(y_s) - iY_1(y_s)] - [Y_0(y_0) + iY_1(y_0)][J_0(y_s) - iJ_1(y_s)]}, \tag{8.3-7}$$

which is the reflection coefficient of an exponential layer, where J_0, J_1 and Y_0, Y_1 are Bessel functions of the first and second kind, respectively, and we recall that the arguments y_0 and y_s are

$$y_0 = \frac{2\pi n_0 L}{\lambda \ln(n_s/n_0)}, \qquad y_s = \frac{2\pi n_s L}{\lambda \ln(n_s/n_0)}.$$

Reflectance is obtained by taking the absolute square of Eq. (8.3-7) and is plotted as a function of L/λ in Fig. 8.2.

We will now examine the asymptotic forms of reflectance for both large wave numbers ($L/\lambda \gg 1$) and small wave numbers ($L/\lambda \ll 1$). This is especially useful for the case of the exponential profile because of the complicated form of Eq. (8.3-7).

Using the asymptotic forms for the various kinds of Bessel functions, we obtain the following asymptotic forms for reflectance:

$$R = \frac{(n_s - n_0)^2}{16\Phi^2 n_0^2 n_s^2}(n_s^2 + n_0^2 - 2n_0 n_s \cos 2\Phi), \qquad \frac{L}{\lambda} \gg 1, \tag{8.3-8}$$

$$R = \left(\frac{n_0 - n_s}{n_0 + n_s}\right)^2 \left[1 - \alpha\left(\frac{4\pi L}{\lambda}\right)^2\right], \qquad \frac{L}{\lambda} \ll 1, \tag{8.3-9}$$

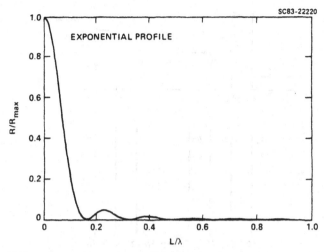

Figure 8.2 Reflectance spectrum of a graded transition layer with an exponential profile. $n_0 = 2.8$, $n_s = 3.5$.

where Φ is the total optical thickness and α is a positive constant. They are given by

$$\Phi = \frac{2\pi}{\lambda} \int_0^L n(z)\, dz = \frac{2\pi L (n_s - n_0)}{\lambda \ln (n_s/n_0)} \tag{8.3-10}$$

and

$$\alpha = \frac{n_0 n_s}{4 \ln^2 (n_s/n_0)} \left[1 - \frac{4 n_0^2 n_s^2}{(n_s^2 - n_0^2)^2} \ln^2 \frac{n_s}{n_0} \right], \tag{8.3-11}$$

respectively.

Note that for large wave numbers ($L/\lambda \gg 1$), the reflectance spectrum decays as Φ^{-2} and oscillates with a period $\Phi = \pi$. For small wave numbers ($L/\lambda \ll 1$), the reflectance decreases from its maximum value by a quadratic function L/λ.

The reflectance for other inhomogeneous layers can be obtained in a similar way. When closed-form solutions are not available, the WKB approximation can be used. When a computer is available, the reflectance and transmittance of any inhomogeneous layer can always be obtained by using an elementary method of integration. In this method, the index profile is replaced by a homogeneous multilayer of sufficiently high order whose refractive index pattern contours the index profile $n(x)$. Such a homogeneous multilayer will

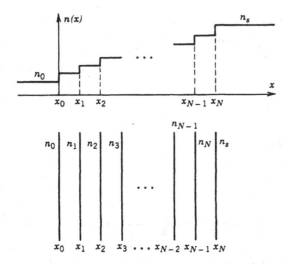

Figure 8.3 Schematic drawing of a graded multilayer dielectric structure.

be called a *graded multilayer*. Using such a subdivision and the matrix method described in Chapter 5, the reflectance and transmittance spectrum can be easily obtained with the help of a computer.

There are many situations when the homogeneous multilayer approximation leads to closed-form expressions for reflectance or transmittance. In the next section, we will discuss the properties of some interesting graded multilayer structures.

8.4 EXPONENTIALLY GRADED MULTILAYERS

Stratified media with graded refractive indices play an important role in the antireflection of broad-band optical waves. This application has benefited greatly from its similarity to electrical transmission line theory.

We now investigate the reflection and transmission of electromagnetic radiation from graded multilayer structures. It is shown that the transmission of an incoherent light wave is maximized when the refractive index is exponentially graded. The coherent reflectance and transmittance are also calculated and compared with those of the continuously varying case.

Referring to Fig. 8.3, we consider the transmission of electromagnetic radiation through a stack of nonabsorbing dielectric layers. If all the layers are of uniform thickness and optically homogeneous, the whole structure may be described by its index of refraction $n(x)$ as

$$
n(x) = \begin{cases}
n_0, & x < x_0, \\
n_1, & x_0 < x < x_1, \\
n_2, & x_1 < x < x_2, \\
\vdots & \vdots \\
n_N, & x_{N-1} < x < x_N, \\
n_s, & x_N < x,
\end{cases} \qquad (8.4\text{-}1)
$$

where n_0, n_s are the indices of refraction of the ambient media and n_i $(i = 1, \ldots, N)$ is the index of refraction of the ith layer. The axis x is chosen along the direction normal to the layer interfaces, and $x_0, x_1, x_2, \ldots, x_N$ are the positions of the interfaces. Note that $x_0 < x_1 < x_2 < x_3 < \cdots < x_{N-1} < x_N$. The layer thicknesses are $t_1 = x_1 - x_0$, $t_2 = x_2 - x_1, \ldots,$ $t_i = x_i - x_{i-1}, \ldots, t_N = x_N - x_{N-1}$. Given such a dielectric multilayer configuration, the reflectance and transmittance of a monochromatic plane optical wave from this structure can be calculated by using the matrix method described in Chapter 5. Generally speaking, the reflectance (or transmittance) spectrum will contain fine structural components that oscillate approximately with a period $\Delta\lambda \sim \lambda^2/2\bar{n}L$, where L is a relevant physical thickness parameter of the stack and $\bar{n}$ is the averaged index of refraction. However, there are many situations where such fine spectral interference oscillations cannot be observed. First, consider the case when the incident wave consists of radiation with a FWHM spectral width greater than the spectral oscillation $\Delta\lambda$. If there is no element to resolve the spectral oscillation, the measured transmission spectrum is spectrally averaged. Second, consider the case when the layers are imperfect and thick compared to the coherence length of the incident radiation (i.e., $L \gg \lambda^2/\Delta\lambda$). The fine structure of the transmittance spectrum will again disappear as a result of thickness averaging. An approximate treatment for calculating the transmission from such a stack is to take the summation of *intensities* (rather than amplitudes) due to all multiple reflections. The resulting transmittance shows no fine structure and is given by

$$
\frac{1}{T} = -N + \sum_{l=0}^{N} \frac{1}{T_l}, \qquad (8.4\text{-}2)
$$

where T_l is the transmittance of the interface at $x = x_l$.

To derive Eq. (8.4-2), we consider first the case of a single layer with two interfaces. Let T_1 and T_2 be the transmittance associated with the interfaces, respectively. Let I_0 be the incident intensity; the transmitted intensity can be obtained by summing the contribution from all the multiple reflections. Thus, the transmitted intensity can be written as

$$I = I_0 T_1 T_2 \{1 + R_1 R_2 + (R_1 R_2)^2 + (R_1 R_2)^3 + \cdots \}, \qquad (8.4\text{-}3)$$

where R_1 and R_2 are the reflectances associated with the interfaces and are given by

$$R_1 = 1 - T_1, \qquad R_2 = 1 - T_2, \qquad (8.4\text{-}4)$$

respectively.

The total resulting transmittance can thus be written [according to Eq. (8.4-3)] as

$$T = \frac{T_1 T_2}{1 - R_1 R_2}. \qquad (8.4\text{-}5)$$

Using Eq. (8.4-4) and inverting Eq. (8.4-5), we obtain

$$\frac{1}{T} = -1 + \frac{1}{T_1} + \frac{1}{T_2}. \qquad (8.4\text{-}6)$$

The general expression for N-layer structures can be easily obtained from Eq. (8.4-6) by mathematical induction.

Let us now turn our attention to the problem of finding the N indices of refraction, $n_1, n_2, n_3, \ldots, n_N$, such that this incoherent transmittance T [Eq. (8.4-2)] is maximized. Mathematically, this corresponds to

$$\frac{\partial T}{\partial n_l} = 0 \quad \text{and} \quad \frac{\partial^2 T}{\partial n_l^2} < 0 \quad \text{for } l = 1, 2, \ldots, N. \qquad (8.4\text{-}7)$$

Using Eq. (8.4-2) and $T_l = 4n_l n_{l+1}/(n_l + n_{l+1})^2$ for the case of normal incidence, that is,

$$T_l = \frac{4n_l n_{l+1}}{(n_l + n_{l+1})^2}, \qquad (8.4\text{-}8)$$

conditions (8.4-7) lead to

$$n_1^2 = n_0 n_2,$$
$$n_2^2 = n_1 n_3,$$
$$\vdots$$
$$n_l^2 = n_{l-1} n_{l+1},$$
$$\vdots$$
$$n_N^2 = n_{N-1} n_s.$$

(8.4-9)

Physically, this means that incoherent transmittance is maximized when the index of refraction of each layer is a geometric average of those of its neighboring layers. The solution is given by

$$n_l = n_0 \left(\frac{n_s}{n_0} \right)^{l/(N+1)}, \qquad l = 1, 2, \ldots, N.$$

(8.4-10)

Substitution of Eqs. (8.4-10) and (8.4-8) into Eq. (8.4-2) leads to a maximized transmittance of

$$T = \left\{ 1 + (N + 1) \sinh^2 \left[\frac{1}{2(N+1)} \ln\left(\frac{n_s}{n_0}\right) \right] \right\}^{-1}.$$

(8.4-11)

A discrete-element dielectric structure with indices of refraction given by Eq. (8.4-10) is called an *exponentially graded-index multilayer stack*. Its transmittance T increases as the number of layers, N, increase. For the case when $\ln(n_s/n_0) \ll N$, transmittance T becomes

$$T \simeq 1 - \frac{1}{4(N+1)} \ln^2 \frac{n_s}{n_0}.$$

(8.4-12)

We now investigate the transmission and reflection of coherent electromagnetic radiation from such an exponentially graded multilayer structure. The matrix method described in Chapter 5 is particularly useful for this purpose. If the layer thicknesses are arbitrary, a computer program may be needed for the calculation. We will, however, consider a special case of particular interest. In this case, all the layers are taken to have exactly the same optical thickness, that is,

$$\frac{2\pi}{\lambda} n_l t_l = \phi, \qquad l = 1, 2, \ldots, N,$$

(8.4-13)

where λ is the wavelength, t_l is the thickness of the lth layer, and ϕ is a constant. For normal incidence, all the matrices linking the amplitudes of left- and right-traveling waves of the neighboring layers are identical, and the product can be simplified by using the Chebyshev identity.

According to the 2 × 2 matrix method described in Section 5.1, the relation between incident amplitude A_0, reflected amplitude B_0, and transmitted amplitude A_s can be written as

$$\begin{pmatrix} A_0 \\ B_0 \end{pmatrix} = [D_0^{-1}D_1P_1][D_1^{-1}D_2P_2] \cdots [D_N^{-1}D_sP_s]\begin{pmatrix} A_s \\ 0 \end{pmatrix}, \quad (8.4\text{-}14)$$

which, aside from a factor of P_s, is identical to Eqs. (5.1-26) and (5.1-27). The D_l's and P_l's are the dynamical and propagation matrices defined in Section 5.1.

As a result of the refractive indices relationship Eq. (8.4-10) and the identical optical thickness [Eq. (8.4-13)], the resultant matrices within each of the square brackets in Eq. (8.4-14) are all identical (for normal incidence only) and are written as

$$D_0^{-1}D_1P_1 = D_1^{-1}D_2P_2 = \cdots = D_N^{-1}D_sP_s$$

$$= \begin{pmatrix} 1 & 1 \\ n_0 & -n_0 \end{pmatrix}^{-1}\begin{pmatrix} 1 & 1 \\ n_1 & -n_1 \end{pmatrix}\begin{pmatrix} e^{i\phi} & 0 \\ 0 & e^{-i\phi} \end{pmatrix} = \sqrt{\beta}\begin{pmatrix} A & B \\ C & D \end{pmatrix},$$

$$(8.4\text{-}15)$$

where

$$\beta = \left(\frac{n_s}{n_0}\right)^{1/(N+1)},$$

$$A = \frac{1+\beta}{2\sqrt{\beta}}e^{i\phi},$$

$$B = \frac{1-\beta}{2\sqrt{\beta}}e^{-i\phi}, \qquad (8.4\text{-}16)$$

$$C = \frac{1-\beta}{2\sqrt{\beta}}e^{i\phi},$$

$$D = \frac{1+\beta}{2\sqrt{\beta}}e^{-i\phi}.$$

The factor $\sqrt{\beta}$ in Eq. (8.4-15) is chosen so that the matrix is unimodular, that is, $AD - BC = 1$.

Equation (8.4-14) can thus be written as

$$\begin{pmatrix} A_0 \\ B_0 \end{pmatrix} = \frac{n_s}{n_0} \begin{pmatrix} A & B \\ C & D \end{pmatrix}^{N+1} \begin{pmatrix} A_s \\ 0 \end{pmatrix}. \tag{8.4-17}$$

Using the Chebyshev identity Eq. (6.3-3) and Eq. (5.2-3), we obtain the following expression for the reflection coefficient:

$$r = \frac{B_0}{A_0} = \frac{C \sin(N+1)y}{A \sin(N+1)y - \sin Ny}, \tag{8.4-18}$$

where A and C are given by Eq. (8.4-16) and

$$y = \cos^{-1}\left(\frac{1+\beta}{2\sqrt{\beta}}\cos\phi\right). \tag{8.4-19}$$

The reflectance is obtained by taking the absolute square of r. Using Eq. (8.4-18) and following a similar procedure that leads to Eq. (6.3-6), we obtain

$$R = |r|^2 = \left(1 + \frac{\sin^2 y}{|C|^2 \sin^2(N+1)y}\right)^{-1}. \tag{8.4-20}$$

The derivation of this reflectance formula is left as an exercise for the student (see Problem 8.4).

From Eqs. (8.4-19) and (8.4-20), we note that reflectance R is a periodic function of ϕ with a period π. This is to be expected because an increment $\Delta\phi = \pi$ merely means that an additional N half-wave layer has been included. Since the addition of any half-wave layer will not affect the transmittance and reflectance at that wavelength, the reflectance formula [Eq. (8.4-20)] is periodic in ϕ with a period π. At $\phi = 0, \pi, 2\pi, \ldots, y = \pm i \ln\sqrt{\beta}$ according to Eq. (8.4-19), and reflectance $R = (n_s - n_0)^2/(n_s + n_0)^2$ according to Eq. (8.4-20), which is the reflectance for the interface between n_0 and n_s. These are the peak reflectances. Between any two neighboring peaks [i.e., $\phi = m\pi, (m+1)\pi$], there are exactly N nodes where reflectance vanishes. These nodes occur at $y = \pi/(N+1), 2\pi/(N+1), \ldots, N\pi/(N+1)$, where $\sin(N+1)y = 0$. For $N \gg 1$, $\sin^2(N+1)y$ is a fast-varying function of y (or ϕ). Therefore, it dominates the oscillation structure of the reflectance spectrum. There are exactly $N-1$ sidelobes under the envelope,

$$\langle R \rangle = \frac{1}{1 + \sin^2 y/|C|^2}. \tag{8.4-21}$$

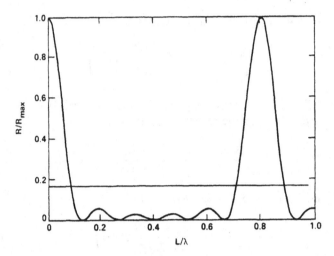

Figure 8.4 Reflectance spectrum of an exponentially graded multilayer structure with $N = 5$, $n_0 = 2.8$, $n_s = 3.5$, and $R_{max} = 0.0125$. The flat horizontal line has been calculated using Eq. (8.4-11).

The reflectance for an exponentially graded multilayer as a function of L/λ is shown in Fig. 8.4, where L is the total thickness of the graded layers.

Next, consider the limiting case when $N \rightarrow \infty$ while the total layer thickness L remains constant. In this limit, the refractive index profile becomes continuous and can be obtained from Eq. (8.4-10) by finding the relationship between l and x. As a result of the constraint of Eq. (8.4-13), x is not linearly proportional to l. Let $x_0 = 0$ and $x_N = L$. The relation x and l can be written as

$$x = \sum_{m=1}^{l} t_m, \qquad (8.4\text{-}22)$$

where m is a dummy index.

Using Eqs. (8.4-10) and (8.4-13), we can write

$$t_m = \frac{\lambda\phi}{2\pi n_m} = \frac{\lambda\phi}{2\pi n_0}\left(\frac{n_s}{n_0}\right)^{-m/(N+1)}. \qquad (8.4\text{-}23)$$

Substituting Eq. (8.4-23) for t_m in Eq. (8.4-12), we obtain

$$x = \frac{\lambda\phi}{2\pi n_0}\frac{\beta^{-l} - 1}{1 - \beta}, \qquad (8.4\text{-}24)$$

where we recall that $\beta = (n_s/n_0)^{1/(N+1)}$.

Since x is the position of the lth layer, the index of refraction n_l can now be written as $n(x)$ for the limiting case. From Eqs. (8.4-10) and (8.4-16), we obtain

$$n(x) = n_0 \left(\frac{n_s}{n_0} \right)^{l/(N+1)} = n_0 \beta^l. \tag{8.4-25}$$

Substituting Eq. (8.4-25) for β^{-l} in Eq. (8.4-24), we obtain

$$n(x) = n_0 \left[\left(1 + \frac{2\pi n_0}{\lambda \phi} (1 - \beta)x \right)^{-1} \right]. \tag{8.4-26}$$

When N approaches infinity, the coefficient of x in the denominator tends to be a constant. In fact, $n_0(1 - \beta)$ is $n_0 - n_1$ and ϕ is $2\pi n_0 t_1/\lambda$. Thus, we have

$$\lim_{N \to \infty} \frac{2\pi n_0}{\lambda \phi} (1 - \beta) = -\frac{1}{n_0} \left(\frac{dn}{dx} \right)_{x=0} = \text{const.}$$

By matching the index of refraction at $x = L$, we obtain

$$n(x) = \left(1 - \frac{n_s - n_0}{n_s} \frac{x}{L} \right)^{-1}, \tag{8.4-27}$$

where we have taken $x_0 = 0$ and $x_N = L$. This is the hyperbolic profile discussed earlier. We will now obtain the reflectance formula for this profile by taking the limit $N \to \infty$ in Eq. (8.4-20).

When $N \to \infty$, each layer thickness approaches zero (as $1/N$), and the total optical thickness is given by

$$\Phi = \lim_{N \to \infty} N\phi = \frac{2\pi}{\lambda} \int_0^L n(x)\, dx = \frac{2\pi n_0 n_s L}{\lambda(n_s - n_0)} \ln \frac{n_s}{n_0}. \tag{8.4-28}$$

The parameters $|C|$ and y also approach zero as $1/N$ and, according to Eq. (8.4-16), are given by

$$|C| \to \frac{1}{2N} \ln \frac{n_s}{n_0}, \tag{8.4-29}$$

$$y \to \frac{1}{N} \left[\Phi^2 - \left(\tfrac{1}{2} \ln \frac{n_s}{n_0} \right)^2 \right]^{1/2}. \tag{8.4-30}$$

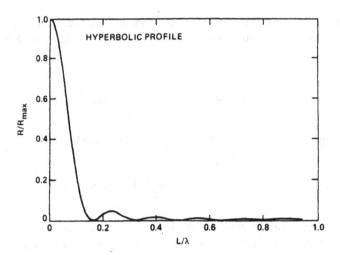

Figure 8.5 Reflectance spectrum of a graded transition layer with a hyperbolic profile, $n_0 = 2.8$, $n_s = 3.5$.

Using Eqs. (8.4-29) and (8.4-30), we obtain the limiting value for reflectance from Eq. (8.4-20):

$$R = \left(1 + \frac{\Phi^2 - \Delta^2}{\Delta^2 \sin^2 \sqrt{\Phi^2 - \Delta^2}}\right)^{-1}, \qquad (8.4.31)$$

where Δ^2 is given by

$$\Delta^2 = \tfrac{1}{4}\ln^2 \frac{n_s}{n_0}. \qquad (8.4\text{-}32)$$

This reflectance is plotted as a function of L/λ in Fig. 8.5. Note that peak reflectance $R_{max} = (n_s - n_0)^2/(n_s + n_0)^2$ occurs at $\Phi = 0$ (or, equivalently, $L/\lambda = 0$). There are sidelobes that occur at $\sin^2 \sqrt{\Phi^2 - \Delta^2} = 1$. In practice, $\Delta^2 = \tfrac{1}{4}\ln^2(n_s/n_0) \ll 1$, so that these sidelobes occur approximately at $\Phi = \tfrac{3}{2}\pi, \tfrac{5}{2}\pi, \ldots$, and under the envelope,

$$\langle R \rangle = \left(1 + \frac{4\Phi^2}{\ln^2(n_s/n_0)}\right)^{-1}. \qquad (8.4\text{-}33)$$

Zero reflectance occurs at approximately $\Phi = \pi, 2\pi, 3\pi, \ldots$.

Consider now the case of large layer thicknesses (i.e., $L \gg \lambda$). In this case, $\Phi \gg \Delta \gg 1$, and Eq. (8.4-31) can be written as

$$R = \tfrac{1}{4} \ln^2 \frac{n_s}{n_0} \frac{\sin^2 \Phi}{\Phi^2}, \qquad \frac{L}{\lambda} \gg 1, \qquad (8.4\text{-}34)$$

where we recall that $\Phi = 2\pi n_0 n_s L \ln (n_s/n_0)/[\lambda(n_s - n_0)]$.

In the preceding discussion on the limiting case when $N \to \infty$, we keep the same optical thickness for each layer of Eq. (8.4-13). This leads to the hyperbolic index profile. If we assume that all the layers have the same physical thickness t, the limiting continuous profile when $N \to \infty$ and $t \to 0$ as L/N becomes the exponential profile

$$n(x) = n_0 \exp\left(\frac{x}{L} \ln \frac{n_s}{n_0} \right), \qquad (8.4\text{-}35)$$

since, in this case, x is proportional to the index l in Eq. (8.4-10).

8.4.1 Exponentially Graded Quarter-Wave Stack for Broad-Band Antireflection Coatings

We have shown that an exponentially graded multilayer maximizes the incoherent transmission of light. From Fig. 8.4, we also note that an exponentially graded quarter-wave stack is ideal for the broad-band anti-reflection of coherent radiation. According to Eq. (8.4-20), reflectance is given approximately by

$$R = |C|^2 \frac{\sin^2(N + 1)y}{\sin^2 y} \qquad (8.4\text{-}36)$$

because $|C| \ll 1$ for most practical cases [see Eq. (8.4-16)]. By carrying out the integration over the region from $y = 0$ to $y = \pi$, we obtain the averaged reflectance in this region,

$$\langle R \rangle = (N + 1)|C|^2. \qquad (8.4\text{-}37)$$

In arriving at Eq. (8.4-37), we have used the integral

$$\int_0^\pi \frac{\sin^2 ax}{\sin^2 x} \, dx = a\pi.$$

Using Eq. (8.4-16), this averaged reflectance can be written as

$$\langle R \rangle = (N + 1) \sinh^2 \left[\frac{1}{2(N + 1)} \ln \frac{n_s}{n_0} \right]. \qquad (8.4\text{-}38)$$

Compared with Eq. (8.4-11), we note that this averaged reflectance is equal to the incoherent reflectance. If we were to average over the sidelobes [i.e., between $y = \pi/(N + 1)$ and $N\pi/(N + 1)$], the averaged reflectance would be even lower. In this region, the phase ϕ is approximately equal to y according to Eq. (8.4-19). Thus, if the layers are quarter-wave at λ_0, the exponentially graded N-layer stack will work as a broad-band antireflection coating covering the spectral regime between $(N + 1)\lambda_0/2$ and $(N + 1)\lambda_0/2N$. For a five-layer graded coating, this spectral region is from $0.6\lambda_0$ to $3\lambda_0$. For $n_s = 3.5$ and $n_0 = 2.8$, such a five-layer graded coating cuts the reflectance from 1.2 to 0.035%.

8.5 SINUSOIDAL LAYERS

We now discuss the properties of an inhomogeneous layer whose index profile is given by

$$n(x) = n_0 + n_1 \cos(Kx), \qquad (8.5\text{-}1)$$

or

$$n(x) = n_0 + n_1 \sin(Kx), \qquad (8.5\text{-}2)$$

where n_0, n_1, and K are constants. The constant n_0 is the averaged index of refraction of the medium, n_1 may be regarded as the depth of the sinusoidal index modulation, and K is related to the period of the index variation Λ by

$$K = \frac{2\pi}{\Lambda}. \qquad (8.5\text{-}3)$$

The depth of modulation is often much smaller than the averaged index of refraction (i.e., $n_1 \ll n_0$). The constant K is also called *the grating momentum* or *the grating wave vector*.

Optical media with sinusoidal index profiles are often called volume gratings. Such index variation can be produced by varying the ratio between the rates of two sources containing substances with different refractive indices. In nonlinear optics and real-time holography, such volume gratings can be produced by two-beam interference inside a nonlinear medium. Spectral filters based on the sinusoidal variation of the refractive index are known as Bragg reflectors. Recently, the term *rugate filter* is also common because of the similarity between the index profile and the wrinkle (*rugate* means wrinkled in Latin).

To investigate the properties of such layers, we substitute Eq. (8.5-1) for n into the wave equation (8.1-1). Using the assumption $n_1 \ll n_0$ and neglecting the small second-order term, we obtain

$$\frac{d^2}{dx^2} E + \frac{\omega^2}{c^2} (n_0^2 + 2n_0 n_1 \cos Kx)E = 0. \qquad (8.5\text{-}4)$$

By introducing a new variable $Kx = 2z$, Eq. (8.5-4) can be transformed into Mathieu's equation, and the solution can be written as Mathieu functions [1, 2]. Although Mathieu functions are tabulated in most mathematical handbooks, many readers may not be familiar with them. Therefore, we will treat the solution of Eq. (8.5-4) by using the so-called coupled-mode analysis [3]. In mathematics, this approach is called the method of variation of constants. The procedure consists of expressing the electric field in terms of the normal modes of the unperturbed structure, where the expansion coefficients evidently depend on x. In other words, the solution of Eq. (8.5-4) is written as

$$E = A(x)\exp(-ik_0 x) + B(x)\exp(ik_0 x), \qquad (8.5\text{-}5)$$

where $\exp(-ik_0 x)$ and $\exp(ik_0 x)$ are the normal modes of the unperturbed structure, with k_0 given by

$$k_0 = \frac{2\pi n_0}{\lambda} = \frac{\omega}{c} n_0. \qquad (8.5\text{-}6)$$

When $A(x)$ and $B(x)$ are constants, Eq. (8.5-5) is the general solution of the unperturbed medium. The dependence on x is necessary when $n_1 \neq 0$ because $\exp(-ik_0 x)$ and $\exp(ik_0 x)$ are no longer the normal modes of the whole structure.

Substituting Eq. (8.5-5) into the wave equation (8.5-4) and using Eq. (8.5-6), we obtain

$$A''e^{-ik_0 x} + B''e^{ik_0 x} - 2ik_0 A'e^{-ik_0 x} + 2ik_0 B'e^{ik_0 x}$$

$$+ 2\frac{\omega^2}{c^2} n_0 n_1 (Ae^{-ik_0 x} + Be^{ik_0 x})\cos Kx = 0, \qquad (8.5\text{-}7)$$

where a prime indicates differentiation with respect to x.

We now assume further that the sinusoidal index variation is "weak" so that the variation of the amplitudes is "slow" and satisfies the condition

$$|A''| \ll |k_0 A'| \quad \text{and} \quad |B''| \ll |k_0 B'|. \qquad (8.5\text{-}8)$$

This condition is known as *parabolic approximation* and is often used when perturbation is small. Thus, neglecting the second-order derivative in Eq. (8.5-7) leads to

$$2ik_0 A'e^{-ik_0 x} - 2ik_0 B'e^{ik_0 x} = 2\frac{\omega^2}{c^2} n_0 n_1 \cos(Kx)(Ae^{-ik_0 x} + Be^{ik_0 x}).$$

$$(8.5\text{-}9)$$

We now multiply both sides of Eq. (8.5-9) by $\exp(ik_0 x)$ and perform spatial averaging. Such spatial averaging is useful in eliminating terms with fast-varying phases. The terms with fast-varying phases do not contribute much in the integration of A and B. Thus, we have

$$2ik_0 A' = 2\frac{\omega^2}{c^2} n_0 n_1 \langle \cos Kx \cdot Be^{2ik_0 x} \rangle, \qquad (8.5\text{-}10)$$

where we used A' for the spatial averaged one, since A is a slowly varying function and we have used the following relations:

$$\langle \cos Kx \rangle = 0, \qquad \langle e^{\pm i2k_0 x} \rangle = 0. \qquad (8.5\text{-}11)$$

The right side of Eq. (8.5-10) is significant only when

$$2k_0 \simeq K. \qquad (8.5\text{-}12)$$

Assuming that Eq. (8.5-12) is satisfied, Eq. (8.5-10) becomes

$$A' = -i\frac{\omega n_1}{2c} Be^{i(2k_0 - K)x}. \qquad (8.5\text{-}13)$$

A similar equation for B' can be obtained and is given by

$$B' = i\frac{\omega n_1}{2c} Ae^{-i(2k_0 - K)x}. \qquad (8.5\text{-}14)$$

We now define a coupling constant

$$\kappa = \frac{\omega n_1}{2c} = \frac{\pi n_1}{\lambda} \qquad (8.5\text{-}15)$$

and a momentum mismatch

$$\Delta k = 2k_0 - K. \qquad (8.5\text{-}16)$$

The coupled equations (8.5-13) and (8.5-14) can thus be written as

$$\frac{d}{dx} A = -i\kappa Be^{i\Delta k x}, \qquad \frac{d}{dx} B = i\kappa Ae^{-i\Delta k x}. \qquad (8.5\text{-}17)$$

Equation (8.5-17) can now be solved for amplitudes $A(x)$ and $B(x)$.

To eliminate $B(x)$ in Eq. (8.5-17), we multiply both sides of the first equation by $\exp(-i\,\Delta k\,x)$ and then differentiate both sides with respect to x. After a few steps of substitution and algebra, we obtain

$$A'' - i\,\Delta k\,A' - \kappa^2 A = 0, \qquad (8.5\text{-}18)$$

where each prime indicates differentiation with respect to x. This is a second-order linear differential equation. The general solution can be written as

$$A(x) = [C_1 \cosh sx + C_2 \sinh sx]e^{i(\Delta k/2)x}, \qquad (8.5\text{-}19)$$

where C_1 and C_2 are arbitrary constants and s is given by

$$s^2 = \kappa^2 - \left(\frac{\Delta k}{2}\right)^2. \qquad (8.5\text{-}20)$$

Amplitude $B(x)$ is obtained from the first equation of (8.5-17) and can be written as

$$B(x) = \frac{i}{\kappa} e^{-i\Delta kx} \frac{d}{dx} A. \qquad (8.5\text{-}21)$$

8.5.1 Reflectance of a Sinusoidal Layer

We now investigate the reflection of light from such a sinusoidally varying layer. Let the light be incident at $x = 0$; the reflection coefficient is given by

$$r = \frac{B(0)}{A(0)}. \qquad (8.5\text{-}22)$$

Since the light is incident from the left at $x = 0$, the boundary condition is

$$B(L) = 0. \qquad (8.5\text{-}23)$$

Using Eqs. (8.5-19)–(8.5-21) and (8.5-23), constants C_1 and C_2 can be expressed in terms of the incident amplitude $A(0)$, and amplitudes $A(x)$ and $B(x)$ can be written as

$$A(x) = \frac{s\cosh s(L-x) + i(\Delta k/2)\sinh s(L-x)}{s\cosh sL + i(\Delta k/2)\sinh sL} A(0)e^{i(\Delta k/2)x},$$

$$B(x) = \frac{-i\kappa\sinh s(L-x)}{s\cosh sL + i(\Delta k/2)\sinh sL} A(0)e^{-i(\Delta k/2)x}. \qquad (8.5\text{-}24)$$

From Eq. (8.5-24), the reflection coefficient is

$$r = \frac{-i\kappa \sinh sL}{s \cosh sL + i(\Delta k/2)\sinh sL}. \qquad (8.5\text{-}25)$$

By taking the absolute square of r, we obtain the following formula for reflectance:

$$R = \frac{\kappa^2 \sinh^2 sL}{s^2 \cosh^2 sL + (\Delta k/2)^2 \sinh^2 sL}. \qquad (8.5\text{-}26)$$

Reflectance may be viewed as the fraction of power transferred to the backward-propagating wave between the region $x = 0$ and $x = L$. We notice that the fractional power transfer decreases as Δk increases according to Eq. (8.5-26). At

$$\Delta k = 2k_0 - K = 0, \qquad (8.5\text{-}27)$$

power transfer is maximum (i.e., maximum reflectance), and reflectance is given by

$$R = \tanh^2 \kappa L. \qquad (8.5\text{-}28)$$

Note that this reflectance approaches unity when κL is large.

The calculated reflectance as a function of ΔkL is plotted in Fig. 8.6. It is seen to be an even function of Δk. The reflectance spectrum consists of a main peak with a sharp cutoff and a series of sidelobes. The bandwidth of the main peak is given approximately by

$$\Delta k = 4|\kappa| \qquad (8.5\text{-}29)$$

because for $|\Delta k| \leqslant 2|\kappa|$, the parameter s is real and the amplitude $A(x)$ decays exponentially as x increases. Thus, the region $\Delta k = 4|\kappa|$ is often called the "stop band."

The reflectance at the band edge $\Delta k = \pm 2|\kappa|$ is

$$R = \frac{\kappa^2 L^2}{1 + \kappa^2 L^2}, \qquad (8.5\text{-}30)$$

which can still be significant because a typical high reflectance requires that $\kappa L \gg 1$, according to Eq. (8.5-28). Using Eqs. (8.5-6), (8.5-15), (8.5-27), and (8.5-29), the bandwidth in terms of wavelength $\Delta\lambda$ is

$$\frac{\Delta\lambda}{\lambda} = 2\left|\frac{n_1}{n_0}\right|. \qquad (8.5\text{-}31)$$

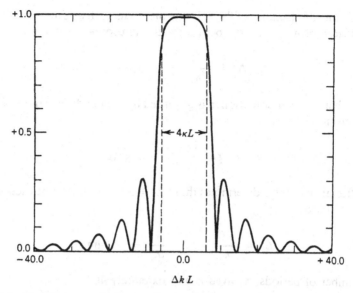

Figure 8.6 The reflectance of a Bragg reflector calculated by using the coupled-mode theory ($|\kappa|L = 3.0$).

Aside from the main peak at $\Delta k = 0$, the reflectance spectrum also consists of a series of sidelobes on both sides of the main peak. These sidelobes peak approximately at $sL = i(p + \frac{1}{2})\pi$ ($p = 1, 2, 3, \ldots$), which corresponds to $\Delta k = \pm 2[\kappa^*\kappa + (p + \frac{1}{2})^2(\pi/L)^2]^{1/2}$. The peak reflectance of these sidelobes is given, according to Eq. (8.5-26), by

$$R = \frac{|\kappa L|^2}{(p + \frac{1}{2})^2\pi^2 + |\kappa L|^2}. \tag{8.5-32}$$

These sidelobes become appreciable when $|\kappa L| > 1/2\pi$. In fact, the peak reflectance of the first sidelobe reaches 10% when $|\kappa L| > 1/2\pi$, at which point the reflectance of the main peak is only 84%. Zero reflectance occurs at $sL = iq\pi$ ($q = 1, 2, 3, \ldots$), which corresponds to $\Delta k = \pm 2(\kappa^*\kappa + q^2\pi^2/L^2)^{1/2}$.

8.5.2 Example: Bragg Reflector Design

To illustrate the use of sinusoidal layers, let us consider the design of a Bragg reflector. Consider the design of a 1.06-μm Bragg reflector with a bandwidth of 100 Å and a reflectance of 99%. Using materials such as ZnS with an index of refraction of $n_0 = 2.3$, the period for the sinusoidal variation is

$$\Lambda = \frac{\lambda}{2n_0} = 0.23\,\mu\text{m}.$$

The depth of index modulation n_1 is governed by the bandwidth [Eq. (8.5-31)] and is given by

$$n_1 = \tfrac{1}{2}n_0 \frac{\Delta\lambda}{\lambda} = 1.1 \times 10^{-2}.$$

The reflectance of 99% determines that $\kappa L = 3.0$. Thus, the thickness of the layer is given by

$$L = \frac{3.0}{\kappa} = 3.0\frac{\lambda}{\pi n_1} = 92\,\mu\text{m}.$$

The number of periods involved is approximately 400.

8.6 RAYS IN INHOMOGENEOUS MEDIUM

So far, we have only considered the situation of normal incidence where the direction of propagation remains constant. This is no longer true at general angles of incidence. We now investigate the path of rays in such optically inhomogeneous media.

We assume that the electromagnetic wave satisfies the scalar wave equation

$$\nabla^2 E + \left(\frac{2\pi}{\lambda} n\right)^2 E = 0. \tag{8.6-1}$$

This equation is the general form of Eq. (8.1-1) and is valid, provided that the index of refraction is a slowly varying function of position. The solution is taken to be

$$E = e^{-i\phi(\mathbf{r})}, \tag{8.6-2}$$

where the phase is now a function of position $\mathbf{r}$. Using an approach similar to that leading to Eq. (8.1-7), the scalar wave equation is reduced to the simple form

$$(\nabla\phi)^2 = \left(\frac{2\pi}{\lambda} n\right)^2. \tag{8.6-3}$$

For the case of homogeneous media, an integration of Eq. (8.6-3) yields
$\phi(r) = \mathbf{k} \cdot \mathbf{x}$, with $k = 2\pi n/\lambda$. Equation (8.6-3) is known as the *eikonal
equation* of geometric optics. The surfaces of constant ϕ determined by this
equation are surfaces of constant optical phase and thus define the wavefronts. The light rays have been defined as the orthogonal trajectories to the
wavefronts $\phi(\mathbf{r}) = $ const and hence are also determined by Eq. (8.6-3). If $\mathbf{r}$
is a position vector of a typical point on a ray trajectory and s is the length
of the ray measured from a fixed point on it, then $d\mathbf{r}/ds$ is a unit vector in the
direction of $\nabla\phi$ that is normal to the wavefronts. Thus, if we take the square
root of Eq. (8.6-3), we obtain

$$\frac{2\pi}{\lambda} n \frac{d\mathbf{r}}{ds} = \nabla\phi. \tag{8.6-4}$$

This equation specifies the rays by means of the *eikonal* $\phi(\mathbf{r})$. Conversely, the
eikonal $\phi(\mathbf{r})$ can be expressed as a ray integral,

$$\phi(\mathbf{r}) = \frac{2\pi}{\lambda} \int n \, ds, \tag{8.6-5}$$

where integration is carried along a ray trajectory. A differential equation
that describes ray propagation directly in terms of the refractive index $n(r)$
can be derived from Eqs. (8.6-3) and (8.6-4) and is given by

$$\frac{d}{ds}\left(n \frac{d\mathbf{r}}{ds}\right) = \nabla n. \tag{8.6-6}$$

According to this equation, the rays always bend toward regions of high
refractive index in an inhomogeneous medium.

Given an index function $n(\mathbf{r})$, Eq. (8.6-6) can be used to trace the optical
path in any inhomogeneous medium. To illustrate this, as an example, we will
treat the case of lenslike medium (also called gradient index fiber) whose
index of refraction in a cylindrical coordinate is given by

$$n^2(r) = \begin{cases} n_0^2[1 - (gr)^2], & r < a, \\ n_c^2, & r > a, \end{cases} \tag{8.6-7}$$

where r is the distance from the z axis and n_0, n_c, and g are constants. The
term n_c is often called the cladding index and is usually chosen such that
$n_c^2 = n_0^2[1 - (ga)^2]$. The parameter a is the core radius of such a fiber.

For paraxial rays (i.e., rays making very small angles with the z axis)
propagating in meridional planes, we may replace d/ds by d/dz and represent

the ray path by a function $r(z)$. Thus, the ray equation can be given approximately as

$$\frac{d^2}{dz^2} r + g^2 r = 0, \tag{8.6-8}$$

where we have neglected higher order terms such as $(gr)^3$. The solution for the ray path can be written, according to Eq. (8.6-8), as

$$r(z) = A \sin(gz) + B \cos(gz), \tag{8.6-9}$$

where A and B are two arbitrary constants that can be determined by the initial conditions, provided $|r(z)| < a$ for all z.

If, at $z = 0$, the ray is at $r = r_0$ and has a slope of r_0',

$$B = r_0, \qquad A = \frac{1}{g} r_0'. \tag{8.6-10}$$

Thus, the ray path becomes

$$r(z) = r_0 \cos(gz) + \frac{1}{g} r_0' \sin(gz). \tag{8.6-11}$$

We note that the ray path is a periodic function of z with a period

$$\Lambda = \frac{2\pi}{g}. \tag{8.6-12}$$

This period Λ is also known as the *pitch*.

As a result of this periodic behavior, sections of such gradient index fibers can be used for imaging purposes. These sections are called *GRIN-rod lenses*. To illustrate this, let us consider the situation when a bundle of parallel rays is incident into such a medium at $z = 0$. All the incident rays are parallel to the z axis. The paths of these rays are represented by

$$r(z) = r_0 \cos(gz), \tag{8.6-13}$$

where r_0 is the point of incidence for each ray. After traversing a distance l such that

$$gl = \tfrac{1}{2}\pi, \tag{8.6-14}$$

all the rays collapse into a point at the axis. If this is called a focal point, the focal length is

$$f = \frac{\pi}{2g} = \frac{\Lambda}{4}. \tag{8.6-15}$$

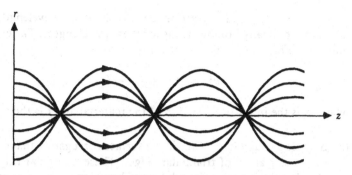

Figure 8.7   Ray paths inside a GRIN-rod.

If the GRIN-rod is long enough, the bundle of rays will diverge and then converge again. In fact, the focusing will repeat itself indefinitely at an interval π/g. Figure 8.7 illustrates the ray paths inside the medium.

The rays described in the preceding are all in meridional planes that contain the z axis. There are other types of rays that propagate out of meridional planes. These are the so-called skew rays. Typical examples of skew rays take a helical path around the z axis.

REFERENCES

1. I. S. Gradshteyn and I. M. Ryzhik, *Tables of Integrals, Series and Products,* Academic Press, New York, 1965, p. 990.

2. J. Mathews and R. L. Walker, *Method of Mathematical Physics,* Benjamin, New York, 1970, p. 198.

3. A. Yariv and P. Yeh, *Optical Waves in Crystals,* Wiley, New York, 1984, Chapter 6.

PROBLEMS

8.1   *Gradient index via surface microstructures.*   Although gradient index layers can be achieved by the codeposition of two species with different rates, it is very difficult to achieve a continuous variation at the air surface. Using microstructures on the surface, it is possible to achieve effective indices of refraction as close to 1 as possible. Consider a microstructure that consists of fine voids (air bubbles) that are very small compared to the wavelengths. In addition, the average spacing between these voids are also small relative to the wavelengths. Since both the size and spacing are much less than the wavelengths, the medium is optically homogeneous for these wavelengths. The effective index of refraction of such a region is

$$n^2 = fn_s^2 + (1 - f),$$

where n_s is the index of refraction of the material and f is the packing density.

(a) *Surface corrugation.* Consider a surface corrugation that consists of a linear array of triangular ridges. These triangular ridges are closely packed with a period a and a height h. Period a is assumed to be much less than the wavelength. The ridges and the substrate are the same material with a refractive index of n_s. The corrugation can be considered a transition region between the air and the substrate. Show that the index of refraction of such a transition layer is

$$n^2(x) = 1 + (n_s^2 - 1)\frac{x}{h}.$$

(b) *Porous surface structures.* Consider a porous microstructure that consists of a two-dimensional array of fine pyramids. These pyramids are all identical with square base $a \times a$ and height h. The pyramids are closely packed so that all the bases are adjacent to each other with no gaps. Show that the index of refraction of such a transition layer is

$$n^2(x) = 1 + (n_s^2 - 1)\left(\frac{x}{h}\right)^2.$$

8.2   *WKB approximation.* Apply the WKB approximation Eq. (8.1-13) to the index profiles studied in Section 8.2 and derive the approximation solutions.

(a) *Linear layer 1.* Show that

$$\frac{\omega}{c}\int n(x)\, dx = \frac{1}{2\alpha}\xi^2 + \text{const},$$

where ξ is defined in Eq. (8.2-2), and then compare the WKB approximation with the asymptotic form of the exact solution Eq. (8.2-5) by using Eq. (8.2-24).

(b) *Exponential layer.* Show that

$$\frac{\omega}{c} \int n(x)\, dx = \frac{\xi}{\gamma} + \text{const},$$

where ξ and γ are defined in Eqs. (8.2-12) and (8.2-13), and then compare the WKB approximation with the asymptotic form of the exact solution Eq. (8.2-16) by using Eq. (8.2-24).

(c) *Hyperbolic layer.* Show that

$$\frac{\omega}{c} \int n(x)\, dx = -m \ln \xi + \text{const},$$

provided that $m \gg \frac{1}{2}$, and then compare the WKB approximation with the exact solution Eq. (8.2-21).

8.3   *Reflectance of exponential layer*

(a) Derive Eq. (8.3-7).

(b) Use the asymptotic forms of the Bessel functions as given by Eqs. (8.2-23) and (8.2-24) and derive the asymptotic forms of the reflectance formulas (8.3-8) and (8.3-9).

8.4   *Reflectance formula.* Derive Eq. (8.4-20) by taking the absolute square of Eq. (8.4-18). *Hint:* Use the unimodular condition $AD - BC = 1$ and $D = A^*$, $B = C^*$. This leads to $|A|^2 - |C|^2 = 1$. Also use $\cos y = \frac{1}{2}(A + A^*)$ and the trigonometric relation $\sin(a + b) = \sin a \cos b - \cos a \sin b$.

8.5   *Rays in atmosphere.* The index of refraction of the atmosphere is a function of the altitude y and is given approximately as

$$n = 1 + \Delta n \exp\left(-\frac{y}{L}\right),$$

where $\Delta n = 0.000292$ and $L = 7.8\,\text{km}$.

(a) Show that the paraxial approximation of the ray equation (8.6-6) is

$$\frac{d^2 y}{dz^2} = -\frac{\Delta n}{L} \exp\left(-\frac{y}{L}\right)$$

Here, we neglect the curvature of the earth.

(b) Show that

$$y = L \ln\left[\frac{(z + a)^2}{2L^2} \Delta n \right]$$

is a solution to the paraxial ray equation, where a is an arbitrary constant.

(c) The solution obtained in (b) can be used to estimate the ray bending due to the gradient in the index of refraction of the atmosphere. Show that the bending of the sun's rays is approximately given by

$$\Delta y' = \sqrt{2 \Delta n}$$

in radians. Using $\Delta n = 0.000292$, estimate this angle. (*Hint:* Calculate the slope of the ray at $y = 0$ and compare with the slope of the ray at $y = \infty$.)

8.6 *Skew rays.* Rays that are not confined to meridional planes exist in GRIN-rods, described by Eq. (8.6-7). These rays are the so-called skew rays.

(a) Starting from the ray equation (8.6-6), show that rays with helical paths exist. In other words, show that the solution

$$r = a, \qquad \theta = bz,$$

satisfies the ray equation, where a and b are constants. Note that the helical pitch is $2\pi/b$.

(b) Show that b is related to the helical radius a by

$$b^2 = \frac{2n_0^2 g^2}{n - 2n_0^2 g^2 a^2},$$

where n is the index of refraction at $r = a$.

(c) Show that circular paths also exist. Show that the radius of such a circular path satisfies the equation

$$n = 2n_0^2 g^2 r^2.$$

Note that the circular path may be considered as a degenerated helix with an infinite b.

8.7   *Snell's law*

(a) Derive Snell's law by using the ray equation (8.6-6).

(b) Consider the ray tracing in a medium with a cylindrically symmetric refractive index profile [i.e., $n = n(r)$]. Let θ be the angle between the ray and the z axis. Show that

$$n \cos \theta = \text{const}$$

provided that the index of refraction is independent of z.

(c) Derive the ray bending in Problem 8.5(c) using Snell's law.

8.8   *Lorentzian GRIN-rod.*   Consider the tracing of helical rays in a medium with a cylindrically symmetric refractive index profile with $dn/dz = 0$.

(a) Let $\theta = gz$; show that the ray equation (8.6-6) reduces to

$$- \frac{nrg^2}{1 + (gr)^2} = \frac{dn}{dr}.$$

(b) Integrate the preceding equation and show that

$$n^2(r) = \frac{n_0^2}{1 + (gr)^2}.$$

Media with such an index profile are often called Lorentzian GRIN-rods.

(c) Show that all helical rays in Lorentzian GRIN-rods have the same pitch regardless of the helical radius. Ray tracing in Lorentzian GRIN-rods was first studied by Kawakami and Nishizawa in 1968 [*IEEE Trans.* **MTT-16**, 814 (1968)].

8.9   *Sech GRIN-rod.*   Consider the tracing of meridional rays in a cylindrically symmetric medium whose refractive index profile is given by

$$n(r) = n_0 \operatorname{sech}(gr).$$

(a) Consider the incidence of a ray parallel to the z axis at $r = r_0$. Using Snell's law [see Problem 8.7(b)], show that

$$n \sin \theta = \sqrt{n^2 - n_1^2}, \qquad n \cos \theta = n_1,$$

where $n_1 = n_0 \operatorname{sech}(gr_0)$.

(b) Using $dr/dz = \tan\theta$, show that

$$\frac{dr}{dz} = \sqrt{(n/n_1)^2 - 1}.$$

(c) Using a substitution of $u = \sinh(gr)$ and integrating the equation in (b), show that the ray path is given by

$$r(z) = \frac{1}{g}\sinh^{-1}[\sinh(gr_0)\cos(gz)].$$

Notice that the ray intersects with the axis periodically with a pitch $2\pi/g$ regardless of the point of incidence r_0. The first intersection occurs at $z = \pi/(2g)$. Thus, Sech GRIN-rods can be used for the focusing of all meridional rays. Ray tracing in such media was first studied in 1953 by Fletcher, Murphy, and Young [*Proc. Roy. Soc. London* **233**, 216 (1954)].

8.10 Show that the refractive index profiles of both the Lorentzian and the Sech GRID-rods reduce to Eq. (8.6-7) for paraxial rays with $(gr) \ll 1$.

9

Optics of Anisotropic Layered Media

So far, we have treated the propagation of optical waves in layers that are isotropic. In this chapter, we consider the propagation of electromagnetic waves in anisotropic layered media. We will first review the propagation of plane waves in homogeneous and anisotropic media. We then take up the subject of refraction and transmission of plane waves and their effect on the polarization state. The Jones matrix will then be introduced to treat the layered media. In the last part of this chapter, we will consider the exact treatment where interface reflection is included in the formulation.

9.1 PLANE WAVES IN HOMOGENEOUS AND ANISOTROPIC MEDIA

In our earlier treatment of reflection and transmission of electromagnetic radiation through layered media, the plane waves are divided into two categories. These are the linearly polarized s and p waves. Both of these waves have the same phase velocity. This is no longer true in anisotropic media such as calcite, quartz, and lithium niobate. Plane-wave propagation in anisotropic media is determined by the dielectric tensor ε_{ij} that links the displacement vector and the electric field vector:

$$\mathbf{D} = \varepsilon \mathbf{E} \tag{9.1-1}$$

or

$$D_i = \varepsilon_{ij} E_j, \tag{9.1-2}$$

where the convention of summation over repeated indices is observed. In nonmagnetic and transparent materials, this tensor is real and symmetric:

$$\varepsilon_{ij} = \varepsilon_{ji}. \tag{9.1-3}$$

The magnitude of these nine tensor elements depends, of course, on the choice of the x, y, and z axes relative to the crystal structure. Because of its real and symmetric nature, it is always possible to find three mutually orthogonal axes in such a way that the off-diagonal elements vanish, leaving

$$
\varepsilon = \varepsilon_0 \begin{pmatrix} n_x^2 & 0 & 0 \\ 0 & n_y^2 & 0 \\ 0 & 0 & n_z^2 \end{pmatrix} = \begin{pmatrix} \varepsilon_x & 0 & 0 \\ 0 & \varepsilon_y & 0 \\ 0 & 0 & \varepsilon_z \end{pmatrix}, \tag{9.1-4}
$$

where ε_x, ε_y, and ε_z are the principal dielectric constants and n_x, n_y, and n_z are the principal indices of refraction. These directions (x, y, and z) are called the principal dielectric axes of the crystal. According to Eqs. (9.1-1) and (9.1-4), a plane wave propagating along the z axis can have two phase velocities, depending on its state of polarization. Specifically, the phase velocity is c/n_x for x-polarized light and c/n_y for y-polarized light. Generally speaking, there are two normal modes of polarization for each direction of propagation.

To study such propagation along a general direction, we assume a monochromatic plane wave with an electric field vector

$$
\mathbf{E} \exp\left[i(\omega t - \mathbf{k} \cdot \mathbf{r})\right] \tag{9.1-5}
$$

and a magnetic field vector

$$
\mathbf{H} \exp\left[i(\omega t - \mathbf{k} \cdot \mathbf{r})\right], \tag{9.1-6}
$$

where $\mathbf{k}$ is the wave vector $\mathbf{k} = (\omega/c)n\mathbf{s}$, with $\mathbf{s}$ as a unit vector in the direction of propagation. The phase velocity c/n, or equivalently n, is to be determined. Substitution for $\mathbf{E}$ and $\mathbf{H}$ from Eqs. (9.1-5) and (9.1-6), respectively, into Maxwell's equations (1.1-1) and (1.1-2) gives

$$
\mathbf{k} \times \mathbf{E} = \omega\mu\mathbf{H}, \tag{9.1-7}
$$

$$
\mathbf{k} \times \mathbf{H} = -\omega\varepsilon\mathbf{E} = -\omega\mathbf{D}. \tag{9.1-8}
$$

By eliminating $\mathbf{H}$ from Eqs. (9.1-7) and (9.1-8), we obtain

$$
\mathbf{k} \times (\mathbf{k} \times \mathbf{E}) + \omega^2\mu\varepsilon\mathbf{E} = 0. \tag{9.1-9}
$$

This equation will now be used to solve for the eigenvectors $\mathbf{E}$ and the corresponding eigenvalue n.

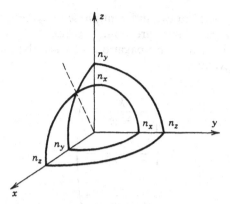

Figure 9.1   One octant of a normal surface

In the principal coordinate system, the dielectric tensor ε is given by Eq. (9.1-4). Equation (9.1-9) can be written as

$$\begin{pmatrix} \omega^2 \mu \varepsilon_x - k_y^2 - k_z^2 & k_x k_y & k_x k_z \\ k_y k_x & \omega^2 \mu \varepsilon_y - k_x^2 - k_z^2 & k_y k_z \\ k_z k_x & k_z k_y & \omega^2 \mu \varepsilon_z - k_x^2 - k_y^2 \end{pmatrix} \begin{pmatrix} E_x \\ E_y \\ E_z \end{pmatrix} = 0,$$

$$(9.1\text{-}10)$$

where we recall that $\varepsilon_x = \varepsilon_0 n_x^2$, $\varepsilon_y = \varepsilon_0 n_y^2$, and $\varepsilon_z = \varepsilon_0 n_z^2$.

For nontrivial solutions to exist, the determinant of the matrix in Eq. (9.1-10) must vanish. This leads to a relation between ω and $\mathbf{k}$:

$$\det \begin{vmatrix} \omega^2 \mu \varepsilon_x - k_y^2 - k_z^2 & k_x k_y & k_x k_z \\ k_y k_x & \omega^2 \mu \varepsilon_y - k_x^2 - k_z^2 & k_y k_z \\ k_z k_x & k_z k_y & \omega^2 \mu \varepsilon_z - k_x^2 - k_y^2 \end{vmatrix} = 0.$$

$$(9.1\text{-}11)$$

This equation can be represented by a three-dimensional surface in $\mathbf{k}$ space (momentum space). This surface is known as the *normal surface* and consists of two shells. These two shells, in general, have four points in common. The two lines that go through the origin and these points are known as the optical axes. Figure 9.1 shows one octant of a general normal surface. Given a direction of propagation, there are in general two k values that are the intersections of the direction of propagation s and the normal surface. These two k values correspond to two different phase velocities (ω/k) of the waves propagating along the chosen direction. The directions of the electric field vector associated with these propagations can also be obtained from Eq. (9.1-10) and are given by

$$\begin{pmatrix} \dfrac{k_x}{k^2 - \omega^2 \mu \varepsilon_x} \\[2ex] \dfrac{k_y}{k^2 - \omega^2 \mu \varepsilon_y} \\[2ex] \dfrac{k_z}{k^2 - \omega^2 \mu \varepsilon_z} \end{pmatrix} \tag{9.1-12}$$

provided that the denominators do not vanish.

For propagation in the direction of the optic axes, there is only one value of k and consequently only one phase velocity. There are, however, two independent directions of polarization.

Equations (9.1-11) and (9.1-12) are often written in terms of the direction cosines of the wave vector. By using the relation $\mathbf{k} = (\omega/c)n\mathbf{s}$ for the plane wave given by Eq. (9.1-5), Eqs. (9.1-11) and (9.1-12) can be written as

$$\frac{s_x^2}{n^2 - n_x^2} + \frac{s_y^2}{n^2 - n_y^2} + \frac{s_z^2}{n^2 - n_z^2} = \frac{1}{n^2} \tag{9.1-13}$$

and

$$\begin{pmatrix} \dfrac{s_x}{n^2 - n_x^2} \\[2ex] \dfrac{s_y}{n^2 - n_y^2} \\[2ex] \dfrac{s_z}{n^2 - n_z^2} \end{pmatrix}, \tag{9.1-14}$$

respectively, where we have used $\varepsilon_x = \varepsilon_0 n_x^2$, $\varepsilon_y = \varepsilon_0 n_y^2$, and $\varepsilon_z = \varepsilon_0 n_z^2$.

Equation (9.1-13) is known as *Fresnel's equation of wave normals* and can be solved for the eigenvalues of refraction, and Eq. (9.1-14) gives the directions of polarization. Equation (9.1-13) is a quadratic equation in n^2. Therefore, for each direction of propagation (a set of s_x, s_y, s_z), two solutions for n^2 can be obtained from Eq. (9.1-13). To complete the solution of the problem, we use the values of n^2, one at a time, in Eq. (9.1-14). This gives us the polarizations (electric field vectors) of these waves. It can be seen that in a nonabsorbing medium these normal modes are linearly polarized since all the components are real in Eq. (9.1-14). Let E_1, E_2 be the electric field vectors and D_1, D_2 be the displacement vectors of the linearly polarized normal modes associated with n_1^2 and n_2^2, respectively. Maxwell's equation $\nabla \cdot D = 0$ requires that D_1, D_2 are orthogonal to s. Since $D_1 \cdot D_2 = 0$ (the proof of this orthogonal relation is left as a problem for students), the three vectors D_1, D_2, and s form an orthogonal triad. According to Eqs. (9.1-7) and (9.1-8), D and H are both perpendicular to the direction of propagation s. Consequently, the direction of energy flow as given by the Poynting vector $E \times H$ is, in general, not collinear with the direction of propagation s. Since D, E, and k are all orthogonal to H, they must lie in the same plane.

To summarize: Along an arbitrary direction of propagation s, there can exist two independent plane-wave, linearly polarized propagation modes. These modes have phase velocities $\pm (c/n_1)$ and $\pm (c/n_2)$, where n_1^2 and n_2^2 are the two solutions of Fresnel's equation (9.1-13). The electric field vectors of these two normal modes are given by Eq. (9.1-12) or (9.1-14).

9.1.1 Classification of Anisotropic Media (Crystals)

We have shown in the preceding that the normal surface contains a good deal of information concerning wave propagation in anisotropic media. The normal surface is uniquely determined by the principal indices of refraction n_x, n_y, n_z. In the general case when the three principal indices n_x, n_y, n_z are all different, there are two optic axes. In this case, the crystal is said to be *biaxial*. In many optical materials, it happens that two of the principal indices are equal, in which case the equation for the normal surface (9.1-11) can be factored according to

$$\left(\frac{k_x^2 + k_y^2}{n_e^2} + \frac{k_z^2}{n_o^2} - \frac{\omega^2}{c^2}\right)\left(\frac{k^2}{n_o^2} - \frac{\omega^2}{c^2}\right) = 0, \qquad (9.1\text{-}15)$$

where $n_o^2 = \varepsilon_x/\varepsilon_0 = \varepsilon_y/\varepsilon_0$ and $n_e^2 = \varepsilon_z/\varepsilon_0$.

The normal surface in this case consists of a sphere and an ellipsoid of revolution. These two sheets of the normal surface touch at two points on the z axis. The z axis is therefore the only optic axis, and the crystal is said to be *uniaxial*. If the three principal indices are equal, the two sheets of normal surface degenerate to a single sphere, and the crystal is *optically isotropic*.

It is obvious that the optical symmetry of the crystal is closely related to the point group of the crystal. For example, in a cubic crystal, the three principal axes are physically equivalent. Therefore, we expect a cubic crystal to be optically isotropic. Table 9.1 lists the optical symmetry of the crystals and the corresponding dielectric tensor.

In a biaxial crystal, the principal coordinate axes are labeled in such a way that the three principal indices are in the order

$$n_x < n_y < n_z. \qquad (9.1\text{-}16)$$

In this convention, the optical axes lie in the xz plane. The cross section of the normal surfaces with the xz plane is shown in Fig. 9.2(a). In a uniaxial crystal, the index of refraction that corresponds to the two equal elements, $n_o^2 = \varepsilon_x/\varepsilon_0 = \varepsilon_y/\varepsilon_0$, is called the *ordinary index* n_o; the other index, corresponding to ε_z, is called the *extraordinary index* n_e. If $n_o < n_e$, the crystal is said to be positive, whereas if $n_o > n_e$, it is called a negative crystal. The intersection of the normal surfaces with the xz plane is again shown in Figs. 9.2(b) and 9.2(c). The optical axis corresponds to the principal axis which has a unique index of refraction. Table 9.2 lists some example of crystals with their indices of refraction.

9.1.2 Gyrotropic Media

In materials such as quartz and sugar solution, where natural optical rotation is present, the dielectric tensor that includes such rotation is no longer symmetric. If absorption is absent, the dielectric tensor is Hermitian:

$$\varepsilon_{ij}^* = \varepsilon_{ji}. \qquad (9.1\text{-}17)$$

For plane-wave propagation in a homogeneous medium, the material equation for such an optically active material is often written as

$$\mathbf{D} = \varepsilon_a \mathbf{E} + i\varepsilon_0 \mathbf{G} \times \mathbf{E}, \qquad (9.1\text{-}18)$$

where ε_a is the dielectric tensor in the absence of optical activity, and $\mathbf{G}$ is a vector parallel to the direction of propagation and is called *the gyration vector*. The typical magnitude of $\mathbf{G}$ is $\sim 10^{-4}$ and is dimensionless (see Table 9.3). The vector product $\mathbf{G} \times \mathbf{E}$ can always be represented by the product of an antisymmetric tensor $[G]$ with $\mathbf{E}$. The matrix elements of this antisymmetric tensor $[G]$ are given by

$$[G]_{23} = -[G]_{32} = -G_x,$$
$$[G]_{31} = -[G]_{13} = -G_y, \qquad (9.1\text{-}19)$$
$$[G]_{12} = -[G]_{21} = -G_z.$$

Table 9.1. Dielectric Tensor and Crystal Symmetry

Optical Symmetry	Crystal System	Point Groups[†]	Dielectric Tensor
Isotropic	Cubic	$\bar{4}3\,m$ 432 m3 23 m3m	$\varepsilon = \varepsilon_0 \begin{pmatrix} n^2 & 0 & 0 \\ 0 & n^2 & 0 \\ 0 & 0 & n^2 \end{pmatrix}$
Uniaxial	Tetragonal	4 $\bar{4}$ 4/m 422 4mm $\bar{4}$2m 4/mm	
	Hexagonal	6 $\bar{6}$ 6/m 622 6mm $\bar{6}$m2 6/mmm	$\varepsilon = \varepsilon_0 \begin{pmatrix} n_o^2 & 0 & 0 \\ 0 & n_o^2 & 0 \\ 0 & 0 & n_e^2 \end{pmatrix}$
	Trigonal	3 $\bar{3}$ 32 3m $\bar{3}$m	
Biaxial	Triclinic	1 $\bar{1}$	$\varepsilon = \varepsilon_0 \begin{pmatrix} n_x^2 & 0 & 0 \\ 0 & n_y^2 & 0 \\ 0 & 0 & n_z^2 \end{pmatrix}$
	Monoclinic	2 m 2/m	
	Orthorhombic	222 mm2 mmm	

[†]For definition of point groups, see for example, M. Tinkham, *Group Theory and Quantum Mechanics* (McGraw-Hill, New York, 1964).

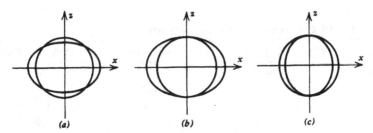

Figure 9.2 Intersection of the normal surface with xz plane for (a) biaxial crystals, (b) positive uniaxial crystals, and (c) negative uniaxial crystals. (Adapted from A. Yariv and P. Yeh, *Optical Waves in Crystals*, Wiley, New York, 1984, p. 84. By permission of John Wiley & Sons, Inc.).

Thus, Eq. (9.1-18) becomes

$$\mathbf{D} = (\varepsilon_a + i\varepsilon_0[G])\mathbf{E}. \tag{9.1-20}$$

It is often convenient to define a new dielectric tensor as

$$\varepsilon = \varepsilon_a + i\varepsilon_0[G]. \tag{9.1-21}$$

This new dielectric tensor is Hermitian, that is, $\varepsilon_{ij} = \varepsilon_{ji}^*$. We can substitute this dielectric tensor ε into Eq. (9.1-9) and solve for the normal modes of propagation.

For plane-wave propagation in isotropic media or along the c axis of uniaxial crystals, the effect of the term $i\varepsilon_0 \mathbf{G} \times \mathbf{E}$ is to cause a rotation of the plane of polarization of a linearly polarized light. If we treat $i\varepsilon_0 \mathbf{G} \times \mathbf{E}$ as a small perturbation (see Problem 9.12) and assume a solution of the form $\mathbf{E} \exp[i(\omega t - kz)]$, we obtain, from Eqs. (1.4-9) and (9.1-18),

$$\frac{d}{dz}\mathbf{E} = \frac{\pi}{\lambda n} \mathbf{G} \times \mathbf{E}, \tag{9.1-22}$$

where n is the index of refraction of the medium (or n_o if the medium is uniaxially birefringent) in the absence of optical activity. According to this equation, the linearly polarized electric field vector $\mathbf{E}$ gyrates (or rotates) as the light travels in the medium. The angle of rotation per unit length of propagation, also called *specific rotatory power* ϱ, is given by

$$\varrho = \frac{\pi}{\lambda n} G, \tag{9.1-23}$$

where G is the magnitude of $\mathbf{G}$ (i.e., $G = |\mathbf{G}|$). Table 9.3 lists the rotatory power and the gyration constant for some materials.

Table 9.2. Typical Refractive Indices of Some Crystals

	Crystal	Refractive Indices		
Isotropic	Fluorite	1.392		
	Sodium chloride, NaCl	1.544		
	Diamond, C	2.417		
	CdTe	2.69		
	GaAs	3.40		
		n_o		n_e
Uniaxial	Ice, H_2O	1.309		1.310
(Positive)	Quartz, SiO_2	1.544		1.553
	Beryllium oxide, BeO	1.717		1.732
	Zircon, $ZrSiO_4$	1.923		1.968
	ZnS	2.354		2.358
	Rutile, TiO_2	2.616		2.903
		n_o		n_e
Uniaxial	ADP, $(NH_4)H_2PO_4$	1.522		1.478
(Negative)	Beryl, $Be_3Al_2(SiO_3)_6$	1.598		1.590
	KDP, KH_2PO_4	1.507		1.467
	Sodium nitrate, $NaNO_3$	1.587		1.366
	Calcite, $CaCO_3$	1.658		1.486
	Tourmaline	1.638		1.618
	Sapphire, Al_2O_3	1.768		1.760
	Lithium niobate $LiNbO_3$	2.300		2.208
	Barium titanate, $BaTiO_3$	2.416		2.364
	Proustite, Ag_3AsS_3	3.019		2.739
		n_x	n_y	n_z
Biaxial	Gypsum	1.520	1.523	1.530
	Feldspar	1.522	1.526	1.530
	Mica	1.552	1.582	1.588
	Topaz	1.619	1.620	1.627
	Sodium nitrite	1.344	1.411	1.651
	$YAlO_3$	1.923	1.938	1.947
	SbSI	2.7	3.2	3.8

In the presence of a magnetic field, some materials also exhibit rotatory power. This is known as the *Faraday effect*. The Faraday effect is a property of transparent substances that causes a rotation of the plane of polarization with distance when the material is placed in a magnetic field for light propagated along the magnetic field. More accurately, the rotation is proportional to the component of the magnetic field along the direction of propagation of

Table 9.3. Rotatory Power and Gyration Constants for Some Materials

Material	λ (μm)	n	ϱ (degrees/mm)	G (10^{-4})
Quartz, SiO_2	0.4	—	50.3	—
	0.45	1.552	339.2	1.52
	0.488	1.550	32.5	1.37
	0.5	1.549	31.0	1.33
	0.5145	1.548	29.1	1.30
	0.55	1.511	25.4	1.17
	0.6	1.509	21.0	1.06
	0.6328	1.507	18.5	0.98
	0.65	—	17.7	—
	0.70	—	15.1	—
Cinnabar, HgS	0.5893	2.94	32.5	3.13
$AgGaS_2$	0.488	2:706	800	58.7
	0.490.	—	700	—
	0.495	—	600	—
	0.500	2.686	500	37.3
	0.505	—	430	—
	0.5145	2.664	300	22.8
TeO_2	0.488	2.332	202	12.8
	0.5	2.322	185	11.9
	0.5145	2.312	165	10.9
	0.55	2.291	132	9.24
	0.6328	2.260	87	6.9
	1.0	2.208	30	3.7

the light. The gyration vector defined by Eq. (9.1-18) is proportional to the external magnetic field

$$\mathbf{G} \;=\; \gamma\mathbf{B}, \tag{9.1-24}$$

in which γ is the *magneto-gyration coefficient* of the medium. In an optically active medium, the direction of rotation bears a fixed relation to the direction of propagation so that if a beam of light is reflected back on itself, the net rotation is zero. In the Faraday effect, however, rotation bears a fixed relation to the magnetic field **B** so that reflection back on itself doubles the rotation. According to Eq. (9.1-23), the specific rotation (i.e., rotation per unit length) is often written as

$$\varrho \;=\; \frac{\pi\gamma}{\lambda n}\, B \;=\; VB, \tag{9.1-25}$$

where V is the *Verdet constant*.

Table 9.4. Values of Verdet Constant and Magneto-Gyration Coefficient at $\lambda = 589.3$ nm

Substance	T (°C)	n	V (degrees/mm G)	γ (G^{-1})
Water	20	1.333	2.18×10^{-5}	9.51×10^{-11}
Fluorite		1.434	1.5×10^{-6}	7.0×10^{-12}
Diamond		2.417	2.0×10^{-5}	1.6×10^{-10}
Glass (crown)	18	1.520	2.68×10^{-5}	1.3×10^{-10}
Glass (flint)		1.650	5.28×10^{-5}	2.8×10^{-10}
Carbon disulfide (CS$_2$)	20	1.625	7.05×10^{-5}	3.7×10^{-10}
Phosphorus	33	2.14	2.21×10^{-4}	1.5×10^{-9}
Sodium chloride		1.544	6.0×10^{-5}	3.0×10^{-10}

The Faraday effect originates from the effect of the static magnetic field on the motion of electrons. In the presence of the electric field of the optical beam, the electrons are displaced from their equilibrium position. This motion coupled with the static magnetic field creates a lateral displacement of electrons due to the Lorentz force $\mathbf{v} \times \mathbf{B}$. As a result, the induced dipole moment involves a term that is proportional to $\mathbf{B} \times \mathbf{E}$. The material relation becomes

$$\mathbf{D} = \varepsilon\mathbf{E} + i\varepsilon_0\gamma\mathbf{B} \times \mathbf{E}. \qquad (9.1\text{-}26)$$

The factor i accounts for the $\frac{1}{2}\pi$ phase delay between the velocity and the electric field. Faraday rotation has been observed in many solids, liquids, and even gases. A few values of the Verdet constant are given in Table 9.4.

9.2 PLANE WAVES IN UNIAXIALLY ANISOTROPIC MEDIA

Many optical devices are made of uniaxially birefringent crystals such as quartz, sapphire, calcite, lithium niobate, and barium titanate. These devices include wave plates, birefringent filters, and prism polarizers. Thus, the propagation of optical waves in uniaxially anisotropic media deserves special attention.

By putting

$$\varepsilon_x = \varepsilon_y = \varepsilon_0 n_o^2, \qquad \varepsilon_z = \varepsilon_0 n_e^2, \qquad (9.2\text{-}1)$$

the normal surface [Eq. (9.1-11)] can be reduced to

$$\left(\frac{k_x^2 + k_y^2}{n_e^2} + \frac{k_z^2}{n_o^2} - \frac{\omega^2}{c^2}\right)\left(\frac{k^2}{n_o^2} - \frac{\omega^2}{c^2}\right) = 0. \qquad (9.2\text{-}2)$$

This is identical to Eq. (9.1-15). The normal surface consists of two parts. The sphere gives the relation between ω and $\mathbf{k}$ of the ordinary (O) wave. The ellipsoid of revolution gives the similar relation for the extraordinary (E) wave. The refractive indices associated with these two modes of propagation are given by

$$O \text{ wave:} \quad n = n_o, \tag{9.2-3}$$

$$E \text{ wave:} \quad \frac{1}{n^2} = \frac{\cos^2 \theta}{n_o^2} + \frac{\sin^2 \theta}{n_e^2}, \tag{9.2-4}$$

where θ is the angle between the direction of propagation and the c axis (the crystal z axis).

The electric field vector of the O wave cannot be obtained directly from Eq. (9.1-12) because of the vanishing denominators. It can be easily obtained from Eq. (9.1-10). By using $\varepsilon_x = \varepsilon_y = \varepsilon_0 n_o^2$, $\varepsilon_z = \varepsilon_0 n_e^2$, and $\mathbf{k} = (\omega/c)n\mathbf{s}$, Eq. (9.1-10) can be written as

$$\begin{pmatrix} s_x^2 & s_x s_y & s_x s_z \\ s_y s_x & s_y^2 & s_y s_z \\ s_z s_x & s_z s_y & \left(\dfrac{n_e}{n_o}\right)^2 - (s_x^2 + s_y^2) \end{pmatrix} \begin{pmatrix} E_x \\ E_y \\ E_z \end{pmatrix} = 0. \tag{9.2-5}$$

A simple inspection of this equation yields the following direction of polarization:

$$O \text{ wave:} \quad \mathbf{E} = \begin{pmatrix} s_y \\ -s_x \\ 0 \end{pmatrix}, \tag{9.2-6}$$

where we recall that s_x, s_y, s_z are directional cosines of the direction of propagation. The electric field vector of the E wave can be obtained from Eq. (9.1-12) and is given by

$$E \text{ wave:} \quad \mathbf{E} = \begin{pmatrix} \dfrac{s_x}{n^2 - n_o^2} \\[2ex] \dfrac{s_y}{n^2 - n_o^2} \\[2ex] \dfrac{s_y}{n^2 - n_e^2} \end{pmatrix}, \tag{9.2-7}$$

where n is given by Eq. (9.2-4).

Notice that the electric field vector of the O wave is perpendicular to the plane formed by the wave vector $\mathbf{k}$ and the c axis, whereas the electric field vector of the E wave is in that plane. The electric field vector of the E wave is not exactly perpendicular to the wave vector $\mathbf{k}$. However, the deviation from 90° is very small. This small angle between the field vectors $\mathbf{E}$ and $\mathbf{D}$ is also the angle between the phase velocity and the group velocity (see Problem 9.3). Therefore, for practical purposes, we may assume that the electric field is transverse to the direction of propagation. The displacement vectors $\mathbf{D}$ of the normal modes are exactly perpendicular to the wave vector $\mathbf{k}$ and can be written as

$$O \text{ wave:} \quad \mathbf{D}_o = \frac{\mathbf{k} \times \mathbf{c}}{|\mathbf{k} \times \mathbf{c}|}, \tag{9.2-8}$$

$$E \text{ wave:} \quad \mathbf{D}_e = \frac{\mathbf{D}_o \times \mathbf{k}}{|\mathbf{D}_o \times \mathbf{k}|}. \tag{9.2-9}$$

If inside the crystal a polarized light is generated that is to propagate along a direction s, the displacement vector of this light can always be written as a linear combination of these two normal modes, that is,

$$\mathbf{D} = C_o \mathbf{D}_o \exp(-i\mathbf{k}_o \cdot \mathbf{r}) + C_e \mathbf{D}_e \exp(-i\mathbf{k}_e \cdot \mathbf{r}), \tag{9.2-10}$$

where C_o and C_e are constants and $\mathbf{k}_o$ and $\mathbf{k}_e$ are the wave vectors that are, in general, different. As the light propagates inside the medium, a phase retardation between these two components is built up due to the difference in their phase velocity. Such a phase retardation leads to a new polarization state. Thus, birefringent crystal plates can be used to alter the polarization state of light.

9.2.1 Example: x-Cut Quartz Plate

A crystal plate is called x-cut (or a-cut) when the front and rear surfaces are perpendicular to the x axis. The indices of refraction of quartz at 589.3 nm are

$$n_o = 1.544, \quad n_e = 1.553$$

which leads to a birefringence of $\Delta n = n_e - n_o = 0.009$. A light with a wavelength of 589.3 nm is propagating along the x axis. The electric field vector of such a light is linearly polarized at the front surface and is along the bisection of the y and z axes. The propagation of such a light inside the crystal can be written as

$$\mathbf{E} = \mathbf{y} \exp(-ik_o x) + \mathbf{z} \exp(-ik_e x), \qquad (9.2\text{-}11)$$

where $\mathbf{y}$ and $\mathbf{z}$ are unit vectors along the axes and k_e and k_o are given by

$$k_e = \frac{2\pi}{\lambda} n_e, \qquad k_o = \frac{2\pi}{\lambda} n_o. \qquad (9.2\text{-}12)$$

The polarization state is linear at the front surface and becomes circular at

$$x = \frac{\lambda}{4(n_e - n_o)}.$$

A plate of birefringent crystal with this thickness is known as a quarter-wave plate (or $\frac{1}{4}\lambda$ plate).

Although most of the materials discussed so far are crystals, there are noncrystalline materials that are optically anisotropic. These include liquid crystals and sheet polarizers. Most of these materials consist of nonspherical molecules (or anisometric molecules) ordered in such a way that a directionality exists between them. These molecules can be either elongated (needlelike) or flat (disclike). Because of the structure, these molecules have very anisotropic polarizabilities. For needlelike molecules, the electrons are pushed further from their equilibrium positions by electric fields that are parallel to the axes of the molecules than by those at right angles to the molecular axis. Thus, the induced electronic polarization in the first case is larger than that of the second case. For disc like molecules, the polarizability for electric fields parallel to the disc is expected to be larger than that of the field that is perpendicular to the disc.

9.2.2 Liquid Crystals

Liquid crystals are, by definition, crystalline solids in which a definite ordered arrangement of molecules exists. Crystalline liquids represent a state of matter that arises under certain conditions in organic substances having sharply anisometric molecules. A direct consequence of the ordering of anisometric molecules is the anisotropy of mechanical, electrical, magnetic, and optical properties. A good example of such a substance is p-methoxybenzylidene-p'-n-butylaniline (MBBA), which shows a liquid crystalline phase of the simplest type (nematic) over the temperature range 21–47 °C. There are three phases of crystalline liquids. In the nematic phase, there is a long-range orientational order of the axes of the molecules, whereas the center of the molecules are randomly distributed. In the smectic phase, both one-dimensional translational order and orientational order exist. In the cholesteric phase, an orientational order exists. However, the molecules are found in rows with the molecular direction in each row twisted at a well-defined angle. A smectic liquid crystal is closest in structure to solid crystals. It is interesting to note that in substances that form both a nematic and a smectic phase, the sequence of phase changes on rising temperature is solid crystal $\rightarrow$ smectic liquid crystal $\rightarrow$ nematic liquid crystal $\rightarrow$ isotropic liquid. Although the smectic phase possesses the highest degree of order, it is the nematic and cholesteric phases that have the greatest number of applications.

Because of the orientational ordering of the anisometric molecules, the smectic and nematic liquid crystals are uniaxially symmetric with the optic axis parallel to the axis of the molecules. The optic axis of the cholesteric liquid crystals is only defined locally. The anisotropy of the refractive index is characterized by $\Delta n = n_e - n_o$. In all known nematic and smectic liquid crystals, $\Delta n > 0$. The dielectric anisotropy, $\Delta \varepsilon = \varepsilon_\| - \varepsilon_\perp$, of liquid crystals can be either positive (up to $+15\varepsilon_0$) or negative (down to $-2\varepsilon_0$); $\varepsilon_\|$ and $\varepsilon_\perp$ refer to the dielectric constants for the electric field parallel and perpendicular to the optic axis (also called director). A positive $\Delta\varepsilon$ is characteristic of molecules having a longitudinal dipole moment. It is the sign as well as the magnitude of $\Delta\varepsilon$ that is most critical in determining how a liquid crystal will respond to an applied electric field.

There are many electro-optic effects in liquid crystals. Most of them are the switching of their optical properties by an external electric field. It is found that the molecules in the liquid crystal tend to rotate in such a way that the direction of the maximum dielectric constant coincides with the direction of the field. Since the liquid crystal itself is very anisotropic, any change in the structure is easily observed optically. The time constant involved in the reorientation of molecules in the liquid crystal is on the order of 10^{-3} s, depending on the viscosity of the liquid.

9.2.3 Anisotropic Absorption and Polarizers

There are materials that exhibit strong anisotropic absorption in addition to the birefringence. Tourmaline, tin oxide crystals, and polaroid sheets are some examples. Polaroid sheets consist of a thin layer of small needlelike crystals of herapathite all aligned with their axes parallel. Such materials are characterized by complex refractive indices. If these materials are made of ordered arrangements of sharply anisometric molecules (e.g., needlelike), they are uniaxially anisotropic (e.g., polaroid). These molecules may be absorptive for light with electric fields parallel to their absorption axes and transparent for light with electric fields perpendicular to these axes. Most sheet polarizers are uniaxially anisotropic in their absorption. These polarizers can be described by the complex ordinary and extraordinary indices of refraction.

These complex refractive indices can be written in their real parts and imaginary parts as

$$\hat{n}_o = n_o - i\kappa_o, \qquad \hat{n}_e = n_e - i\kappa_e. \qquad (9.2\text{-}13)$$

A good polarizer should have a large difference between the imaginary parts of the complex refractive indices so that one of the modes suffers strong attenuation in the bulk of the polarizer. Thus, in this model, polarizers can be classified into two types. The O-type polarizer, which transmits ordinary waves and attenuates extraordinary waves, has a real ordinary refractive index (i.e., $\kappa_o = 0$) and a complex extraordinary refractive index (i.e., $0 < \kappa_e$). The E-type polarizer, which transmits extraordinary waves and attenuates ordinary waves, has a real extraordinary refractive index n_e (i.e., $\kappa_e = 0$) and a complex ordinary refractive index n_o (i.e., $0 < \kappa_o$). Most of the commercially produced sheet-type polarizers are made by processes involving the stretching of large plastic sheets. A commonly used plastic is the polymer polyvinyl alcohol (PVA). The stretching creates a unidirectional parallelism for the anisometric molecules. If the absorption axis is parallel to the long axis, these stretched PVA sheet polarizers are of the O type (i.e., $0 \simeq \kappa_o \leqslant \kappa_e$). Typical examples of these polarizers are HN-22 and HN-38. If the polarizers are prepared by a unidirectional compression, these anisometric molecules will have a high degree of uniplanar parallelism. These compressed sheet polarizers are of the E type, provided that the absorption axis of the molecules is parallel to the long axis. An ideal O-type polarizer would have a real ordinary refractive index and a complex extraordinary refractive index such that $|\exp(-ik_o d)| = 1$ and $|\exp(-ik_e d)| \simeq 0$, where d is the thickness of the plate.

9.3   JONES MATRIX FORMULATION

Many birefringent optical systems are made of anisotropic layered media that consist of a train of polarizers and crystal plates. These include electro-optic modulators, Lyot filters, and Solc filters [1]. The effect of each individual element, either polarizer or crystal plate, on the polarization state of the light beam can be easily pictured without the aid of any matrix algebra. However, when an optical system consists of many crystal plates and polarizers, each oriented at a different azimuth angle, the calculation of the overall

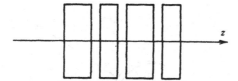

Figure 9.3 Schematic drawing of an anisotropic layered medium.

transmission becomes complicated and is greatly facilitated by a systematic approach.

Jones matrix algebra is a powerful 2 × 2 matrix method in which the electric field vector (or state of polarization) is represented by a two-element column vector, while each crystal plate or polarizer is represented by a 2 × 2 matrix. The whole optical system is represented by a 2 × 2 matrix obtained by the multiplication of all the matrices in sequence. The transmission of light is thus described by the multiplication of the input vector with this matrix.

Referring to Fig. 9.3, let us consider the transmission of light through an anisotropic layered media. For the sake of clarity in introducing the Jones matrix algebra, we will limit ourselves to the case of normal incidence. In addition, we assume that all the crystals are uniaxial and cut such that the c axis lies in the plane of the plate surfaces. Furthermore, we assume that all the surfaces of crystals or polarizers are antireflection coated so that reflection of light can be neglected and light is totally transmitted.

Since there are many crystal plates each oriented at a different azimuth angle, it is essential that we select a fixed coordinate system to describe the polarization state of light. The orientation of each crystal can thus be described by the direction of the c axis relative to this fixed coordinate system (x, y, z). Let the z axis be the direction of propagation that is also normal to the plate surfaces. The azimuth angle of a plate is defined as the angle between the c axis and the x axis. The x and y axes are chosen so that the x axis is vertical and the y axis is horizontal (see Fig. 9.4).

Referring to Fig. 9.4, we consider an incident light beam with an electric field described by the Jones vector

$$E = \begin{pmatrix} A_x \\ A_y \end{pmatrix}, \tag{9.3-1}$$

where A_x and A_y are two complex numbers that represent the components of the electric field vector. The x and y axes are the fixed laboratory axes and are, in general, different from the principal axes of the crystal. To determine how the light propagates in the crystal plate, we need to find the amplitudes of the normal modes. Since this is normal incidence and the light is totally

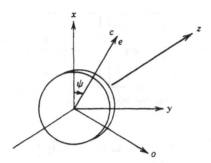

Figure 9.4 Coordinate axes and the azimuthal angle.

transmitted, the amplitudes of normal modes are obtained by a simple geometric projection and are given by

$$C_e = \mathbf{e} \cdot \mathbf{x} A_x + \mathbf{e} \cdot \mathbf{y} A_y, \qquad C_o = \mathbf{o} \cdot \mathbf{x} A_x + \mathbf{o} \cdot \mathbf{y} A_y, \quad (9.3\text{-}2)$$

where C_e and C_o are the amplitudes of the normal modes, $\mathbf{x}$ and $\mathbf{y}$ are unit vectors along the x and y axes, and $\mathbf{e}$ and $\mathbf{o}$ are unit vectors along the direction of the normal modes. Using the definition of azimuth angles in Fig. 9.4, these two unit vectors are given by

$$\mathbf{e} = \mathbf{x} \cos \psi + \mathbf{y} \sin \psi, \qquad \mathbf{o} = -\mathbf{x} \sin \psi + \mathbf{y} \cos \psi. \quad (9.3\text{-}3)$$

Notice that the unit vector $\mathbf{e}$ is parallel to the c axis and the unit vector $\mathbf{o}$ is perpendicular to the c axis.

Using matrix representation, Eq. (9.3-2) can be written as

$$\begin{pmatrix} C_e \\ C_o \end{pmatrix} = \begin{pmatrix} \mathbf{e} \cdot \mathbf{x} & \mathbf{e} \cdot \mathbf{y} \\ \mathbf{o} \cdot \mathbf{x} & \mathbf{o} \cdot \mathbf{y} \end{pmatrix} \begin{pmatrix} A_x \\ A_y \end{pmatrix}. \quad (9.3\text{-}4)$$

This is a general relationship and can be viewed simply as a rotation of the coordinate. Using Eq. (9.3-3), such a coordinate transformation can be written as

$$\begin{pmatrix} C_e \\ C_o \end{pmatrix} = \begin{pmatrix} \cos \psi & \sin \psi \\ -\sin \psi & \cos \psi \end{pmatrix} \begin{pmatrix} A_x \\ A_y \end{pmatrix} = R(\psi) \begin{pmatrix} A_x \\ A_y \end{pmatrix}, \quad (9.3\text{-}5)$$

where $R(\psi)$ represents the rotation matrix.

Here C_e is the extraordinary component of the incident electric field vector, whereas C_o is the ordinary component. These two components are normal modes of the crystal and will propagate with their own phase velocities. Because of the difference in phase velocity, one component is retarded relative to the other. This phase retardation changes the polarization state of the emerging beam.

Let k_e and k_o be the wave numbers of the extraordinary and ordinary waves, respectively. The electric field of the emerging beam can be written as

$$\begin{pmatrix} C'_e \\ C'_o \end{pmatrix} = \begin{pmatrix} \exp(-ik_e l) & 0 \\ 0 & \exp(-ik_o l) \end{pmatrix} \begin{pmatrix} C_e \\ C_o \end{pmatrix} \tag{9.3-6}$$

where l is the thickness of the crystal plate.

Phase retardation is given by the difference of the exponents in (9.3-6) and is equal to

$$\Gamma = (k_e - k_o)l = \frac{2\pi l}{\lambda}(n_e - n_o). \tag{9.3-7}$$

Notice that the phase retardation Γ is a measure of the relative change in phase, not the absolute change. The birefringence of a typical crystal retardation plate is small, that is, $|n_e - n_o| \ll n_e$ or $|n_e - n_o| \ll n_o$. Consequently, the absolute change in phase caused by the plate may be hundreds of times greater than the phase retardation. Let ϕ be the mean absolute phase change:

$$\phi = \tfrac{1}{2}(n_e + n_o)\frac{2\pi l}{\lambda}. \tag{9.3-8}$$

Then, Eq. (9.3-6) can be written in terms of ϕ and Γ as

$$\begin{pmatrix} C'_e \\ C'_o \end{pmatrix} = e^{-i\phi}\begin{pmatrix} e^{-i\Gamma/2} & 0 \\ 0 & e^{i\Gamma/2} \end{pmatrix} \begin{pmatrix} C_e \\ C_o \end{pmatrix}. \tag{9.3-9}$$

The electric field vector (or Jones vector) of the polarization state of the emerging beam in the xy coordinate is given by transforming back from the crystal eo coordinate system:

$$\begin{pmatrix} A'_x \\ A'_y \end{pmatrix} = \begin{pmatrix} \cos\psi & -\sin\psi \\ \sin\psi & \cos\psi \end{pmatrix} \begin{pmatrix} C'_e \\ C'_o \end{pmatrix}. \tag{9.3-10}$$

By combining Eqs. (9.3-5), (9.3-9), and (9.3-10), we can write the transformation due to the retardation plate as

$$\begin{pmatrix} A'_x \\ A'_y \end{pmatrix} = R(-\psi)W_0 R(\psi)\begin{pmatrix} A_x \\ A_y \end{pmatrix}, \tag{9.3-11}$$

where $R(\psi)$ is the rotation matrix and W_0 is the Jones matrix for the retardation plate. These are given, respectively, by

$$R(\psi) = \begin{pmatrix} \cos\psi & \sin\psi \\ -\sin\psi & \cos\psi \end{pmatrix} \qquad (9.3\text{-}12)$$

and

$$W_0 = e^{-i\phi}\begin{pmatrix} e^{-i\Gamma/2} & 0 \\ 0 & e^{i\Gamma/2} \end{pmatrix}. \qquad (9.3\text{-}13)$$

The phase factor $\exp(-i\phi)$ can be neglected provided interference effects are not important or not observable. A retardation plate is characterized by its phase retardation Γ and its azimuth angle ψ and is represented by the product of three matrices (9.3-11):

$$W = R(-\psi)W_0 R(\psi). \qquad (9.3\text{-}14)$$

Note that the Jones matrix of a wave plate is a unitary matrix:

$$W^\dagger W = 1, \qquad (9.3\text{-}15)$$

where the dagger (†) means Hermitian conjugate. The passage of a polarized light beam through a wave plate is mathematically described as a unitary transformation. Many physical properties are invariant under the unitary transformation; these include the orthogonal relation between the Jones vectors and the magnitude of the vectors. Thus, if the polarization states of two beams are mutually orthogonal, they will remain orthogonal after passing through an arbitrary wave plate.

The Jones matrix of an ideal homogeneous linear sheet polarizer oriented with its transmission axis parallel to the laboratory x axis is

$$P_0 = e^{-i\phi}\begin{pmatrix} 1 & 0 \\ 0 & 0 \end{pmatrix}, \qquad (9.3\text{-}16)$$

where ϕ is the absolute phase accumulated due to the finite optical thickness of the polarizer. The Jones matrix of a polarizer rotated by an angle ψ about z is given by

$$P = R(-\psi)P_0 R(\psi). \qquad (9.3\text{-}17)$$

Thus, if we neglect the absolute phase ϕ, the Jones matrix representation of the polarizers transmitting light with electric field vectors parallel to the x and y axes, respectively, are given by

$$P_x = \begin{pmatrix} 1 & 0 \\ 0 & 0 \end{pmatrix} \quad \text{and} \quad P_y = \begin{pmatrix} 0 & 0 \\ 0 & 1 \end{pmatrix}. \tag{9.3-18}$$

To find the effect of a train of retardation plates and polarizers on the polarization state of a polarized light, we first write the electric field vector of the incident beam and then write the Jones matrices of the various elements. The electric field vector of the emerging beam is obtained by carrying out the matrix multiplication in sequence.

9.3.1 Example: A Half-Wave Retardation Plate

A half-wave plate has a phase retardation $\Gamma = \pi$. According to Eq. (9.3-7), an a-cut (or x-cut) uniaxial crystal will act as a half-wave plate provided the thickness is $t = \lambda/2(n_e - n_o)$ (or odd multiples thereof). We will determine the effect of a half-wave plate on the polarization state of a transmitted light beam. The azimuth angle of the wave plate is taken as 45°, and the incident beam is horizontally polarized. The Jones vector for the incident beam can be written as

$$E = \begin{pmatrix} 0 \\ 1 \end{pmatrix}, \tag{9.3-19}$$

and the Jones matrix for the half-wave plate is obtained by using Eqs. (9.3-12)–(9.3-14):

$$W = \frac{1}{\sqrt{2}}\begin{pmatrix} 1 & -1 \\ 1 & 1 \end{pmatrix}\begin{pmatrix} -i & 0 \\ 0 & i \end{pmatrix}\frac{1}{\sqrt{2}}\begin{pmatrix} 1 & 1 \\ -1 & 1 \end{pmatrix} = \begin{pmatrix} 0 & -i \\ -i & 0 \end{pmatrix}. \tag{9.3-20}$$

The Jones vector for the emerging beam is obtained by multiplying Eqs. (9.3-20) and (9.3-19); the result is

$$E' = \begin{pmatrix} -i \\ 0 \end{pmatrix} = -i\begin{pmatrix} 1 \\ 0 \end{pmatrix}. \tag{9.3-21}$$

This is a vertically polarized light. The effect of this half-wave plate is to rotate the polarization by 90°. It can be shown that for a general azimuth

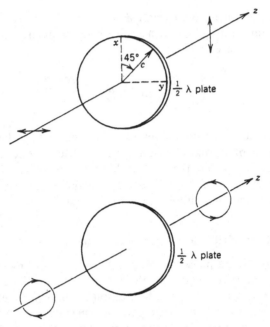

Figure 9.5 Effect of a half-wave plate on the polarization state of a beam. (Adapted from A. Yariv and P. Yeh, *Optical Waves in Crystals*, Wiley, New York, 1984, p. 127. Copyright © 1984. By permission of John Wiley & Sons, Inc.)

angle ψ, the half-wave plate will rotate the polarization by an angle 2ψ (see Problem 9.7). In other words, a linearly polarized light remains linearly polarized, except that the plane of polarization is rotated by an angle of 2ψ.

When the incident light is circularly polarized, a half-wave plate will convert a right-handed circularly polarized light to a left-handed circularly polarized light and vice versa, regardless of the azimuth angle. The proof is left as an exercise for the student (see Problem 9.7). Figure 9.5 illustrates the effect of a half-wave plate.

9.3.2 Example: A Quarter-Wave Plate

A quarter-wave plate has phase retardation $\Gamma = \frac{1}{2}\pi$. If the plate is made of an x-cut (or y-cut) uniaxially anisotropic crystal, thickness $t = \lambda/4(n_e - n_o)$ (or odd multiples thereof). Suppose again that the azimuth angle of the plate is $\psi = 45°$ and the incident beam is horizontally polarized. The normalized

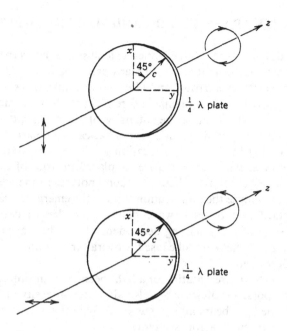

Figure 9.6 The effect of a quarter-wave plate on the polarization state of linearly polarized beam. (Adapted from A. Yariv and P. Yeh, *Optical Waves in Crystals*, Wiley, New York, 1984, p. 128. Copyright © 1984. By permission of John Wiley & Sons, Inc.)

electric field vector for the incident beam is given again by Eq. (9.3-19). The Jones matrix for this quarter-wave plate, according to Eq. (9.3-14), is

$$W = \frac{1}{\sqrt{2}}\begin{pmatrix} 1 & -1 \\ 1 & 1 \end{pmatrix}\begin{pmatrix} e^{-i\pi/4} & 0 \\ 0 & e^{i\pi/4} \end{pmatrix}\frac{1}{\sqrt{2}}\begin{pmatrix} 1 & 1 \\ -1 & 1 \end{pmatrix} = \frac{1}{\sqrt{2}}\begin{pmatrix} 1 & -i \\ -i & 1 \end{pmatrix}.$$

$$(9.3\text{-}22)$$

The Jones vectors for the emerging beam is obtained by multiplying Eqs. (9.3-22) and (9.3-19) and is given by

$$E' = \frac{1}{\sqrt{2}}\begin{pmatrix} -i \\ 1 \end{pmatrix} = \frac{-i}{\sqrt{2}}\begin{pmatrix} 1 \\ i \end{pmatrix}.$$

$$(9.3\text{-}23)$$

This is a left-handed circularly polarized light [see Eq. (1.5-23)]. The effect of a 45°-oriented quarter-wave plate is to convert a horizontally polarized light to a left-handed circularly polarized light. If the incident beam is vertically polarized, the emerging beam will be right-handed circularly polarized. The effect of this quarter-wave plate is illustrated in Fig. 9.6.

9.4 INTENSITY TRANSMISSION AND SOME EXAMPLES

So far, our development of the Jones calculus was concerned with the polarization state of the light beam. In many cases, we need to determine the transmitted intensity. A narrow-band filter, for example, transmits radiation only around a small spectral regime and rejects (or absorbs) radiation with other wavelengths. To change the intensity of the transmitted beam, an analyzer is usually required. An analyzer is basically a polarizer. It is called an analyzer simply because of its location in the optical system. In most birefringent optical systems, a polarizer is placed in front of the system in order to "prepare" a polarized light. A second polarizer (analyzer) is placed at the output to analyze the polarization state of the emerging beam. Because the phase retardation of each wave plate is wavelength dependent, the polarization state of the emerging beam depends on the wavelength of the light. A polarizer at the rear will cause the overall transmitted intensity to be wavelength dependent.

The Jones vector representation of a light beam contains information not only about the polarization state but also about the intensity of light. Let us now consider the light beam after it passes through the polarizer. Its electric vector can be written as a Jones vector,

$$E = \begin{pmatrix} A_x \\ A_y \end{pmatrix}. \tag{9.4-1}$$

The intensity is calculated as follows:

$$I = E^\dagger E = |A_x|^2 + |A_y|^2. \tag{9.4-2}$$

where the dagger (†) indicates the Hermitian conjugate. If the Jones vector of the emerging beam after it passes through the analyzer is written as

$$E' = \begin{pmatrix} A'_x \\ A'_y \end{pmatrix}, \tag{9.4-3}$$

the transmittance of the birefringent optical system is calculated as

$$T = \frac{|A'_x|^2 + |A'_y|^2}{|A_x|^2 + |A_y|^2}. \tag{9.4-4}$$

9.4.1 Example: A Birefringent Plate Sandwiched Between Parallel Polarizers

Referring to Fig. 9.7, we consider a birefringent plate sandwiched between a pair of parallel polarizers. The plate is oriented such that the c axis is at 45°

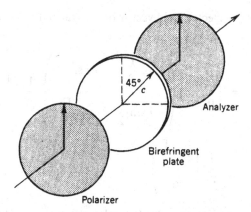

Figure 9.7 A birefringent plate sandwiched between a pair of parallel polarizers. (Adapted from A. Yariv and P. Yeh, *Optical Waves in Crystals*, Wiley, New York, 1984, p. 130. Copyright © 1984. By permission of John Wiley & Sons, Inc.)

with respect to the polarizers. Let the birefringence be $n_e - n_o$ and the plate thickness be d. The phase retardation is then given by

$$\Gamma = \frac{2\pi(n_e - n_o)d}{\lambda} \tag{9.4-5}$$

and the corresponding Jones matrix is, according to Eq. (9.3-14),

$$W = \begin{pmatrix} \cos\dfrac{\Gamma}{2} & -i\sin\dfrac{\Gamma}{2} \\[3mm] -i\sin\dfrac{\Gamma}{2} & \cos\dfrac{\Gamma}{2} \end{pmatrix}. \tag{9.4-6}$$

Let the incident beam be unpolarized so that after it passes through the front polarizer, the electric field vector can be represented by the Jones vector

$$\frac{1}{\sqrt{2}}\begin{pmatrix} 1 \\ 0 \end{pmatrix}, \tag{9.4-7}$$

where we assumed that the intensity of the incident beam is unity and only half of the intensity passes through the polarizer. The Jones vector representation of the electric field vector of the transmitted beam is obtained as follows:

$$E' = \begin{pmatrix} 1 & 0 \\ 0 & 0 \end{pmatrix} \begin{pmatrix} \cos\dfrac{\Gamma}{2} & -i\sin\dfrac{\Gamma}{2} \\ -i\sin\dfrac{\Gamma}{2} & \cos\dfrac{\Gamma}{2} \end{pmatrix} \dfrac{1}{\sqrt{2}} \begin{pmatrix} 1 \\ 0 \end{pmatrix}$$

$$= \dfrac{1}{\sqrt{2}} \begin{pmatrix} \cos\dfrac{\Gamma}{2} \\ 0 \end{pmatrix}. \tag{9.4-8}$$

The transmitted beam is vertically (x) polarized with an intensity given by

$$I = \tfrac{1}{2}\cos^2\dfrac{\Gamma}{2} = \tfrac{1}{2}\cos^2\dfrac{\pi(n_e - n_o)d}{\lambda}. \tag{9.4-9}$$

It can be seen from Eq. (9.4-9) that the transmitted intensity is a sinusoidal function of the wave number and peaks at $\lambda = (n_e - n_o)d$, $\tfrac{1}{2}(n_e - n_o)d$, $\tfrac{1}{3}(n_e - n_o)d$, ... The wave number separation between transmission maxima increases with decreasing plate thickness.

9.4.2 Example: A Birefringent Plate Sandwiched with a Pair of Crossed Polarizers

If we rotate the analyzer shown in Fig. 9.7 by 90°, the input and output polarizers are crossed. The transmitted beam for this case is obtained as follows:

$$E' = \begin{pmatrix} 0 & 0 \\ 0 & 1 \end{pmatrix} \begin{pmatrix} \cos\dfrac{\Gamma}{2} & -i\sin\dfrac{\Gamma}{2} \\ -i\sin\dfrac{\Gamma}{2} & \cos\dfrac{\Gamma}{2} \end{pmatrix} \dfrac{1}{\sqrt{2}} \begin{pmatrix} 1 \\ 0 \end{pmatrix}$$

$$= \dfrac{-i}{\sqrt{2}} \begin{pmatrix} 0 \\ \sin\dfrac{\Gamma}{2} \end{pmatrix}. \tag{9.4-10}$$

The transmitted beam is horizontally (y) polarized with an intensity given by

$$I = \tfrac{1}{2}\sin^2\dfrac{\Gamma}{2} = \tfrac{1}{2}\sin^2\dfrac{\pi(n_e - n_o)d}{\lambda}. \tag{9.4-11}$$

This is again a sinusoidal function of the wave number. The transmission spectrum consists of a series of maxima at $\lambda = 2(n_e - n_o)d$, $\frac{2}{3}(n_e - n_o)d$, ... These wavelengths correspond to phase retardations of π, 3π, 5π, ..., that is, when the wave plate becomes a half-wave plate or an odd integral multiple of the half-wave plate.

9.5 DOUBLE REFRACTION AT A BOUNDARY

Consider a plane wave incident on the surface of an anisotropic crystal. The refracted wave, in general, is a mixture of the two normal modes. In a uniaxial crystal, the refracted wave, in general, is a mixture of the ordinary wave and the extraordinary wave. According to the arguments used in connection with reflection and refraction at a plane boundary, the boundary condition requires that all the wave vectors lies in the plane of incidence and that the tangential components of these wave vectors along the boundary be the same. This kinematic condition remains true for refraction at a boundary of an anisotropic crystal.

Let $\mathbf{k}$ be the propagation vector of the incident wave and $\mathbf{k}_1$, $\mathbf{k}_2$ the propagation vectors of the refracted waves. Given a prescribed value of the projection of the propagation vector $\mathbf{k}$ on the boundary, the two shells of the normal surface in general yields two propagation vectors, thus giving rise to two refracted waves, as shown in Fig. 9.8. The kinematic condition requires that

$$k \sin \theta = k_1 \sin \theta_1 = k_2 \sin \theta_2, \qquad (9.5\text{-}1)$$

where θ is the incident angle and θ_1, θ_2 are the refraction angles.

Equation (9.5-1) looks like Snell's law. However, it is important to remember that k_1, k_2 are not, in general, constant; rather, they vary with the directions of the vectors k_1, k_2. The problem of determining θ_1 and θ_2 involves solving an algebraic quartic equation. The graphic method is relatively easier and is shown in Fig. 9.8.

In the case of uniaxial crystals, one shell of the normal surface is a sphere. The corresponding wave number k is therefore a constant for all directions of propagation. This wave is the ordinary wave and obeys Snell's law, that is,

$$n \sin \theta = n_o \sin \theta_o, \qquad (9.5\text{-}2)$$

where n is the index of refraction of the incident medium, θ_o is the refraction angle, and n_o is the ordinary refractive index of the crystal. The other shell

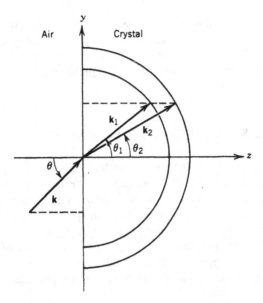

Figure 9.8 Double refraction at a boundary of an anisotropic medium and the graphic method of determining θ_1 and θ_2.

of the normal surface is an ellipsoid of revolution. Therefore, the corresponding wave number k_2 depends on the direction of propagation. This wave is the extraordinary wave.

9.5.1 An a-Cut Uniaxially Crystal Surface

For the case of an a-cut uniaxially crystal surface, expressions for the double refraction can be easily obtained by using the explicit form of the normal surface Eq. (9.2-2). For the sake of clarity in distinguishing from the laboratory coordinates x, y, and z, we will use a, b, and c to designate the principal dielectric axes of the crystal. The laboratory coordinate system is chosen such that the plane of incidence coincides with the yz plane. Since the crystal surface is perpendicular to the a axis, the orientation of the crystal relative to the laboratory coordinate can be described by the angle between the x axis and the c axis. Let ϕ be such an angle between the x and c axes so that the unit vector c representing the c axis can be written (see Fig. 9.9) as

$$\mathbf{c} = \mathbf{x} \cos \phi + \mathbf{y} \sin \phi, \tag{9.5-3}$$

where $\mathbf{x}$ and $\mathbf{y}$ are unit vectors along the x and y axes.

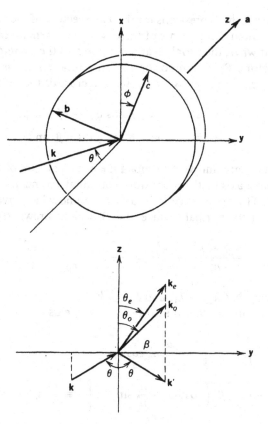

Figure 9.9 (a) xyz and abc coordinate systems. Incident wave vector $\mathbf{k}$ lies in yz-plane and the c axis of crystal lies in xy plane; (b) reflection and refraction at the crystal surface.

Let β be the tangential component of the incident wave so that the wave vector can be written as

$$\mathbf{k} = \mathbf{y}\beta + \mathbf{z}k_z, \qquad (9.5\text{-}4)$$

where $\mathbf{z}$ is a unit vector along the z axis and k_z is the z component of the incident wave vector. These two components are related to the incident angle θ by

$$\beta = k \sin \theta, \qquad k_z = k \cos \theta, \qquad (9.5\text{-}5)$$

where $k = (\omega/c)n$.

We now derive the expressions for the wave vectors of the refracted waves. Since the tangential components of these two waves remain the same as that of the incident wave, the problem at hand is to find the z component of these two wave vectors. By decomposing the tangential component β along the crystal axes c and b, we can write the wave vectors of the refracted waves as

$$
\begin{aligned}
\mathbf{k}_o &= \mathbf{a}k_{oz} + \mathbf{b}(-\beta\cos\phi) + \mathbf{c}\beta\sin\phi, \\
\mathbf{k}_e &= \mathbf{a}k_{ez} + \mathbf{b}(-\beta\cos\phi) + \mathbf{c}\beta\sin\phi,
\end{aligned}
\tag{9.5-6}
$$

where $\mathbf{a}$, $\mathbf{b}$, and $\mathbf{c}$ are unit vectors along the principal axes of the crystal. We notice that the z axis of the laboratory coordinate coincides with the a axis of the crystal. The z components k_{oz} and k_{ez} can now be derived by using the explicit form of the normal surface (9.2-2), which is rewritten as

$$
\frac{k_{ea}^2 + k_{eb}^2}{n_e^2} + \frac{k_{ec}^2}{n_o^2} = \frac{\omega^2}{c^2}, \qquad \frac{k_o^2}{n_o^2} = \frac{\omega^2}{c^2},
\tag{9.5-7}
$$

where we recall that $k_{ea} = k_{ez}$ and $k_{oa} = k_{oz}$.

Substitution of Eq. (9.5-6) into Eq. (9.5-7) yields

$$
\begin{aligned}
k_{oz} &= \left[\left(\frac{n_o\omega}{c}\right)^2 - \beta^2\right]^{1/2} \equiv \frac{\omega}{c}n_o\cos\theta_o, \\
k_{ez} &= \left[\left(\frac{n_e\omega}{c}\right)^2 - \beta^2\left(\cos^2\phi + \frac{n_e^2}{n_o^2}\sin^2\phi\right)\right]^{1/2} \equiv \frac{\omega}{c}n_e(\theta_e)\cos\theta_e,
\end{aligned}
\tag{9.5-8}
$$

where θ_o, θ_e are the refraction angles and $n_e(\theta_e)$ is the refractive index associated with the propagation of the E wave, and we recall that the angle between the c axis and the plane of incidence is $(\frac{1}{2}\pi - \phi)$. The refraction angle θ_e can also be obtained directly from

$$
\theta_e = \tan^{-1}\frac{\beta}{k_{ez}}.
\tag{9.5-9}
$$

The electric field vectors of these two waves can be obtained by using Eqs. (9.2-6) and (9.2-7). In the principal coordinate of the crystal (a, b, c), these electric field vectors are given by

$$
O \text{ wave:} \qquad \mathbf{E} = \begin{pmatrix} \beta\cos\phi \\ k_{oz} \\ 0 \end{pmatrix}
\tag{9.5-10}
$$

and

$$
\text{E wave:} \quad \mathbf{E} = \begin{vmatrix} \dfrac{k_{ez}}{n_e^2(\theta_e) - n_o^2} \\[2mm] \dfrac{-\beta \cos \phi}{n_e^2(\theta_e) - n_o^2} \\[2mm] \dfrac{\beta \sin \phi}{n_e^2(\theta_e) - n_e^2} \end{vmatrix}, \tag{9.5-11}
$$

respectively.

These two vectors can also be written in terms of their components in the laboratory coordinate system by using the transformation

$$
\begin{aligned}
\mathbf{x} &= \mathbf{b} \sin \phi + \mathbf{c} \cos \phi, \\
\mathbf{y} &= -\mathbf{b} \cos \phi + \mathbf{c} \sin \phi, \\
\mathbf{z} &= \mathbf{a}.
\end{aligned} \tag{9.5-12}
$$

From Eqs. (9.5-8), (9.5-6), and (9.5-5) and using $[n_e(\theta_e)\omega/c]^2 = \beta^2 + k_{ez}^2$, we obtain

$$
n_e^2(\theta_e) = n_e^2 - \left(\frac{\beta c}{\omega}\right)^2 \left(\frac{n_e^2}{n_o^2} - 1\right) \sin^2 \phi. \tag{9.5-13}
$$

Substituting Eq. (9.5-13) for $n_e^2(\theta_e)$ in Eqs. (9.5-10) and (9.5-11) and using Eq. (9.5-12), we obtain the following expressions for the electric field vectors in the xyz coordinate system:

$$
\begin{aligned}
\mathbf{o} &= N_o(\mathbf{x}k_{oz} \sin \phi - \mathbf{y}k_{oz} \cos \phi + \mathbf{z}\beta \cos \phi), \\
\mathbf{e} &= N_e(\mathbf{x}k_o^2 \cos \phi + \mathbf{y}k_{oz}^2 \sin \phi - \mathbf{z}\beta k_{ez} \sin \phi),
\end{aligned} \tag{9.5-14}
$$

where N_o and N_e are normalization constants that make

$$
\mathbf{o} \cdot \mathbf{o} = \mathbf{e} \cdot \mathbf{e} = 1, \tag{9.5-15}
$$

and

$$
k_o^2 = \left(\frac{n_o \omega}{c}\right)^2, \qquad k_{oz}^2 = k_o^2 - \beta^2.
$$

Note that $\mathbf{o}$ and $\mathbf{e}$ are, in general, not orthogonal because they correspond to normal modes of two different directions of propagation. However, $\mathbf{o} \cdot \mathbf{e}$ does tend to zero when anisotropy disappears.

So far, we have only discussed the kinematic property of double refraction. To find the amplitudes of the ordinary and extraordinary waves, we need to match the continuity conditions at the boundary. In addition, we need to include the presence of reflected waves.

9.6 REFLECTION AND REFRACTION OF ELECTROMAGNETIC RADIATION AT A CRYSTAL SURFACE

We have so far neglected the reflection of electromagnetic radiation from the surfaces of crystal plates. This is only legitimate when the surfaces are treated with antireflection coatings. It is interesting to study the reflection and refraction of electromagnetic radiation at an interface between air and the crystal. Unlike the Fresnel reflection and refraction of a dielectric interface between two isotropic media, s and p waves are no longer independent. They are coupled because of the optical anisotropy of the crystal. In other words, an incidence wave polarized along the s direction will generate a reflected wave that is a mixture of s and p waves. In addition, an O wave incident from the inside of the crystal will generate a reflected wave that is a mixture of O and E waves.

Referring to Fig. 9.9, we consider the incidence of a plane wave from the lower half-space ($z < 0$). Let $\mathbf{k}$, $\mathbf{k}'$ be the wave vectors of the incident and reflected waves, respectively, and $\mathbf{k}_o$, $\mathbf{k}_e$ be the wave vectors of the ordinary and extraordinary refracted waves, respectively. The electric fields of incident, reflected, and refracted waves are

$$\text{Incident:} \quad \mathbf{E} = (A_s\mathbf{s} + A_p\mathbf{p}) \exp\left[i(\omega t - \mathbf{k}\cdot\mathbf{r})\right], \qquad (9.6\text{-}1)$$

$$\text{Reflected:} \quad \mathbf{E} = (B_s\mathbf{s} + B_p\mathbf{p}') \exp\left[i(\omega t - \mathbf{k}'\cdot\mathbf{r})\right], \qquad (9.6\text{-}2)$$

$$\text{Refracted:} \quad \mathbf{E} = (C_o\mathbf{o}e^{-i\mathbf{k}_o\cdot\mathbf{r}} + C_e\mathbf{e}e^{-i\mathbf{k}_e\cdot\mathbf{r}}) \exp(i\omega t), \qquad (9.6\text{-}3)$$

where $\mathbf{s}$ is a unit vector perpendicular to the plane of incidence and is given by $\mathbf{s} = \mathbf{x}$ in the chosen coordinate (see Fig. 9.9) and $\mathbf{p}$ and $\mathbf{p}'$ are unit vectors parallel to the plane of incidence and are given by

$$\mathbf{p} = \frac{\mathbf{k} \times \mathbf{s}}{|\mathbf{k}|}, \qquad (9.6\text{-}4)$$

$$\mathbf{p}' = \frac{-\mathbf{k}' \times \mathbf{s}}{|\mathbf{k}|}. \qquad (9.6\text{-}5)$$

In Eq. (9.6-5) we put a minus sign to conform with our earlier definition of the reflected p wave (see Section 3.1).

The terms o and e represent unit vectors parallel to the electric field vector of the ordinary mode and extraordinary mode, respectively, in the uniaxially anisotropic medium. The wave vectors can be written as

$$\mathbf{k} = \beta \mathbf{y} + k_z \mathbf{z}, \tag{9.6-6}$$

$$\mathbf{k}' = \beta \mathbf{y} - k_z \mathbf{z}, \tag{9.6-7}$$

$$\mathbf{k}_o = \beta \mathbf{y} + k_{oz} \mathbf{z}, \tag{9.6-8}$$

$$\mathbf{k}_e = \beta \mathbf{y} + k_{ez} \mathbf{z}, \tag{9.6-9}$$

where β, which remains constant throughout the media, is the tangential component of all the wave vectors. These components of wave vectors are given by Eqs. (9.5-5) and (9.5-8). The unit vectors o and e are given by Eq. (9.5-14).

In Eqs. (9.6-1)–(9.6-3), A_s, A_p, B_s, B_p, C_o, and C_e are constants, where A_s and A_p are the amplitudes of the incident s and p waves, respectively; B_s and B_p are those of the reflected wave; and C_o and C_e are the amplitudes of the transmitted O and E waves, respectively.

The magnetic fields of the incident, reflected, and refracted waves can be derived from Eqs. (9.6-1)–(9.6-3) and by Maxwell's equation,

$$\mathbf{H} = \frac{i}{\omega\mu} \nabla \times \mathbf{E}, \tag{9.6-10}$$

and are given, respectively, by

Incident: $$\mathbf{H} = \frac{1}{\omega\mu} \mathbf{k} \times (A_s \mathbf{s} + A_p \mathbf{p}) \exp[i(\omega t - \mathbf{k} \cdot \mathbf{r})], \tag{9.6-11}$$

Reflected: $$\mathbf{H} = \frac{1}{\omega\mu} \mathbf{k}' \times (B_s \mathbf{s} + B_p \mathbf{p}') \exp[i(\omega t - \mathbf{k}' \cdot \mathbf{r})], \tag{9.6-12}$$

Refracted: $$\mathbf{H} = \frac{1}{\omega\mu} [C_o \mathbf{k}_o \times \mathbf{o} e^{-i\mathbf{k}_o \cdot \mathbf{r}} + C_e \mathbf{k}_e \times \mathbf{e} e^{-i\mathbf{k}_e \cdot \mathbf{r}}] \exp(i\omega t). \tag{9.6-13}$$

The tangential component of $\mathbf{E}$ and $\mathbf{H}$ must be continuous at the boundary $z = 0$. In terms of the fields (9.6-1)–(9.6-3) and (9.6-11)–(9.6-13), these boundary conditions at $z = 0$ are

$$A_s + B_s = \mathbf{x} \cdot \mathbf{o} C_o + \mathbf{x} \cdot \mathbf{e} C_e, \tag{9.6-14}$$

$$\mathbf{y} \cdot \mathbf{p} A_p + \mathbf{y} \cdot \mathbf{p}' B_p = \mathbf{y} \cdot \mathbf{o} C_o + \mathbf{y} \cdot \mathbf{e} C_e, \tag{9.6-15}$$

$$\mathbf{x} \cdot (\mathbf{k} \times \mathbf{p})A_p + \mathbf{x} \cdot (\mathbf{k'} \times \mathbf{p'})B_p = \mathbf{x} \cdot (\mathbf{k}_o \times \mathbf{o})C_o + \mathbf{x} \cdot (\mathbf{k}_e \times \mathbf{e})C_e, \tag{9.6-16}$$

$$\mathbf{y} \cdot (\mathbf{k} \times \mathbf{s})A_s + \mathbf{y} \cdot (\mathbf{k'} \times \mathbf{s})B_s = \mathbf{y} \cdot (\mathbf{k}_o \times \mathbf{o})C_o + \mathbf{y} \cdot (\mathbf{k}_e \times \mathbf{e})C_e. \tag{9.6-17}$$

These four equations can be used to solve for the four unknowns B_s, B_p, C_o, and C_e in terms of the amplitudes A_s and A_p of the incident wave. Using the expressions for $\mathbf{k}$, $\mathbf{k'}$, $\mathbf{p}$ and $\mathbf{p'}$ from Eqs. (9.6-4)–(9.6-7), these equations can be written as

$$k(A_s + B_s) = \mathbf{x} \cdot \mathbf{o}kC_o + \mathbf{x} \cdot \mathbf{e}kC_e, \tag{9.6-18}$$

$$k_z(A_p + B_p) = \mathbf{y} \cdot \mathbf{o}kC_o + \mathbf{y} \cdot \mathbf{e}kC_e, \tag{9.6-19}$$

$$-k(A_p - B_p) = \mathbf{x} \cdot (\mathbf{k}_o \times \mathbf{o})C_o + \mathbf{x} \cdot (\mathbf{k}_e \times \mathbf{e})C_e, \tag{9.6-20}$$

$$k_z(A_s - B_s) = \mathbf{y} \cdot (\mathbf{k}_o \times \mathbf{o})C_o + \mathbf{y} \cdot (\mathbf{k}_e \times \mathbf{e})C_e, \tag{9.6-21}$$

where $k_z = k \cos \theta$.

We now eliminate B_s and B_p and obtain

$$AC_o + BC_e = 2k_z A_s, \tag{9.6-22}$$

$$CC_o + DC_e = 2k_z A_p, \tag{9.6-23}$$

where A, B, C and D are constants given by

$$A = \mathbf{o} \cdot (\mathbf{y} \times \mathbf{k}) + \mathbf{o} \cdot (\mathbf{y} \times \mathbf{k}_o), \tag{9.6-24}$$

$$B = \mathbf{e} \cdot (\mathbf{y} \times \mathbf{k}) + \mathbf{e} \cdot (\mathbf{y} \times \mathbf{k}_e), \tag{9.6-25}$$

$$C = \mathbf{o} \cdot \mathbf{y}k - \frac{(\mathbf{y} \times \mathbf{k}) \cdot (\mathbf{k}_o \times \mathbf{o})}{k}, \tag{9.6-26}$$

$$D = \mathbf{e} \cdot \mathbf{y}k - \frac{(\mathbf{y} \times \mathbf{k}) \cdot (\mathbf{k}_e \times \mathbf{e})}{k}. \tag{9.6-27}$$

In arriving at these expressions, we have used $\mathbf{x}k_z = (\mathbf{y} \times \mathbf{k})$. Equations (9.6-22) and (9.6-23) can now be solved for C_o and C_e in terms of A_s and A_p. This leads to the following linear relations:

$$C_o = A_s t_{so} + A_p t_{po}, \tag{9.6-28}$$

$$C_e = A_s t_{se} + A_p t_{pe}, \tag{9.6-29}$$

where t_{so}, t_{po}, t_{se}, and t_{pe} are the Fresnel transmission coefficients given by

$$t_{so} = \frac{2k_z D}{AD - BC},$$ (9.6-30)

$$t_{po} = \frac{-2k_z B}{AD - BC},$$ (9.6-31)

$$t_{se} = \frac{-2k_z C}{AD - BC},$$ (9.6-32)

$$t_{pe} = \frac{2k_z A}{AD - BC}.$$ (9.6-33)

Notice that we now have four transmission coefficients. Here t_{so} is the transmission coefficient for the case of an s-polarized incident wave and a transmitted O wave. The other three coefficients have their similar physical meaning according to their subscripts.

The other two unknowns, B_s and B_p, which are the amplitudes of the reflected waves, can now be obtained by substituting Eqs. (9.6-28) and (9.6-29) into Eqs. (9.6-18)–(9.6-21). Similar linear relations are obtained:

$$B_s = A_s r_{ss} + A_p r_{ps},$$ (9.6-34)

$$B_p = A_s r_{sp} + A_p r_{pp},$$ (9.6-35)

where r_{ss}, r_{sp}, r_{ps}, and r_{pp} are the reflection coefficients given by

$$r_{ss} = \frac{A'D - B'C}{AD - BC},$$ (9.6-36)

$$r_{ps} = \frac{AB' - A'B}{AD - BC},$$ (9.6-37)

$$r_{sp} = -\frac{CD' - C'D}{AD - BC},$$ (9.6-38)

$$r_{pp} = \frac{AD' - BC'}{AD - BC},$$ (9.6-39)

with A', B', C', and D' given by

$$A' = \mathbf{o} \cdot (\mathbf{y} \times \mathbf{k}) - \mathbf{o} \cdot (\mathbf{y} \times \mathbf{k}_o) = 2\mathbf{x} \cdot \mathbf{o} k_z - A,$$ (9.6-40)

$$B' = \mathbf{e} \cdot (\mathbf{y} \times \mathbf{k}) - \mathbf{e} \cdot (\mathbf{y} \times \mathbf{k}_e) = 2\mathbf{x} \cdot \mathbf{e} k_z - B,$$ (9.6-41)

$$C' = \mathbf{o} \cdot \mathbf{y}k + \frac{(\mathbf{y} \times \mathbf{k}) \cdot (\mathbf{k}_o \times \mathbf{o})}{k} = 2\mathbf{o} \cdot \mathbf{y}k - C, \quad (9.6\text{-}42)$$

$$D' = \mathbf{e} \cdot \mathbf{y}k + \frac{(\mathbf{y} \times \mathbf{k}) \cdot (\mathbf{k}_c \times \mathbf{e})}{k} = 2\mathbf{e} \cdot \mathbf{y}k - D. \quad (9.6\text{-}43)$$

Among these four reflection coefficients, r_{ss} and r_{pp} are the direct reflection coefficients, whereas r_{sp} and r_{ps} can be viewed as the cross-reflection coefficients.

The cross-reflection coefficients r_{sp} and r_{ps} vanish when the anisotropy disappears (i.e., when $n_e = n_o$). The proof is left as an exercise for the student (see Problem 9.11).

We are now ready to introduce some important concepts. According to Eqs. (9.6-28) and (9.6-29), both ordinary and extraordinary waves are, in general, excited by the incidence of a polarized wave. It is interesting to note that there exist two input polarization states of the incident wave that will excite only normal modes (either ordinary or extraordinary). According to Eqs. (9.6-28) and (9.6-29), these two polarization states are given by

$$O\text{-wave excitation:} \quad \left(\frac{A_p}{A_s}\right)_{C_e = 0} = -\frac{t_{se}}{t_{pe}}, \quad (9.6\text{-}44)$$

$$E\text{-wave excitation:} \quad \left(\frac{A_p}{A_s}\right)_{C_o = 0} = -\frac{t_{so}}{t_{po}}. \quad (9.6\text{-}45)$$

The expressions for the four reflection coefficients reduce to simple forms at normal incidence. When the incidence angle θ is zero, all the wave vectors are parallel to the z axis, and the unit vectors $\mathbf{o}$ and $\mathbf{e}$ are given by

$$\mathbf{o} = -\mathbf{x} \sin \phi + \mathbf{y} \cos \phi, \quad (9.6\text{-}46)$$

$$\mathbf{e} = \mathbf{x} \cos \phi + \mathbf{y} \sin \phi. \quad (9.6\text{-}47)$$

In addition, we note that $\mathbf{s} = \mathbf{x}$, $\mathbf{p} = \mathbf{y}$ at normal incidence.

Substituting Eqs. (9.6-46) and (9.6-47) into Eqs. (9.6-22)–(9.6-27) and Eqs. (9.6-40)–(9.6-43) and according to Eqs. (9.6-36)–(9.6-39), we obtain

$$r_{ss} = \frac{(n^2 - n_o n_e) - n(n_e - n_o) \cos 2\phi}{(n + n_o)(n + n_e)}, \quad (9.6\text{-}48)$$

$$r_{ps} = -\frac{(n_e - n_o)n \sin 2\phi}{(n + n_o)(n + n_e)}, \quad (9.6\text{-}49)$$

$$r_{sp} = -\frac{(n_e - n_o)n \sin 2\phi}{(n + n_o)(n + n_e)}, \tag{9.6-50}$$

$$r_{pp} = \frac{(n^2 - n_o n_e) - n(n_e - n_o) \cos 2\phi}{(n + n_o)(n + n_e)}, \tag{9.6-51}$$

where n is the index of refraction of the incident medium.

Notice that the cross-reflection coefficients vanish when the c axis is either perpendicular ($\phi = 0$) or parallel ($\phi = \frac{1}{2}\pi$) to the plane of incidence. At these special angles, the s and p waves are not coupled.

The transmission coefficients for normal incidence are obtained in a similar fashion and are given by

$$t_{so} = \frac{-2n}{n + n_o} \sin \phi, \tag{9.6-52}$$

$$t_{po} = \frac{2n}{n + n_o} \cos \phi, \tag{9.6-53}$$

$$t_{se} = \frac{2n}{n + n_e} \cos \phi, \tag{9.6-54}$$

$$t_{pe} = \frac{2n}{n + n_e} \sin \phi. \tag{9.6-55}$$

According to Eqs. (9.6-46), (9.6-47), and (9.6-4), these expressions of transmission coefficients at normal incidence can also be written as

$$t_{so} = \frac{2n}{n + n_o} \mathbf{s} \cdot \mathbf{o}, \tag{9.6-56}$$

$$t_{po} = \frac{2n}{n + n_o} \mathbf{p} \cdot \mathbf{o}, \tag{9.6-57}$$

$$t_{se} = \frac{2n}{n + n_e} \mathbf{s} \cdot \mathbf{e}, \tag{9.6-58}$$

$$t_{pe} = \frac{2n}{n + n_e} \mathbf{p} \cdot \mathbf{e}. \tag{9.6-59}$$

We notice that the transmission coefficients are mostly dependent on the scalar product between the electric field of the normal modes in the incident medium and the crystal.

9.6.1 Small Birefringence Approximation

As we have seen, the general expression for the reflection and transmission coefficients are complicated, especially for off-axis light with the c axis of the crystal oriented at an arbitrary angle ϕ. It is desirable to have approximate expressions for these eight coefficients.

In most birefringent filters, the crystals or polarizers required must have small birefringence (i.e., $|n_e - n_o| \ll 1$). Currently, the most widely used materials in conventional birefringent filters are quartz crystals ($n_e - n_o = 0.009$) and oriented sheets of PVA ($|n_e - n_o| \simeq 0.006$) [2], whereas the most widely used crystals in dispersive birefringent filters is CdS ($|n_e - n_o| \simeq 0.01$) [3]. Under these conditions, the eight coefficients can be greatly simplified by an approximation. In this approximation, we derive the expressions for the reflection and transmission coefficients, disregarding the anisotropy of the elements. This is legitimate because the wave vectors $\mathbf{k}_o$ and $\mathbf{k}_e$ are almost equal (i.e., $\mathbf{k}_o \simeq \mathbf{k}_e$) according to Eqs. (9.5-8) and (9.5-6), provided that $|n_e - n_o| \ll n_o, n_e$. Also, the refraction angles θ_o and θ_e are almost equal (i.e., $\theta_e \simeq \theta_o$), and the polarization vectors $\mathbf{o}$ and $\mathbf{e}$ can be given approximately by

$$\mathbf{o} = \frac{\mathbf{c} \times \mathbf{k}_o}{|\mathbf{c} \times \mathbf{k}_o|} \qquad (9.6\text{-}60)$$

and

$$\mathbf{e} = \frac{\mathbf{k}_o \times \mathbf{o}}{|\mathbf{k}_o \times \mathbf{o}|}, \qquad (9.6\text{-}61)$$

respectively [note that Eq. (9.6-60) is exact]. Thus, an s wave will retain its s polarization on refraction at the interfaces. The same thing happens to the p wave. Consequently, these eight coefficients can be given approximately by

$$r_{ss} = r_s, \qquad (9.6\text{-}62)$$

$$r_{sp} = r_{ps} = 0, \qquad (9.6\text{-}63)$$

$$r_{pp} = r_p, \qquad (9.6\text{-}64)$$

and

$$t_{so} = \mathbf{s} \cdot \mathbf{o} t_s, \qquad (9.6\text{-}65)$$

$$t_{se} = \mathbf{s} \cdot \mathbf{e} t_s, \qquad (9.6\text{-}66)$$

$$t_{po} = \mathbf{p} \cdot \mathbf{o} t_p, \qquad (9.6\text{-}67)$$

$$t_{pe} = \mathbf{p} \cdot \mathbf{e} t_p, \qquad (9.6\text{-}68)$$

where o and e are unit vectors of the normal modes in crystals and are given by Eqs. (9.6-60) and (9.6-61) and r_s, r_p, t_s, and t_p are the Fresnel reflection and transmission coefficients given by

$$r_s = \frac{n \cos \theta - n_o \cos \theta_o}{n \cos \theta + n_o \cos \theta_o}, \tag{9.6-69}$$

$$r_p = \frac{n \cos \theta_o - n_o \cos \theta}{n_o \cos \theta + n \cos \theta_o}, \tag{9.6-70}$$

and

$$t_s = \frac{2n \cos \theta}{n \cos \theta + n_o \cos \theta_o}, \tag{9.6-71}$$

$$t_p = \frac{2n \cos \theta}{n \cos \theta_o + n_o \cos \theta}, \tag{9.6-72}$$

where we recall that n is the index of refraction of the incident medium (usually air), n_o is the index of refraction of the birefringent elements (since $n_e \simeq n_o$), θ is the incident angle, and θ_o is the refraction angle ($\theta_e \simeq \theta_o$). Here t_s and t_p are the transmission coefficients for the s and p waves, respectively, on entering the birefringent element disregarding the anisotropy. These approximate expressions may be used to study the wide-field property of birefringent optical systems.

9.7   4 × 4 MATRIX FORMULATION

The Jones matrix method introduced earlier in this chapter neglects the reflection of light completely. This is, of course, only an approximation. In practice, the reflection of light occurs at crystal surfaces where dielectric discontinuity is present.

In the case of isotropic layered media, the electromagnetic radiation can be divided into two independent (uncoupled) modes: s modes (with electric field vector E perpendicular to the plane of incidence) and p modes (with electric field vector E parallel to the plane of incidence). Since they are uncoupled, the matrix method involves the manipulation of 2 × 2 matrices only. In the case of birefringent layered media, the electromagnetic radiation consists of four partial waves. Mode coupling takes place at the interface where an incident plane wave produces waves with different polarization states due to the anisotropy of the layers. As a result, 4 × 4 matrices are needed in the matrix method.

We will now formulate this problem for the general birefringent layered media, which consists of anisotropic materials with their principal axes oriented at arbitrary directions. The laboratory coordinate system is again chosen such that the z axis is normal to the interfaces.

Since the medium is not isotropic, the propagation characteristics depend on the direction of propagation. The orientations of the crystal axes are described by the Euler angles θ, ϕ, and ψ with respect to a fixed xyz coordinate system. The dielectric tensor in the xyz coordinate system is given by

$$\varepsilon = A \begin{pmatrix} \varepsilon_1 & 0 & 0 \\ 0 & \varepsilon_2 & 0 \\ 0 & 0 & \varepsilon_3 \end{pmatrix} A^{-1}, \tag{9.7-1}$$

where ε_1, ε_2, and ε_3 are the principal dielectric constants and A is the coordinate rotation matrix given by [4]

$$A =$$

$$\begin{pmatrix} \cos\psi\cos\phi - \cos\theta\sin\phi\sin\psi & -\sin\psi\cos\phi - \cos\theta\sin\phi\cos\psi & \sin\theta\sin\phi \\ \cos\psi\sin\phi + \cos\theta\cos\phi\sin\psi & -\sin\psi\sin\phi + \cos\theta\cos\phi\cos\psi & -\sin\theta\cos\phi \\ \sin\theta\sin\psi & \sin\theta\cos\psi & \cos\theta \end{pmatrix}.$$

$$\tag{9.7-2}$$

Since A is orthogonal, the dielectric tensor ε in the xyz coordinate must be symmetric, that is, $\varepsilon_{ij} = \varepsilon_{ji}$. The electric field can be assumed to have $\exp[i(\omega t - \alpha x - \beta y - \gamma z)]$ dependence on each crystal layer, which is assumed to be homogeneous. Since the whole birefringent layered medium is homogeneous in the xy plane, α and β remain the same throughout the layered medium. Therefore, the two components (α, β) of the propagation vector are chosen as the dynamical variables characterizing the electromagnetic waves propagating in the layered media. Given α and β, the z component γ is determined directly from the wave equation in momentum space:

$$\mathbf{k} \times (\mathbf{k} \times \mathbf{E}) + \omega^2 \mu \varepsilon \mathbf{E} = 0, \tag{9.7-3}$$

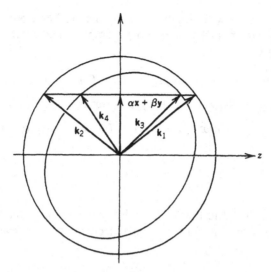

Figure 9.10 Graphic method to determine the propagation constants from the normal surface.

or equivalently,

$$\begin{pmatrix} \omega^2 \mu \varepsilon_{xx} - \beta^2 - \gamma^2 & \omega^2 \mu \varepsilon_{xy} + \alpha\beta & \omega^2 \mu \varepsilon_{xz} + \alpha\gamma \\ \omega^2 \mu \varepsilon_{yx} + \alpha\beta & \omega^2 \mu \varepsilon_{yy} - \alpha^2 - \gamma^2 & \omega^2 \mu \varepsilon_{yz} + \beta\gamma \\ \omega^2 \mu \varepsilon_{zx} + \alpha\gamma & \omega^2 \mu \varepsilon_{zy} + \beta\gamma & \omega^2 \mu \varepsilon_{zz} - \alpha^2 - \beta^2 \end{pmatrix} \begin{pmatrix} E_x \\ E_y \\ E_z \end{pmatrix} = 0.$$

$$(9.7\text{-}4)$$

To have nontrivial plane-wave solutions, the determinant of the matrix in Eq. (9.7-4) must vanish. This gives a quartic equation in γ that yields four roots γ_σ, $\sigma = 1, 2, 3, 4$. These roots may be either real or complex. Since all the coefficients of this quartic equation are real, complex roots are always in conjugate pairs. These four roots can also be obtained graphically from Fig. 9.10 if they are real. The plane of incidence is defined as the plane formed by $\alpha x + \beta y$ and z. The intersection of this plane with the normal surface yields two closed curves that are symmetric with respect to the origin of the axes. Drawing a line from the tip of the vector $\alpha x + \beta y$ parallel to the z direction yields, in general, four points of intersection. The four wave vectors $\mathbf{k}_\sigma = \alpha x + \beta y + \gamma_\sigma z$ all lie in the plane of incidence, which also remains the same throughout the layered medium because α and β are constant. However, the four group velocities associated with these partial waves are, in general, not lying in the plane of incidence. If all the four wave vectors $\mathbf{k}_\sigma$ are real, two of them have group velocities with positive z components, and the other two have group velocities with negative z components. The z component of the group velocity vanishes when γ_σ becomes complex. The polarization of these waves is given by

$$
\mathbf{p} = N_\sigma \begin{pmatrix} (\omega^2\mu\varepsilon_{yy} - \alpha^2 - \gamma_\sigma^2)(\omega^2\mu\varepsilon_{zz} - \alpha^2 - \beta^2) - (\omega^2\mu\varepsilon_{yz} + \beta\gamma_\sigma)^2 \\ (\omega^2\mu\varepsilon_{yz} + \beta\gamma_\sigma)(\omega^2\mu\varepsilon_{zx} + \alpha\gamma_\sigma) - (\omega^2\mu\varepsilon_{xy} + \alpha\beta)(\omega^2\mu\varepsilon_{zz} - \alpha^2 - \beta^2) \\ (\omega^2\mu\varepsilon_{xy} + \alpha\beta)(\omega^2\mu\varepsilon_{yz} + \beta\gamma_\sigma) - (\omega^2\mu\varepsilon_{xz} + \alpha\gamma_\sigma)(\omega^2\mu\varepsilon_{yy} - \alpha^2 - \gamma_\sigma^2) \end{pmatrix}
$$

$$(9.7\text{-}5)$$

where $\sigma = 1, 2, 3, 4$ and the N_σ's are the normalization constant such that $\mathbf{p} \cdot \mathbf{p} = 1$. The electric field of the plane electromagnetic waves can thus be written as

$$
\mathbf{E} = \sum_{\sigma=1}^{4} A_\sigma \mathbf{p}_\sigma \exp\left[i(\omega t - \alpha x - \beta y - \gamma_\sigma z)\right]. \qquad (9.7\text{-}6)
$$

Partial waves with complex propagation vectors cannot exist in an infinite homogeneous birefringent medium. If the medium is semi-infinite, the exponentially damped partial waves are legitimate solutions near the interface, and the field envelope decays exponentially as a function of z, where z is the distance from the interface. These exponentially damped partial waves are called evanescent waves. The evanescent waves in birefringent media in general have complex γ's, that is $\gamma = \gamma_R + i\gamma_I$. In a uniaxially birefringent medium, the ordinary evanescent wave has a purely imaginary γ. If the three principal dielectric constants are all real, these partial waves with complex γ's can be shown to have their Poynting vectors parallel to the interface. In other words, the energy is flowing parallel to the interface, and the propagation is lossless, as it should be. A mathematical proof is given in Ref. 5 for the special case of extraordinary evanescent waves in a uniaxially birefringent medium.

9.7.1 4 × 4 Matrix Method

The 4 × 4 matrix algebra, which analyzes the propagation of monochromatic plane waves in birefringent layered media, can now be introduced. The approach is general so that the results will be used later for many special cases of propagation in anisotropic layered media. The materials are assumed to be nonmagnetic so that μ = constant throughout the whole layered medium. The dielectric permittivity tensor ε in the xyz coordinates is given by

$$\varepsilon = \begin{cases} \varepsilon(0), & z < z_0, \\ \varepsilon(1), & z_0 < z < z_1, \\ \varepsilon(2), & z_1 < z < z_2, \\ \vdots & \\ \varepsilon(N), & z_{N-1} < z < z_N, \\ \varepsilon(s), & z_N < z. \end{cases} \qquad (9.7\text{-}7)$$

The electric field distribution within each homogeneous anisotropic layer can be expressed as a sum of those four partial waves [Eq. (9.7-6)]. The complex amplitudes of these four partial waves constitute the components of a column vector. The electromagnetic field in the nth layer of the anisotropic layered medium can thus be represented by a column vector $A_\sigma(n)$, $\sigma = 1, 2, 3, 4$.

As a result, the electric field distribution in the same nth layer can be written as

$$\mathbf{E} = \sum_{\sigma=1}^{4} A_\sigma(n)\mathbf{p}_\sigma(n) \exp\{i[\omega t - \alpha x - \beta y - \gamma_\sigma(n)(z - z_n)]\}. \quad (9.7\text{-}8)$$

The column vectors are not independent of each other. They are related through the continuity conditions at the interfaces. As a matter of fact, only one vector (or any four components of four different vectors) can be arbitrarily chosen. The magnetic field distribution is obtained using Maxwell's equations and is given by

$$\mathbf{H} = \sum_{\sigma=1}^{4} A_\sigma(n)\mathbf{q}_\sigma(n) \exp\{i[\omega t - \alpha x - \beta y - \gamma_\sigma(n)(z - z_n)]\}, \quad (9.7\text{-}9)$$

$$\mathbf{q}_\sigma(n) = \frac{ck_\sigma(n)}{\omega\mu} \times \mathbf{p}_\sigma(n), \qquad (9.7\text{-}10)$$

$$\mathbf{k}_\sigma(n) = \alpha\mathbf{x} + \beta\mathbf{y} + \gamma_\sigma(n)\mathbf{z}. \qquad (9.7\text{-}11)$$

Note that the $\mathbf{q}_\sigma(n)$'s are not unit vectors.

Imposing the continuity of E_x, E_y, H_x, and H_y at the interface $z = z_{n-1}$ leads to

$$\sum_{\sigma=1}^{4} A_\sigma(n-1)\mathbf{p}_\sigma(n-1) \cdot \mathbf{x} = \sum_{\sigma=1}^{4} A_\sigma(n)\mathbf{p}_\sigma(n) \cdot \mathbf{x} \exp\left[i\gamma_\sigma(n)t_n\right], \quad (9.7\text{-}12)$$

$$\sum_{\sigma=1}^{4} A_\sigma(n-1)\mathbf{p}_\sigma(n-1) \cdot \mathbf{y} = \sum_{\sigma=1}^{4} A_\sigma(n)\mathbf{p}_\sigma(n) \cdot \mathbf{y} \exp\left[i\gamma_\sigma(n)t_n\right], \quad (9.7\text{-}13)$$

$$\sum_{\sigma=1}^{4} A_\sigma(n-1)\mathbf{q}_\sigma(n-1) \cdot \mathbf{x} = \sum_{\sigma=1}^{4} A_\sigma(n)\mathbf{q}_\sigma(n) \cdot \mathbf{x} \exp\left[i\gamma_\sigma(n)t_n\right], \quad (9.7\text{-}14)$$

$$\sum_{\sigma=1}^{4} A_\sigma(n-1)\mathbf{q}_\sigma(n-1) \cdot \mathbf{y} = \sum_{\sigma=1}^{4} A_\sigma(n)\mathbf{q}_\sigma(n) \cdot \mathbf{y} \exp\left[i\gamma_\sigma(n)t_n\right], \quad (9.7\text{-}15)$$

where $t_n = z_n - z_{n-1}$, $n = 1, 2, \ldots, N$.

These four equations can be rewritten as a matrix equation:

$$\begin{pmatrix} A_1(n-1) \\ A_2(n-1) \\ A_3(n-1) \\ A_4(n-1) \end{pmatrix} = D^{-1}(n-1)D(n)P(n) \begin{pmatrix} A_1(n) \\ A_2(n) \\ A_3(n) \\ A_4(n) \end{pmatrix}, \quad (9.7\text{-}16)$$

where

$$D(n) = \begin{pmatrix} \mathbf{x} \cdot \mathbf{p}_1(n) & \mathbf{x} \cdot \mathbf{p}_2(n) & \mathbf{x} \cdot \mathbf{p}_3(n) & \mathbf{x} \cdot \mathbf{p}_4(n) \\ \mathbf{y} \cdot \mathbf{q}_1(n) & \mathbf{y} \cdot \mathbf{q}_2(n) & \mathbf{y} \cdot \mathbf{q}_3(n) & \mathbf{y} \cdot \mathbf{q}_4(n) \\ \mathbf{y} \cdot \mathbf{p}_1(n) & \mathbf{y} \cdot \mathbf{p}_2(n) & \mathbf{y} \cdot \mathbf{p}_3(n) & \mathbf{y} \cdot \mathbf{p}_4(n) \\ \mathbf{x} \cdot \mathbf{q}_1(n) & \mathbf{x} \cdot \mathbf{q}_2(n) & \mathbf{x} \cdot \mathbf{q}_3(n) & \mathbf{x} \cdot \mathbf{q}_4(n) \end{pmatrix}, \quad (9.7\text{-}17)$$

$$P(n) = \begin{pmatrix} \exp\left[i\gamma_1(n)t_n\right] & 0 & 0 & 0 \\ 0 & \exp\left[i\gamma_2(n)t_n\right] & 0 & 0 \\ 0 & 0 & \exp\left[i\gamma_3(n)t_n\right] & 0 \\ 0 & 0 & 0 & \exp\left[i\gamma_4(n)t_n\right] \end{pmatrix}.$$

$$(9.7\text{-}18)$$

The matrices $D(n)$ are called *dynamical matrices* because they depend only on the direction of polarization of those four partial waves. The dynamical matrices are defined in a way such that they are block diagonalized when the mode coupling disappears. This requires that A_1 and A_2 are the amplitudes of the plane waves of the same mode (polarization) such that the plane wave with amplitude A_1 propagates to the right, whereas the plane wave with amplitude A_2 propagates to the left. Likewise, A_3 and A_4 are the amplitudes of the plane waves of the same mode, propagating to the right or left, respectively. The matrices $P(n)$ are called *propagation matrices* and depend only on the phase shift of these four partial waves as they traverse through the layer. The transfer matrix is defined as

$$T_{n-1,n} = D^{-1}(n - 1)D(n)P(n). \tag{9.7-19}$$

Equation (9.7-16) can thus be written as

$$
\begin{pmatrix} A_1(n - 1) \\ A_2(n - 1) \\ A_3(n - 1) \\ A_4(n - 1) \end{pmatrix} = T_{n-1,n} \begin{pmatrix} A_1(n) \\ A_2(n) \\ A_3(n) \\ A_4(n) \end{pmatrix}. \tag{9.7-20}
$$

The matrix equation that related $A(0)$ and $A(s)$ is therefore given by

$$
\begin{pmatrix} A_1(0) \\ A_2(0) \\ A_3(0) \\ A_4(0) \end{pmatrix} = T_{0,1} T_{1,2} T_{2,3} \cdots T_{N-1,N} T_{N,s} \begin{pmatrix} A_1(s) \\ A_2(s) \\ A_3(s) \\ A_4(s) \end{pmatrix}, \tag{9.7-21}
$$

where $s \equiv N + 1$ and $t_{N+1} \equiv 0$.

Equations (9.7-16) and (9.7-21) show how systematic the matrix method is for treating electromagnetic propagation in anisotropic layered media. If Eq. (9.7-21) is represented graphically in Fig. 9.11, two dynamical matrices can be seen to be associated with each interface and one propagation matrix is associated with the bulk of each layer. The overall transfer matrix is the product of all these matrices from left to right. This completes the theoretical formulation of the 4 × 4 matrix method.

The matrix method just described is an exact approach to the propagation of electromagnetic radiation in anisotropic layered media. Both birefringent phase retardation and thin-film interference are considered. This differs from the traditional 2 × 2 Jones matrix method, which neglects the reflection from each interface. Therefore, in calculating the transmission and reflection

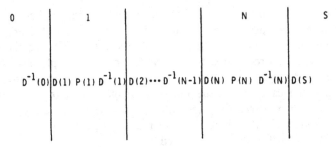

Figure 9.11 Diagram representation of the matrix method.

properties of some birefringent filters, the results obtained from these methods are expected to be different.

In addition, there are several interesting optical phenomena in periodic birefringent layered media that have been analyzed by using this 4×4 matrix algebra. These include the indirect optical bandgap, exchange Bragg reflection, and exchange Solc–Bragg transmission. Interested readers are referred to Ref. 5.

9.7.2 Reflection and Transmission

The matrix method just discussed is very useful in the calculation of the reflectance and transmittance of an anisotropic layered medium. Because of the anisotropy of the medium, mode coupling appears at the interfaces. Therefore, there are four complex amplitudes associated with the reflection and another four associated with the transmission. These eight complex amplitudes can be expressed in terms of the matrix elements of the overall transfer matrix. To illustrate this, one considers, without loss of generality, the case of an anisotropic layered medium sandwiched between two isotropic ambient and substrate media. Assume that the light is incident from the left side of the structure, and let A_s, A_p, B_s, B_p, and C_s, C_p be the incident, reflected, and transmitted electric field amplitudes, respectively. By using the matrix method described in the preceding, a transfer matrix can be found for any given anisotropic layered structure such that

$$
\begin{pmatrix} A_s \\ B_s \\ A_p \\ B_p \end{pmatrix} = \begin{pmatrix} M_{11} & M_{12} & M_{13} & M_{14} \\ M_{21} & M_{22} & M_{23} & M_{24} \\ M_{31} & M_{32} & M_{33} & M_{34} \\ M_{41} & M_{42} & M_{43} & M_{44} \end{pmatrix} \begin{pmatrix} C_s \\ 0 \\ C_p \\ 0 \end{pmatrix}.
\tag{9.7-22}
$$

The reflection and transmission coefficients are defined and expressed in terms of the matrix elements as follows:

$$r_{ss} = \left(\frac{B_s}{A_s}\right)_{A_p=0} = \frac{M_{21}M_{33} - M_{23}M_{31}}{M_{11}M_{33} - M_{13}M_{31}}, \qquad (9.7\text{-}23)$$

$$r_{sp} = \left(\frac{B_p}{A_s}\right)_{A_p=0} = \frac{M_{41}M_{33} - M_{43}M_{31}}{M_{11}M_{33} - M_{13}M_{31}}, \qquad (9.7\text{-}24)$$

$$r_{ps} = \left(\frac{B_s}{A_p}\right)_{A_s=0} = \frac{M_{11}M_{23} - M_{21}M_{13}}{M_{11}M_{33} - M_{13}M_{31}}, \qquad (9.7\text{-}25)$$

$$r_{pp} = \left(\frac{B_p}{A_p}\right)_{A_s=0} = \frac{M_{11}M_{43} - M_{41}M_{13}}{M_{11}M_{33} - M_{13}M_{31}}, \qquad (9.7\text{-}26)$$

$$t_{ss} = \left(\frac{C_s}{A_s}\right)_{A_p=0} = \frac{M_{33}}{M_{11}M_{33} - M_{13}M_{31}}, \qquad (9.7\text{-}27)$$

$$t_{sp} = \left(\frac{C_p}{A_s}\right)_{A_p=0} = \frac{-M_{31}}{M_{11}M_{33} - M_{13}M_{31}}, \qquad (9.7\text{-}28)$$

$$t_{ps} = \left(\frac{C_s}{A_p}\right)_{A_s=0} = \frac{-M_{13}}{M_{11}M_{33} - M_{13}M_{31}}, \qquad (9.7\text{-}29)$$

$$t_{pp} = \left(\frac{C_p}{A_p}\right)_{A_s=0} = \frac{M_{11}}{M_{11}M_{33} - M_{13}M_{31}}. \qquad (9.7\text{-}30)$$

These reflection and transmission formulas are extremely useful in the calculation of the spectral response of an anisotropic layered structure. The matrix elements are obtained by carrying out the matrix multiplication in Eq. (9.7-21). The general explicit forms are normally not available. For fast results, a computer program is in general required. Even for the special case of periodic layered medium, closed forms for the reflectance and transmittance are too complicated to derive. These eight complex amplitudes are spectrally correlated (see Ref. 5).

REFERENCES

1. A. Yariv and P. Yeh, *Optical Waves in Crystals*, Wiley, New York, 1984.
2. A. M. Title, Recent advances in birefringent filters, *Proc. Soc. Photo-Opt. Instrum. Eng.* **88**, 23 (1976).
3. P. Yeh, Dispersive birefringent filters, *Opt. Comm.* **37**, 153 (1981).

4. H. Goldstein, *Classical Mechanics*, Addison-Wesley, Reading, MA, 1965.

5. P. Yeh, Electromagnetic propagation in birefringent layered media, *J. Opt. Soc. Am.* **69**, 742 (1979).

PROBLEMS

9.1   *Eigenvectors for* **E** *and* **D**.

   (a) Derive the expression for the eigenpolarization of the electric field vector (9.1-12) and (9.1-14).

   (b) Use the relation $\mathbf{D} = \varepsilon\mathbf{E}$ and obtain an expression for the corresponding eigenpolarization of the displacement vector **D**.

   (c) Let n_1 and n_2 be the solution of the Fresnel equation (9.1-13) and $\mathbf{E}_1$, $\mathbf{E}_2$, $\mathbf{D}_1$, and $\mathbf{D}_2$ the corresponding eigenvectors. Evaluate $\mathbf{E}_1 \cdot \mathbf{E}_2$ and $\mathbf{D}_1 \cdot \mathbf{D}_2$ and show that $\mathbf{D}_1$ and $\mathbf{D}_2$ are always mutually orthogonal, whereas $\mathbf{E}_1$ and $\mathbf{E}_2$ are only mutually orthogonal in a uniaxial medium and an isotropic medium.

   (d) Show that $\mathbf{E}_1 \cdot \mathbf{D}_2 = 0$ and $\mathbf{E}_2 \cdot \mathbf{D}_1 = 0$

9.2   *Fresnel's equation*

   (a) Derive the Fresnel equation (9.1-13) directly from Eq. (9.1-11).

   (b) Show that the Fresnel equation (9.1-13) is a quadratic equation in n^2, that is,

$$An^4 + Bn^2 + C = 0,$$

   and obtain expressions for A, B, and C.

   (c) Show that $B^2 - 4AC > 0$ for the case of pure dielectrics with real ε_x, ε_y, ε_z.

   (d) Derive Eq. (9.1-15) from Eq. (9.1-11) for the case of uniaxial crystals.

   (e) Show that, in an isotropic medium, Eq. (9.1-11) reduces to

$$k^2 - \frac{\varepsilon}{\varepsilon_0}\left(\frac{\omega}{c}\right)^2 = 0.$$

9.3   *Group velocity and phase velocity.*   The phase velocity of a wave is parallel to its wave vector, whereas the group velocity is given by $\partial\omega/\partial\mathbf{k}$ and is perpendicular to the normal surface.

   (a) Derive an expression for the group velocity of the extraordinary wave in a uniaxial crystal as a function of the polar angle θ of the propagation vector.

(b) Derive an expression for the angle α between the phase velocity and the group velocity. This angle is also the angle between the field vectors $\mathbf{E}$ and $\mathbf{D}$.

(c) Show that $\alpha = 0$ when $\theta = 0, \frac{1}{2}\pi$. Find the angle θ at which α is maximized, and obtain an expression for α_{max}. Calculate angle α_{max} for quartz with $n_o = 1.544$ and $n_e = 1.553$.

(d) Show that for $n_o \simeq n_e$ the maximum angular separation α_{max} occurs at $\theta \sim 45°$, and show that α_{max} is proportional to $|n_o - n_e|$.

9.4 *Prism polarizers.* The phenomenon of double refraction in an anisotropic crystal may be utilized to produce polarized light. Consider a light beam incident on the plane boundary from the inside of a calcite crystal ($n_o = 1.658$, $n_e = 1.486$). Suppose that the c axis of the crystal is normal to the plane of incidence.

(a) Find the range of the internal angle of incidence such that the ordinary wave is totally reflected. The transmitted wave is thus completely polarized.

(b) *Glan prism.* Use the basic principle described in (a) to design the calcite Glan prism shown in the following figure*:

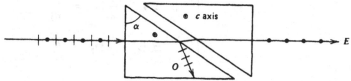

Find the range of the apex angle α.

9.5 *Quarter-wave plate.* A linearly polarized electromagnetic wave at $\lambda = 6328\,\text{Å}$ is incident normally at $x = 0$ on the yz face of a quartz crystal plate (x-cut) so that it propagates along the x axis. If the wave is polarized initially so that it has equal components along y and z.

(a) What is the state of polarization at the plane x, where

$$(k_z - k_y)x = \tfrac{1}{2}\pi.$$

Plot the position of the electric field vector in this plane at times $t = 0, \pi/6\omega, \pi/3\omega, \pi/2\omega, 2\pi/3\omega, 5\pi/6\omega$.

(b) A plate that satisfies the condition in (a) is known as a quarter-wave plate because the difference in the phase shift for the two orthogonal polarization states is a quarter of 2π. Find the thickness of a quartz quarter-wave plate at $\lambda = 6328\,\text{Å}$.

(c) *Half-wave plate.* What is the state of polarization at the plane x, where

$$\frac{2\pi}{\lambda}(n_c - n_a)x = \pi.$$

9.6   *Displacement eigenmodes.*   The wave equation (9.1-9) can be written as

$$\mathbf{s} \times \left[\mathbf{s} \times \frac{\varepsilon_0}{\varepsilon}\mathbf{D}\right] = -\frac{1}{n^2}\mathbf{D},$$

where $\mathbf{s}$ is the unit vector in the direction of propagation. Let $\mathbf{D}_1$ and $\mathbf{D}_2$ be the normalized eigenvectors with eigenvalues $1/n_1^2$ and $1/n_2^2$, respectively. We assume that $\varepsilon_0/\varepsilon$ is a Hermitian tensor.

(a) Show that

$$\left(\frac{\varepsilon_0}{\varepsilon}\right)_{11} \equiv \mathbf{D}_1^* \cdot \frac{\varepsilon_0}{\varepsilon}\mathbf{D}_1 = \frac{1}{n_1^2},$$

$$\left(\frac{\varepsilon_0}{\varepsilon}\right)_{22} \equiv \mathbf{D}_2^* \cdot \frac{\varepsilon_0}{\varepsilon}\mathbf{D}_2 = \frac{1}{n_2^2},$$

$$\left(\frac{\varepsilon_0}{\varepsilon}\right)_{12} \equiv \mathbf{D}_1^* \cdot \frac{\varepsilon_0}{\varepsilon}\mathbf{D}_2 = 0.$$

(b) Show that

$$\mathbf{D}_1^* \cdot \mathbf{D}_2 = 0.$$

9.7   *Half-wave plate.*   A half-wave plate has a phase retardation $\Gamma = \pi$. Assume that the plate is oriented such that the azimuth angle (i.e., the angle between the x axis and the slow axis of the plate) is ψ.

(a) Find the polarization state of the transmitted beam assuming that the incident beam is linearly polarized in the y direction.

(b) Show that a half-wave plate will convert a right-handed circularly polarized light into a left-handed circularly polarized light, and vice versa, regardless of the azimuth angle of the plate.

(c) Lithium tantalate (LiTaO$_3$) is a uniaxial crystal with $n_o = 2.1391$ and $n_e = 2.1432$ at $\lambda = 1 \ \mu m$. Find the half-wave plate thickness at this wavelength, assuming the plate is cut in such a way that the surfaces are perpendicular to the x axis of the principal coordinate (i.e., x-cut).

9.8   *Quarter-wave plate.* A quarter-wave plate has a phase retardation of
      $\Gamma = \frac{1}{2}\pi$. Assume that the plate is oriented in a direction with azimuth
      angle ψ.

   (a) Find the polarization state of the transmitted beam, assuming that
       the incident beam is polarized in the y direction.

   (b) If the resulting polarization state from (a) is represented by a
       complex number on the complex plane, show that the locus of
       these points as ψ varies from 0 to $\frac{1}{2}\pi$ is a branch of a hyperbola.
       Obtain the equation of the hyperbola.

   (c) Quartz (α-SiO$_2$) is a uniaixial crystal with $n_o = 1.53283$ and
       $n_e = 1.54152$ at $\lambda = 1.592\,\mu m$. Find the thickness of an x-cut
       quartz quarter-wave plate at this wavelength.

9.9   *Polarization transformation by wave plate.* A wave plate is charac-
      terized by its phase retardation Γ and azimuth angle ψ. Carry out the
      matrix multiplication in Eq. (9.3-14) and obtain an explicit expression
      for the Jones matrix.

   (a) Find the polarization state of the emerging beam, assuming that
       the incident beam is polarized in the x direction.

   (b) Use a complex number to represent the resulting polarization state
       obtained in (a).

   (c) The polarization state of the incident x-polarized beam is
       represented by a point at the origin of the complex plane. Show
       that the transformed polarized state can be anywhere on the
       complex plane, provided Γ can be varied from 0 to 2π and ψ can
       be varied from 0 to $\frac{1}{2}\pi$. Physically, this means that any polar-
       ization state can be produced from a linearly polarized light,
       provided a proper wave plate is available.

   (d) Show that the locus of these points on the complex plane obtained
       by rotating a wave plate from $\psi = 0$ to $\psi = \frac{1}{2}\pi$ is a hyperbola.
       Derive the equation of this hyperbola.

   (e) Show that the Jones matrix W of a wave plate is unitary, that is,

$$W^{\dagger}W = 1,$$

      where the dagger indicates Hermitian conjugation.

   (f) Let V_1' and V_2' be the transformed Jones vectors from V_1 and V_2,
       respectively. Show that if V_1 and V_2 are orthogonal, so are V_1' and
       V_2'.

9.10 *Polarizers and projection operators.* An ideal polarizer can be considered as a projection operator that acts on the incident polarization state and results in the projection of the polarization vector along the transmission axis of the polarizer.

(a) If we neglect the absolute phase factor in Eq. (9.3-16), show that

$$P_0^2 = P_0 \quad \text{and} \quad P^2 = P.$$

Operators satisfying these conditions are called projection operators in linear algebra.

(b) Show that if $\mathbf{E}_1$ is the amplitude of the relative field of the electric field, the amplitude of the beam after it passes through the polarizer is given by $\mathbf{p}(\mathbf{p} \cdot \mathbf{E}_1)$, where $\mathbf{p}$ is the unit vector along the transmission axis of the polarizer.

(c) If the incident beam is vertically polarized (i.e., $\mathbf{E}_1 = \mathbf{y}E_0$), the polarizer transmission axis is in the x direction (i.e., $P = \mathbf{p} = \mathbf{x}$). The transmitted beam has zero amplitude since $\mathbf{x} \cdot \mathbf{y} = 0$. However, if a second polarizer is placed in front of the first polarizer and is oriented 45° with respect to the first polarizer, the transmitted amplitude is not zero. Find this amplitude.

(d) Consider a series of N polarizers with the first one oriented at $\psi_1 = \pi/2N$, the second one at $\psi_2 = 2(\pi/2N)$, the third one at $\psi_3 = 3(\pi/2N)$, and the Nth one at $N(\pi/2N)$. Let the incident beam be horizontally polarized. Show that the transmitted beam is vertically polarized with an amplitude of $[\cos(\pi/2N)]^N$. Evaluate the amplitude for $N = 1, 2, 3, \ldots, 10$. Show that in the limit of $N \to \infty$, the amplitude becomes 1. In other words, a series of polarizers oriented similar to a fan can rotate the polarization of the light without attenuation.

9.11 *Cross-reflection coefficients.* Show that r_{sp} and r_{ps} vanish when $n_a = n_c$ for an arbitrary angle of incidence.

9.12 *Equation of gyration.* The gyration of the electric field vector in isotropic media or along the c axis of uniaxial crystals can be described by using the wave equation (1.4-9) and Eq. (9.1-20).

(a) Show that for monochromatic waves, the wave equation can be written as

$$\frac{d^2}{dz^2}\mathbf{E} + \left(\frac{\omega}{c}n\right)^2 \mathbf{E} + i\left(\frac{\omega}{c}\right)^2 \mathbf{G} \times \mathbf{E} = 0,$$

where $n^2 = \varepsilon_a/\varepsilon_0$.

(b) Show that substitution of $E(z) \exp(-ikz)$ for E in the wave equation in (a), with $k = n\omega/c$, yields

$$E''(z) - 2ikE'(z) + i\left(\frac{\omega}{c}\right)^2 G \times E = 0,$$

where a prime indicates differentiation with respect to z.

(c) Derive the equation of gyration (9.1-22) by neglecting the second-order derivative in (b).

9.13 *Faraday isolators.* A faraday rotator that rotates the plane of polarization of an incident light by 45° can be used as an isolator. Any light that is reflected will undergo another rotation of 45° and can thus be eliminated by polarizers.

(a) Show that such an isolater can be obtained by placing the Faraday rotator between two polarizers whose transmission axes are oriented at an angle 45° relative to each other.

(b) Using a 1-cm-thick flint glass as the material, find the magnetic field required for the isolation of light with wavelength $\lambda = 589.3$ nm (use the data in Table 9.4).

10

Some Applications of Anisotropic Layered Media

Layered media that consist of anisotropic materials play an important role in many optical systems. One of the most important applications is in polarization interference filters. These filters can accommodate simultaneously an extremely narrow bandwidth and a wide field of view. Such narrow-band and wide-field spectral filters are essential in many situations when weak signals from an extended source are to be detected against a noisy background. For example, in the area of solar physics, the distribution of hydrogen may be measured by photographing the solar corona in the light of the H_z ($\lambda = 6563$ Å) line. In view of the large amount of light present in neighboring wavelengths, a filter of extremely narrow bandwidth (~ 1 Å) is required if reasonable discrimination is to be attained. In the area of laser communication when the signal is transmitted through a random medium such as an atmosphere or seawater, the received signal is carried by a scattered component of the laser radiation, which will appear to come from a wide field of view, often up to several steradians. Optical communications under these severe circumstances require a filter with not only an extremely narrow bandwidth to reject the unwanted background light and hence increase the signal-to-noise ratio, but also a large angular aperture to receive as much signal as possible.

In this chapter, we will first describe the basic principle of polarization interference filters. We then describe the wide-field element that makes the field of view several orders of magnitude larger than that of Fabry–Perot thin-film filters. Polarization interference filters that employ the dispersion of birefringence will also be discussed.

10.1 LYOT–OHMAN FILTERS

Polarization interference filters consist of an anisotropic layered medium, which can be viewed simply as a series of crystal plates and polarizers. Since the crystal plates must be optically birefringent, these filters are also called

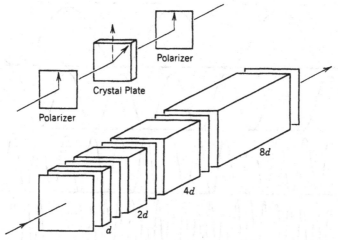

Figure 10.1 Schematic drawing of a 4-stage Lyot–Ohman filter. The upper left drawing shows a typical stage.

birefringent filters. The first birefringent filter was invented by Lyot in 1933 [1, 2] and independently by Ohman in 1938 [3, 4]. A different filter configuration using fewer polarizers was invented by Solc in 1953 [5, 6]. All these earlier filters were designed for the purpose of the investigation in solar physics.

The birefringent filter devised by Lyot and Ohman consists of a set of birefringent crystal plates separated by parallel polarizers (see Fig. 10.1). All the crystal plates have the same orientation of their principal axes, which are oriented at an azimuth angle of 45° relative to the polarizer axis. In addition, the plate thicknesses are in geometric progression, that is, $d, 2d, 4d, 8d, \ldots$. The transmission of light through such a system can be easily obtained by multiplying the transmittance of each individual stage. From Eq. (9.4-9), the transmittance of the first stage for unpolarized light can be written as

$$T = \tfrac{1}{2} \cos^2 \frac{\Gamma}{2}, \tag{10.1-1}$$

with

$$\Gamma = \frac{2\pi}{\lambda}(n_e - n_o)d, \tag{10.1-2}$$

where d is the thickness, n_o, n_e are the indices of refraction, λ is the wavelength, and Γ is the phase retardation. Here, for the sake of clarity in

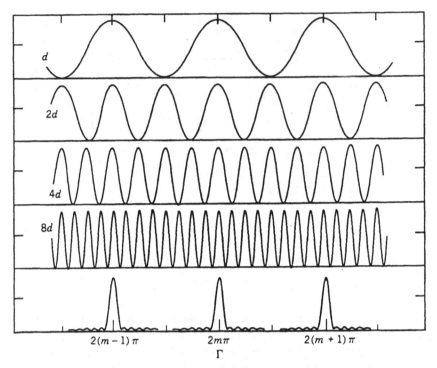

Figure 10.2 Transmission curves of birefringent crystals separated by polarizers.

introducing the concept, we assume that the crystals are uniaxial with their
c axes parallel to the plate surfaces. The factor $\frac{1}{2}$ accounts for the loss of
50% energy at the front polarizer.

As illustrated in Fig. 10.2, when such birefringent plates separated by
parallel polarizers are in series, the transmittance is

$$T = \tfrac{1}{2} \cos^2\left(\frac{\Gamma}{2}\right) \cos^2\left(2\,\frac{\Gamma}{2}\right) \cos^2\left(4\,\frac{\Gamma}{2}\right) \cdots \cos^2\left(2^{N-1}\,\frac{\Gamma}{2}\right), \quad (10.1\text{-}3)$$

where N is the number of stages. The bottom curve in Fig. 10.2 shows the
result of the multiplication. Each of the other curves shows the transmittance
of one of the stages. Since the phase retardation depends on the wavelength,
such a birefringent layered medium can be used as a spectral filter. We notice
that the bandwidth of such a filter is determined by the thickness of the
largest plates, while the free spectral range is determined by the thickness of
the thinnest plate.

To study in detail the transmittance property of such a filter, it is useful to simplify the transmittance formula [Eq. (10.1-3)]. Using the mathematical identity

$$\cos x = \tfrac{1}{2}(e^{ix} + e^{-ix}) \tag{10.1-4}$$

and carrying out the multiplication, we obtain

$$\cos x \cos 2x \cos 4x \cdots \cos 2^{N-1}x = 2^{-N}[e^{i(2^N-1)x} + e^{i(2^N-3)x} \cdots + e^{-i(2^N-1)x}]. \tag{10.1-5}$$

The series on the right side can be summed into a single form. This leads to

$$\cos x \cos 2x \cos 4x \cdots \cos 2^{N-1}x = \frac{\sin 2^N x}{2^N \sin x}. \tag{10.1-6}$$

Thus, the transmittance [Eq. (10.1-3)] is

$$T = \frac{1}{2}\left(\frac{\sin 2^N x}{2^N \sin x}\right)^2, \quad x = \frac{\Gamma}{2} = \frac{\pi}{\lambda}(n_e - n_o)d. \tag{10.1-7}$$

The function on the right side of this equation is very familiar and also appears in diffraction gratings. We thus expect that the spectral revolving power of such a filter will be the same as that of a diffraction grating.

According to Eq. (10.1-7), maximum transmittance occurs when

$$\Gamma = 2m\pi, \quad m = \text{integer}. \tag{10.1-8}$$

In other words, maximum transmittance occurs when the thinnest plate is an integral multiple of full-wave plates. As the phase retardation Γ deviates from Eq. (10.1-8), this function starts to drop sharply. The first time it reaches zero is when $2^N\Delta\Gamma = 2\pi$ because $\sin\pi = 0$. In other words,

$$\Delta\Gamma = \Gamma - 2m\pi = 2^{-N}(2\pi) \tag{10.1-9}$$

corresponds to the first minimum of the transmittance curve. From Fig. 10.2, it is obvious that this minimum corresponds to a zero transmittance of the thickest stage.

There are a series of sidelobes on each side of a maximum transmittance. These sidelobes are interlaced with zero transmittances. The first sidelobe occurs at approximately

$$\Delta\Gamma = \Gamma - 2m\pi = 2^{-N}(3\pi) \tag{10.1-10}$$

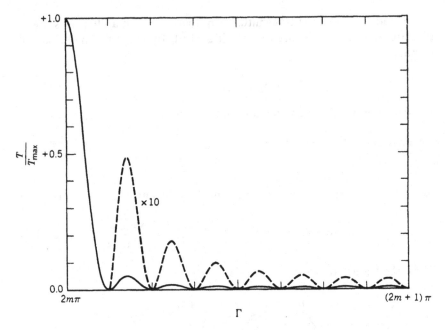

Figure 10.3 Calculated transmittance spectrum of a four-stage Lyot–Ohman filter.

with a transmittance $T = (4/9\pi^2)T_{max}$. In other words, the first sidelobe is only 0.047, less than 5% of the maximum transmittance. The other sidelobes are even smaller. So, we have a very sharp transmission maximum with very weak subsidiary maxima on the sides. In fact, there are exactly $2^N - 1$ minima (zero transmittance) and $2^N - 2$ sidelobes. The transmittance between $\Gamma = 2m\pi$ and $\Gamma = (2m + 1)\pi$ is shown in Fig. 10.3.

Maximum transmittance is $T_{max} = 0.5$ according to Eq. (10.1-7). Half-maximum transmittance (i.e., $T = 0.5T_{max}$) occurs at approximately

$$\Delta\Gamma = 2^{-N}(0.886\pi), \qquad (10.1\text{-}11)$$

where the factor 0.886 is due to the presence of those thinner stages. This factor is 1 when there is only one stage. It tends to 0.886 as the number of stages N increases.

According to Eq. (10.1-8), the passband of the Lyot–Ohman filter occurs at

$$\lambda = \frac{\Delta n}{m}d = \frac{n_e - n_o}{m}d, \qquad m = \text{nonzero integer}, \quad (10.1\text{-}12)$$

where $\Delta n = n_e - n_o$ is the birefringence. The bandwidth (FWHM) $\Delta\lambda_{1/2}$ is given by [according to Eqs. (10.1-11) and (10.1-12)]

$$\Delta\lambda_{1/2} = 2^{-N}\frac{0.886}{m}\lambda = \frac{0.886\lambda^2}{2^N|n_e - n_o|d}. \qquad (10.1\text{-}13)$$

The free spectral range (FSR) of this filter is determined by the thickness of the thinnest plate and is given by [according to Eqs. (10.1-8) and (10.1-2)]

$$\Delta\lambda = \frac{\lambda}{m} = \frac{\lambda^2}{\Delta nd}. \qquad (10.1\text{-}14)$$

The finesse of this filter, defined as the ratio of the FSR to the bandwidth, is thus given by

$$F = 1.13 \times 2^N. \qquad (10.1\text{-}15)$$

Equations (10.1-13) and (10.1-14) for bandwidth and FSR are derived by assuming that the birefringence is relatively constant over the range of spectral regime.

Let D be the thickness of the largest plate, that is,

$$D = 2^{N-1}d. \qquad (10.1\text{-}16)$$

The bandwidth $\Delta\lambda_{1/2}$ can be written in terms of D as

$$\Delta\lambda_{1/2} = \frac{0.443\lambda^2}{|n_e - n_o|D}. \qquad (10.1\text{-}17)$$

10.1.1 Example

Consider a Lyot–Ohman filter that consists of five quartz crystal plates with thickness 1, 2, 4, 8, and 16 mm. Using a birefringence of $\Delta n = n_e - n_o = 0.00900$, find the passbands and the bandwidths.

According to Eq. (10.1-12), passbands occur at

$$\lambda = (1/m)(9\,\mu m), \qquad m = 1, 2, 3, \ldots.$$

Take $m = 18$, which corresponds to a passband at $\lambda = 0.5\,\mu m$. The bandwidth for the passband at $\lambda = 0.5\,\mu m$ is [according to Eq. (10.1-13)]

$$\Delta\lambda_{1/2} = 7.7\,\text{Å}.$$

By adding three more stages with plate thicknesses 32, 64, and 128 mm, the bandwidth can be reduced to 1 Å.

10.1.2 Example

Design a Lyot–Ohman filter with a bandwidth of 1 Å at $\lambda = 0.5\,\mu m$ using calcite. Assume that $n_o = 1.658$ and $n_e = 1.486$, that is,

$$\Delta n = n_e - n_o = -0.172.$$

According to Eq. (10.1-17), the thickness of the largest plate must be

$$D = 6.4\,mm.$$

If a FSR of 500 Å is to be attained, the number of stages needs to be 9. This means that the thinner plates are of thicknesses 3.2, 1.6, 0.8, 0.4, 0.2, 0.1, 0.05, and 0.025 mm. Notice that the thinnest plate is only $25\,\mu m$ thick.

The transmission of light through Lyot–Ohman filters can be understood by examining the polarization state of light as it propagates through the crystal plates. At the passband, all the crystal plates are exactly an integral number of full waves. The light passes through each plate without any change in its polarization state and transmits through the polarizers without any loss except due to the imperfection of the polarizers. For light with other wavelengths, the polarization state is altered by the crystal plates, and the light suffers loss at each polarizer. This leads to the low transmittance at out-of-band wavelengths.

Most sheet polarizers have some residual absorptions even if the polarization is parallel to the transmission axis. This leads to a problem in high-finesse Lyot–Ohman filters. Because of the large number of polarizers involved, the total transmittance is greatly reduced by the residual absorption. In the next section, we will discuss a different type of birefringent filter that requires only two polarizers.

10.2 SOLC FILTERS

There are two basic configurations of Solc filters: folded and fan filters. The folded Solc filter works between two crossed polarizers, whereas the fan Solc filter works between a pair of parallel polarizers. These two configurations will be described separately.

10.2.1 Folded Solc Filters

The folded Solc filter consists of a periodic series of thin birefringent crystal plates. The simplest configuration consists of a series of identical crystal

Table 10.1. The Folded Solc Filter

Element	Azimuth
Front polarizer	$0°$
Plate 1	ϱ
Plate 2	$-\varrho$
Plate 3	ϱ
$\vdots$	$\vdots$
Plate N	$(-1)^{N-1}\varrho$
Rear polarizer	$90°$

plates with c axes that rock back and forth. The azimuthal angles of the individual plates are described in Table 10.1. The geometric arrangement of an eight-plate Solc filter is sketched in Fig. 10.4. As described in Table 10.1, the front polarizer has its transmission axis parallel to the x axis (vertical), and the rear polarizer is parallel to the y axis (horizontal). The overall Jones matrix for these N plates is given by

$$M = [R(\varrho)W_0 R(-\varrho)R(-\varrho)W_0 R(\varrho)]^m, \qquad (10.2\text{-}1)$$

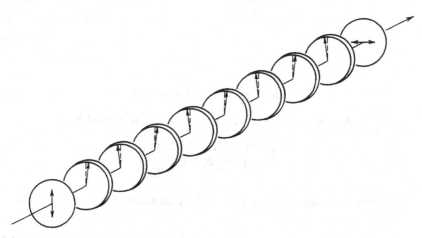

Figure 10.4 Schematic drawing of an 8-stage folded Solc filter.

where we assume that the number of plates is an even number, $N = 2m$. Substituting Eqs. (9.3-12) and (9.3-13) into Eq. (10.2-1) and carrying out the matrix multiplication result in

$$M = \begin{pmatrix} A & B \\ C & D \end{pmatrix}^m, \qquad (10.2\text{-}2)$$

with

$$A = \left(\cos \frac{\Gamma}{2} - i \cos 2\varrho \sin \frac{\Gamma}{2} \right)^2 + \sin^2 2\varrho \sin^2 \frac{\Gamma}{2},$$

$$B = \sin 4\varrho \sin^2 \frac{\Gamma}{2},$$

$$C = -B, \qquad (10.2\text{-}3)$$

$$D = \left(\cos \frac{\Gamma}{2} + i \cos 2\varrho \sin \frac{\Gamma}{2} \right)^2 + \sin^2 2\varrho \sin^2 \frac{\Gamma}{2},$$

where Γ is the phase retardation of each plate. Note that this matrix is unimodular, that is, $AD - BC = 1$, since all the matrices in Eq. (10.2-1) are unimodular. Equation (10.2-2) can be simplified by using Chebyshev's identity equation (6.3-3) to

$$\begin{pmatrix} A & B \\ C & D \end{pmatrix}^m = \begin{pmatrix} \dfrac{A \sin mK\Lambda - \sin (m-1)K\Lambda}{\sin K\Lambda} & \dfrac{B \sin mK\Lambda}{\sin K\Lambda} \\[2ex] \dfrac{C \sin mK\Lambda}{\sin K\Lambda} & \dfrac{D \sin mK\Lambda - \sin (m-1)K\Lambda}{\sin K\Lambda} \end{pmatrix},$$

$$(10.2\text{-}4)$$

with

$$K\Lambda = \cos^{-1}[\tfrac{1}{2}(A + D)]. \qquad (10.2\text{-}5)$$

The incident wave and the transmitted wave are related by

$$\begin{pmatrix} E'_x \\ E'_y \end{pmatrix} = P_y M P_x \begin{pmatrix} E_x \\ E_y \end{pmatrix}. \qquad (10.2\text{-}6)$$

The transmitted beam is polarized in the y direction, with a field amplitude given by

$$E'_y = M_{21} E_x. \qquad (10.2\text{-}7)$$

If the incident light is linearly polarized in the x direction, the transmittance of this filter is

$$T = |M_{21}|^2. \qquad (10.2\text{-}8)$$

From Eqs. (10.2-2) and (10.2-3), we obtain

$$T = \left| \sin 4\varrho \sin^2 \frac{\Gamma}{2} \left(\frac{\sin mK\Lambda}{\sin K\Lambda} \right) \right|^2, \qquad (10.2\text{-}9)$$

with

$$\cos K\Lambda = 1 - 2\cos^2 2\varrho \sin^2 \frac{\Gamma}{2}. \qquad (10.2\text{-}10)$$

The transmittance T is often expressed in terms of a new variable χ, defined as

$$K\Lambda = \pi - 2\chi. \qquad (10.2\text{-}11)$$

In terms of this new variable χ, the transmittance becomes

$$T = \left| \tan 2\varrho \cos \chi \, \frac{\sin N\chi}{\sin \chi} \right|^2, \qquad (10.2\text{-}12)$$

with

$$\cos \chi = \cos 2\varrho \sin \frac{\Gamma}{2}. \qquad (10.2\text{-}13)$$

According to Eqs. (10.2-12) and (10.2-13), transmittance $T = \sin^2 2N\varrho$ when the phase retardation of each plate is $\Gamma = \pi, 3\pi, 5\pi, \ldots$, that is, when each plate becomes a half-wave plate. This transmittance is 100% if the azimuth angle ϱ is such that

$$\varrho = \frac{\pi}{4N}. \qquad (10.2\text{-}14)$$

Transmission under these conditions can be easily understood if we examine the polarization state after passing through each plate within the Solc filter. We recall that in passing through a "half-wave" ($\Gamma = \pi, 3\pi, \ldots$) plate, the azimuthal angle between the polarization vector and the c axis of the crystal changes sign. Past the front polarizer, the light is linearly polarized in the x direction (azimuth $\psi = 0$). Since the first plate is at the azimuth angle ϱ, the emerging beam after passage through the first plate is linearly polarized at $\psi = 2\varrho$. The second plate is oriented at the azimuth angle $-\varrho$, making an angle of 3ϱ with respect to the polarization direction of light incident on it. The polarization direction at its output face will be rotated by 6ϱ and oriented at the azimuth angle -4ϱ. The plates are oriented successively at $+\varrho$, $-\varrho$, $+\varrho$, $-\varrho$, ... , while the polarization direction at the exit of the plates assumes the values 2ϱ, -4ϱ, 6ϱ, -8ϱ, ... The final azimuthal angle after N plates is thus $2N\varrho$. If this final azimuth angle is 90° (i.e., $2N\varrho = \frac{1}{2}\pi$), the light passes through the rear polarizer without any loss of intensity. Light at other wavelengths, where the plates are not half-wave plates, does not experience a 90° rotation of polarization and suffers loss at the rear polarizer.

The Solc filter can also be viewed as a periodic medium as far as wave propagation is concerned. The alternating azimuth angles of the crystal axes constitute a periodic perturbation to the propagation of both waves. This perturbation couples the ordinary and extraordinary waves. Because these waves propagate at different phase velocities, complete exchange of electromagnetic energy is possible only when the perturbation is periodic so as to maintain the relative relations necessary to cause continuous power transfer from the E wave to the O wave and vice versa. The basic physical explanation is as follows: If power is to be transferred gradually with distance from a mode A to a mode B by a static perturbation, it is necessary that both waves travel with the same phase velocity. If the phase velocities are not equal, the incident wave A gets progressively out of phase with the wave into which it couples. This limits the total fraction of power that can be exchanged. Such a situation can be avoided if the sign of the perturbation is reversed whenever the phase mismatch (between the coupled field and the field into which it couples) is equal to π. This reverses the sign of the coupled power, which thus maintain the proper phase for continuous power transfer.

The transmission characteristics around the peak and its sidelobes of a Solc filter are interesting and deserve some investigation. Assume that each plate is characterized by refractive indices n_e and n_o and thickness d. Let λ_m denote the wavelength at which the phase retardation is $(2m + 1)\pi$. The phase retardation at a general wavelength is

$$\Gamma = \frac{2\pi}{\lambda}(n_e - n_o)d. \qquad (10.2\text{-}15)$$

If λ is slightly away from λ_m (i.e., $\lambda - \lambda_m \ll \lambda_m$), Γ can be approximately given by

$$\Gamma = (2m + 1)\pi + \Delta\Gamma,$$

where

$$\Delta\Gamma = -\frac{(2m + 1)\pi}{\lambda}(\lambda - \lambda_m).$$ (10.2-16)

We assume further that the azimuth angle of the plate obeys condition (10.2-14), and N is much larger than 1. Under these conditions, the trigonometric functions in Eq. (10.2-13) can be expanded to yield

$$\chi \simeq \frac{\pi}{2N}\left[1 + \left(\frac{N\,\Delta\Gamma}{\pi}\right)^2\right]^{1/2}.$$ (10.2-17)

Substituting χ into Eq. (10.2-12), we obtain

$$T = \left(\frac{\sin\left[(\pi/2)\sqrt{1 + (N\,\Delta\Gamma/\pi)^2}\right]}{\sqrt{1 + (N\,\Delta\Gamma/\pi)^2}}\right)^2.$$ (10.2-18)

This approximate expression for the transmittance is valid provided $N \gg 1$ and $(\lambda - \lambda_m) \ll \lambda_m$. From Eq. (10.2-18), the FWHM of the main transmission peak is given approximately by $\Delta\Gamma_{1/2} = 1.60\pi/N$, which in terms of wavelength is

$$\Delta\lambda_{1/2} \simeq 1.60\frac{\lambda}{(2m + 1)N} = 0.80\frac{\lambda^2}{Nd|n_e - n_o|},$$ (10.2-19)

where we recall that d is the thickness of the plates and N is the number of plates. Note that this bandwidth is inversely proportional to the total plate thickness.

The FSR of such a filter is determined in $\Delta\Gamma = 2\pi$ and is given by

$$\Delta\lambda = \frac{\lambda^2}{d|n_e - n_o|}.$$ (10.2-20)

In a manner very similar to Lyot–Ohman filters, the FSR is determined by the thickness of the thinnest plate (in this case, individual plate thickness d), whereas the bandwidth is determined by the total thickness (Nd).

The finesse of such a Solc filter is given by [according to Eqs. (10.2-19) and (10.2-20)]

$$F = \frac{\Delta\lambda}{\Delta\lambda_{1/2}} = 1.25N,$$ (10.2-21)

which is proportional to the total number of plates.

We now examine the transmission spectrum, including all the sidelobes and minima. According to Eqs. (10.2-12) and (10.2-13), the transmission is a periodic function of Γ with a period of 2π. Thus, it is sufficient to examine the region between $\Gamma = \pi$ and $\Gamma = 3\pi$. According to Eq. (10.2-13), the parameter χ varies from $\chi = 2\varrho$ to $\chi = \pi - 2\varrho$ in this region. The transmission spectrum, as given by Eq. (10.2-12), is dominated by the fast variation of the factor $|\sin N\chi|$ because N is a large integer. This factor varies between zero and 1. There are $N - 1$ zeros in the spectral regime between $\Gamma = \pi$ and $\Gamma = 3\pi$. Thus, there are $N - 1$ minima (zero transmittance) between the passbands. The transmittance minima are interlaced with sidelobes. There are exactly $N - 2$ sidelobes. Those sidelobes near the passband can be estimated by using Eq. (10.2-18). They occur approximately at

$$\sqrt{1 + (N\,\Delta\Gamma/\pi)^2} \simeq (2l + 1), \qquad l = 1, 2, 3, \ldots, \qquad (10.2\text{-}22)$$

with transmittance given approximately by

$$T \simeq \frac{1}{(2l + 1)^2}. \qquad (10.2\text{-}23)$$

The first sidelobe ($l = 1$) has an approximate transmittance of $\frac{1}{9}$, or 11%. Recall that the first sidelobe in the Lyot–Ohman filters is 4.7%. A calculated transmission spectrum of a Solc filter is shown in Fig. 10.5.

Example 1. Consider a folded Solc filter that consists of 90 identical quartz plates, each 1 mm thick. Assuming a birefringence $\Delta n = n_e - n_o = 0.009$, find the passbands and the bandwidths.

The passbands are determined by $\Gamma = (2m + 1)\pi$, or

$$\lambda = \frac{2\,\Delta n\,d}{2m + 1} = \frac{18\,\mu m}{2m + 1}.$$

Take $m = 20$; a passband at $\lambda = 0.44\,\mu m$ is obtained. The bandwidth of this passband is given by [according to Eq. (10.2-19)]

$$\Delta\lambda_{1/2} = 1.9\,\text{Å}.$$

Example 2. Design a 1-Å folded Solc filter at $\lambda = 0.5\,\mu m$ by using calcite crystals. Take $|n_e - n_o| = 0.172$. Eq. (10.2-19) yields

$$Nd = 1.16\,\text{cm}.$$

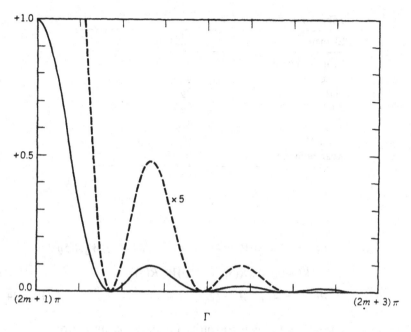

Figure 10.5 Calculated transmittance spectrum of an 8-stage Solc filter.

If a FSR of 100 Å is desired, the number of plates must be 80, according to Eq. (10.2-21). The calcite plate thickness is thus

$$d = 145\,\mu\text{m}.$$

The rocking angle ϱ must be maintained at

$$\varrho = \frac{\varrho}{4N} = 0.56°.$$

10.2.2 Fan Solc Filters

A fan Solc filter also consists of a stack of identical birefringent plates each oriented at a prescribed azimuth. A brief summary of the basic type of fan Solc filter is given in Table 10.2. The geometric arrangement of the fan Solc filter is sketched in Fig. 10.6. According to the Jones matrix method formulated in the previous chapter, the overall matrix for these N plates is

Table 10.2. Fan Solc Filter

Element	Azimuth
Front polarizer	$0°$
Plate 1	ϱ
Plate 2	3ϱ
Plate 3	5ϱ
$\vdots$	$\vdots$
Plate N	$(2N - 1)\varrho = \frac{1}{2}\pi - \varrho$
Rear polarizer	$0°$

given by

$$
\begin{aligned}
M &= R(-\tfrac{1}{2}\pi + \varrho)W_0 R(\tfrac{1}{2}\pi - \varrho) \cdots R(-5\varrho)W_0 R(5\varrho) \\
&\quad \times\ R(-3\varrho)W_0 R(3\varrho)R(-\varrho)W_0 R(\varrho) \\
&= R(-\tfrac{1}{2}\pi + \varrho)[W_0 R(2\varrho)]^N R(-\varrho),
\end{aligned}
\tag{10.2-24}
$$

where we have used the following identity for the rotation matrix:

$$
R(\varrho_1)R(\varrho_2) = R(\varrho_1 + \varrho_2).
\tag{10.2-25}
$$

Notice that the last plate always appears first in the product (10.2-24).

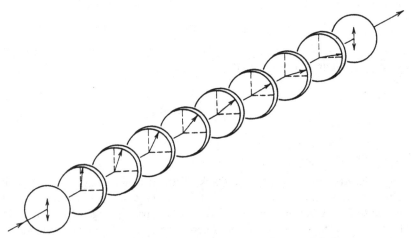

Figure 10.6   Schematic drawing of an 8-stage fan-type Solc filter.

By using the Chebyshev identity (10.2-4) and carrying out the matrix multiplication in Eq. (10.2-24), we obtain

$$M_{11} = \sin 2\varrho \cos \frac{\Gamma}{2} \frac{\sin N\chi}{\sin \chi},$$

$$M_{12} = -\cos Nx - i\sin\frac{\Gamma}{2}\frac{\sin N\chi}{\sin \chi},$$

$$M_{21} = \cos Nx - i\sin\frac{\Gamma}{2}\frac{\sin N\chi}{\sin \chi},$$

$$M_{22} = M_{11},$$

(10.2-26)

with

$$\cos \chi = \cos 2\varrho \cos\frac{\Gamma}{2}. \tag{10.2-27}$$

These are the matrix elements of the overall Jones matrix not including the polarizers.

The incident wave E and the transmitted wave E' are thus related

$$\begin{pmatrix} E'_x \\ E'_y \end{pmatrix} = \begin{pmatrix} 1 & 0 \\ 0 & 0 \end{pmatrix} \begin{pmatrix} M_{11} & M_{12} \\ M_{21} & M_{22} \end{pmatrix} \begin{pmatrix} 1 & 0 \\ 0 & 0 \end{pmatrix} \begin{pmatrix} E_x \\ E_y \end{pmatrix}. \tag{10.2-28}$$

The transmitted beam is vertically polarized (x) with the amplitude given by

$$E'_x = M_{11} E_x. \tag{10.2-29}$$

If the incident wave is linearly polarized in the x direction, the transmittance is given by

$$T = |M_{11}|^2. \tag{10.2-30}$$

From (10.2-26), we have the following expression for the transmittance:

$$T = \left| \tan 2\varrho \cos \chi \frac{\sin N\chi}{\sin \chi} \right|^2, \tag{10.2-31}$$

with

$$\cos \chi = \cos 2\varrho \cos\frac{\Gamma}{2}. \tag{10.2-32}$$

Notice that transmission formula (10.2-31) is formally identical to Eq. (10.2-12), provided that Γ is replaced by $\Gamma + \pi$. Maximum transmittance $(T = 1)$ occurs when $\Gamma = 0, 2\pi, 4\pi, \ldots$, and $\varrho = \pi/4N$. This unity transmission results simply from the fact that at these wavelengths the plates are full-wave plates. The light will remain linearly polarized in the x direction after transmitting through each plate and will suffer no loss at the rear polarizer. Light at other wavelengths, where the plates are not full-wave plates, does not remain linearly polarized in the x direction and suffers loss at the rear polarizer. Let λ_m be the wavelength at which the phase retardation $\Gamma = 2m\pi$. If λ differs slightly away from λ_m, that is, $\lambda - \lambda_m \ll \lambda$, Γ can be given approximately by

$$\Gamma = 2m\pi + \Delta\Gamma = 2m\pi - \frac{2m\pi}{\lambda}(\lambda - \lambda_m), \qquad (10.2\text{-}33)$$

where $m = 1, 2, 3, \ldots$. If we now further assume that N is much larger than 1 and follow the same procedure as Eq. (10.2-17), we obtain the following approximate expression for the transmittance

$$T = \left(\frac{\sin\left[(\pi/2)\sqrt{1 + (N\,\Delta\Gamma/\pi)^2}\,\right]}{\sqrt{1 + (N\,\Delta\Gamma/\pi)^2}} \right)^2, \qquad (10.2\text{-}34)$$

which is identical to Eq. (10.2-18). The FWHM $\Delta\lambda_{1/2}$ of the transmission maxima is again given approximately by

$$\Delta\lambda_{1/2} \simeq 1.60\,\frac{\lambda}{2mN} = 0.80\,\frac{\lambda^2}{Nd|n_e - n_o|}, \qquad m = 1, 2, 3, \ldots . \qquad (10.2\text{-}35)$$

The transmission spectrum of the fan Solc filter is identical to that of the folded Solc filter except that the curves are shifted by $\Gamma = \pi$. In other words, the transmittance of the fan Solc filter at phase retardation Γ is identical to that of the folded Solc filter at phase retardation $\Gamma + \pi$. This can also be seen from the expression for transmittance [(10.2-12) and (10.2-31)].

10.3 ANGULAR PROPERTIES OF BIREFRINGENT FILTERS

We have shown that birefringent layered structures such as Lyot–Ohman and Solc filters can provide extremely narrow bandwidths. Such narrow

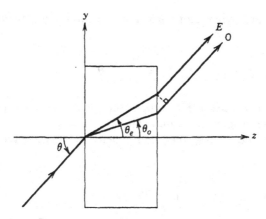

Figure 10.7   Transmission of off-axis light.

bandwidths can also be easily obtained by using Fabry–Perot interference filters. The great advantage of the birefringent filters comes from their capability of accommodating a very large field of view. As we mentioned before, such wide fields of view are essential in many applications. In this section, we will examine the angular properties of these filters and then illustrate how these wide fields of view are achieved.

In the previous two sections, we studied the spectral transmittance of Lyot–Ohman and Solc filters and derived the passbands and their bandwidths for normally incident light. This was done by examining the variation of the phase retardation with respect to the wavelength, since the transmittance is completely determined by the phase retardation of the plates [see, e.g., Eq. (10.1-7)]. To study the angular properties of these filters, we must examine the variation of the phase retardation of the crystal plates with respect to the angle of incidence. For the sake of clarity in introducing the concepts, we will again limit ourselves to the case when the c axes of the crystals are all parallel to the plate surfaces.

Referring to Fig. 10.7, we consider the incidence of an off-axis light onto a uniaxially birefringent crystal plate. Following the notation used in Section 9.6, the incident and refracted wave vectors can be written as

$$\mathbf{k} = \beta\mathbf{y} + k_z\mathbf{z}, \tag{10.3-1}$$

$$\mathbf{k}_o = \beta\mathbf{y} + k_{oz}\mathbf{z}, \tag{10.3-2}$$

$$\mathbf{k}_e = \beta\mathbf{y} + k_{ez}\mathbf{z}, \tag{10.3-3}$$

where **y** and **z** are unit vectors, β is the tangential component of the wave vectors and is given by

$$\beta = \frac{\omega}{c} \sin \theta \qquad (10.3\text{-}4)$$

and k_z, k_{ez}, and k_{oz} are the z components of the wave vectors. Here, we assume that the incident medium is air, with $n = 1$.

According to Eq. (9.5-8), the z components of the wave vectors k_o, k_e are given by

$$k_{oz} = \left[\left(\frac{n_o \omega}{c} \right)^2 - \beta^2 \right]^{1/2}, \qquad (10.3\text{-}5)$$

$$k_{ez} = \left[\left(\frac{n_e \omega}{c} \right)^2 - \beta^2 \left(\cos^2 \phi + \frac{n_e^2}{n_o^2} \sin^2 \phi \right) \right]^{1/2}, \qquad (10.3\text{-}6)$$

respectively, where we recall that θ is the angle between the incident wave vector and the normal of the crystal surface, and $(\frac{1}{2}\pi - \phi)$ is the angle between the c axis and the plane of incidence (see Fig. 9.9). Specifically, θ and ϕ can be written as

$$\theta = \cos^{-1} \frac{\mathbf{k} \cdot \mathbf{z}}{|\mathbf{k}|}, \qquad (10.3\text{-}7)$$

$$\phi = \cos^{-1} \frac{\mathbf{c} \cdot (\mathbf{k} \times \mathbf{z})}{|\mathbf{c} \cdot (\mathbf{k} \times \mathbf{z})|}, \qquad (10.3\text{-}8)$$

where **c** is a unit vector along the c axis and **z** is a unit vector along the z axis.

The phase retardation between the E wave and the O wave, after traversing from $z = 0$ to $z = d$ in the crystal, is given by

$$\Gamma = (k_{ez} - k_{oz})d, \qquad (10.3\text{-}9)$$

with k_{ez} and k_{oz} given by Eqs. (10.3-6) and (10.3-5), respectively. We notice that for light with nonnormal incidence, the phase retardation is no longer given by the simple expression Eq. (10.1-2). It is a function of both θ and ϕ.

If the angle of incidence θ is such that

$$\sin^2 \theta \ll n_o^2, n_e^2, \qquad (10.3\text{-}10)$$

the z component of the wave vectors $\mathbf{k}_o$, $\mathbf{k}_e$ can be written approximately as

$$k_{oz} = \frac{2\pi}{\lambda} n_o \left[1 - \frac{1}{2} \frac{\sin^2 \theta}{n_o^2} \right] \qquad (10.3\text{-}11)$$

and

$$k_{ez} = \frac{2\pi}{\lambda} n_e \left[1 - \tfrac{1}{2} \sin^2\theta \left(\frac{\sin^2\phi}{n_o^2} + \frac{\cos^2\phi}{n_e^2} \right) \right], \qquad (10.3\text{-}12)$$

respectively. These two equations are obtained from Eqs. (10.3-5) and (10.3-6) by the series expansion of the square roots and by neglecting the higher order terms. We have also used $\omega/c = 2\pi/\lambda$.

Substitution of Eqs. (10.3-11) and (10.3-12) for k_{oz} and k_{ez}, respectively, into Eq. (10.3-9) yields

$$\Gamma = \frac{2\pi}{\lambda} (n_e - n_o)d \left[1 - \frac{\sin^2\theta}{2n_o^2} \left(1 - \frac{n_e + n_o}{n_e} \cos^2\phi \right) \right]. \quad (10.3\text{-}13)$$

For the case of small birefringence (i.e., $n_o \simeq n_e$), the phase retardation becomes

$$\Gamma = \frac{2\pi d}{\lambda} \Delta n \left[1 - \frac{\sin^2\theta}{2n_o^2} (1 - 2\cos^2\phi) \right], \qquad (10.3\text{-}14)$$

where Δn is the birefringence ($n_e - n_o$). Since the phase retardation Γ of each plate is a function of both θ and ϕ, the transmission maximum of a filter is expected to be dependent on the angle of incidence (θ, ϕ). For the case of the Lyot–Ohman birefringent filters, the spectral shift in wavelength is given by

$$\Delta \lambda = - \frac{\lambda}{2n_o^2} \sin^2\theta (1 - 2\cos^2\phi). \qquad (10.3\text{-}15)$$

Notice that this spectral shift of the transmission maxima is a function of both θ and ϕ. For a given θ, maximum shift occurs at $\phi = 0$ and $\tfrac{1}{2}\pi$. Let us now examine the field of view at these adverse angles of incidence. Assuming $\phi = 0$, the spectral shift [Eq. (10.3-15)] becomes

$$\Delta \lambda = - \frac{\lambda}{2n_o^2} \sin^2\theta. \qquad (10.3\text{-}16)$$

If the field of view is defined as the angle of incidence θ when the spectral shift is one-half of the bandwidth, then according to Eq. (10.3-16), the field of view of a Lyot–Ohman birefringent filter is

$$\theta = \pm n_o \left| \frac{\Delta \lambda_{1/2}}{\lambda} \right|^{1/2}, \qquad (10.3\text{-}17)$$

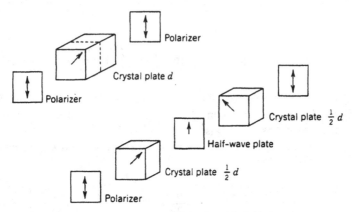

Figure 10.8 Schematic drawing of a wide-field element.

where $\Delta\lambda_{1/2}$ is the bandwidth given by Eq. (10.1-13). This expression is not much different from that of a Fabry–Perot filter as given by Eq. (7.6-5). In other words, the field of view of a simple Lyot–Ohman filter is also very narrow. For example, for the simple Lyot–Ohman filter described in the second example of Section 10.1, the field of view for a 1-Å bandwidth at $\lambda = 0.5\,\mu m$ is only $\theta = \pm 1.25°$.

There is, however, a big difference between the Lyot–Ohman filter and the Fabry–Perot filter. In birefringent crystals, the phase retardation [Eq. (10.3-13)] is also a function of the azimuthal angle ϕ. This dependence allows the possibility of arranging crystal plates at different azimuth angles to minimize their dependence on angles. This is the basic principle of a wide-field element.

10.3.1 Wide-Field Element

Figure 10.8 illustrates a wide-field element invented by Lyot. It differs from the simple element in that the crystal plate is split in half and the second half is rotated 90° with respect to the first. In addition, an achromatic half-wave plate is inserted between these two halves. This half-wave plate is oriented such that its c axis is parallel to the transmission axis of the polarizers.

Let Γ_1 and Γ_2 be the phase retardation for the two split halves, respectively, for a given angle of incidence. Using the Jones matrix formulation described in Section 9.3, the Jones matrix for each of the three parts of the wide-field element between the polarizers can be written as

$$W_1 = \begin{pmatrix} \cos\frac{1}{2}\Gamma_1 & -i\sin\frac{1}{2}\Gamma_1 \\ -i\sin\frac{1}{2}\Gamma_1 & \cos\frac{1}{2}\Gamma_1 \end{pmatrix}, \tag{10.3-18}$$

$$W_{\lambda/2} = \begin{pmatrix} -i & 0 \\ 0 & i \end{pmatrix}, \tag{10.3-19}$$

$$W_2 = \begin{pmatrix} \cos\frac{1}{2}\Gamma_2 & i\sin\frac{1}{2}\Gamma_2 \\ i\sin\frac{1}{2}\Gamma_2 & \cos\frac{1}{2}\Gamma_2 \end{pmatrix}, \tag{10.3-20}$$

respectively. The Jones matrix representing the whole wide-field element is simply the product of these three matrices, that is,

$$W = W_2 W_{\lambda/2} W_1. \tag{10.3-21}$$

By carrying out the matrix multiplication, we obtain

$$W = \begin{pmatrix} -i\cos\frac{1}{2}(\Gamma_1 + \Gamma_2) & -\sin\frac{1}{2}(\Gamma_1 + \Gamma_2) \\ \sin\frac{1}{2}(\Gamma_1 + \Gamma_2) & i\cos\frac{1}{2}(\Gamma_1 + \Gamma_2) \end{pmatrix}. \tag{10.3-22}$$

When sandwiched between parallel polarizers oriented along the x axis, the transmittance for unpolarized light is

$$T = \frac{1}{2}\cos^2\frac{1}{2}(\Gamma_1 + \Gamma_2). \tag{10.3-23}$$

According to this equation, the three-component wide-field element is, so far as transmittance is concerned, equivalent to a wave plate oriented at an azimuth of 45° and with a phase retardation of

$$\Gamma = \Gamma_1 + \Gamma_2. \tag{10.3-24}$$

Let $\Gamma_0(\theta, \phi)$ be the phase retardation of the crystal plate before it was split. The phase retardation Γ_1 and Γ_2 for the split elements are

$$\Gamma_1(\theta, \phi) = \frac{1}{2}\Gamma_0(\theta, \phi) \tag{10.3-25}$$

and

$$\Gamma_2(\theta, \phi) = \frac{1}{2}\Gamma_0(\theta, \phi + \frac{1}{2}\pi), \tag{10.3-26}$$

respectively. Thus, the equivalent phase retardation becomes

$$\Gamma(\theta, \phi) = \frac{1}{2}[\Gamma_0(\theta, \phi) + \Gamma_0(\theta, \phi + \frac{1}{2}\pi)]. \tag{10.3-27}$$

Substitution of Eq. (10.3-13) for $\Gamma_0(\theta, \phi)$ in Eq. (10.3-7) yields

$$\Gamma(\theta, \phi) = \frac{2\pi}{\lambda}(n_e - n_o)d\left(1 - \frac{\sin^2\theta}{2n_o^2}\frac{n_e - n_o}{2n_e}\right). \qquad (10.3\text{-}28)$$

Notice that, indeed, the effective phase retardation of the wide-field element is much less dependent on the angle of incidence compared with the simple crystal plate. The shift of the transmittance maximum with respect to the angle of incidence is [according to Eq. (10.3-28)]

$$\Delta\lambda = -\frac{\lambda}{2n_o^2}\frac{\Delta n}{2n_e}\sin^2\theta \qquad (10.3\text{-}29)$$

for the wide-field element. Compared with Eq. (10.3-16) for the simple element, we find that the spectral shift with respect to the angle of incidence is reduced by a factor of $\Delta n/2n_e$. The field of view of a Lyot–Ohman filter made of those wide-field elements is thus given by

$$\theta = \pm n_o\left|\frac{2n_e\,\Delta\lambda_{1/2}}{\lambda\,\Delta n}\right|^{1/2}. \qquad (10.3\text{-}30)$$

Notice that the field of view is increased by a factor of $(2n_e/\Delta n)^{1/2}$ in θ. In terms of solid angles, which are proportional to θ^2, the increase in the field of view is by a factor of $2n_e/\Delta n$. For quartz crystals, with $n_e = 1.553$ and $\Delta n = 0.009$, the field of view is increased by a factor of 13 in θ. For the 1-Å filter at 0.5 μm, this field of view is increased from $\theta = 1.25°$ to $\theta = \pm 16.4°$.

Lyot–Ohman birefringent filters that use the wide-field elements described are called *Lyot-1 filters*. Figure 10.9 compares the spectral transmittance of a simple Lyot–Ohman filter and its wide-field version. The filter is designed to have a passband at 0.5 μm with a bandwidth of 0.5 Å. The birefringent materials are quartz crystal plates. We notice that the passband of the simple Lyot–Ohman filter is shifted by more than 1 Å at an incidence angle of only 2°. The spectral shift for the wide-field version (i.e., the Lyot-1 filter) at an incidence angle of 16° is less than 0.5 Å. We see that, indeed, Lyot-1 filters can offer simultaneously a very large field of view and an extremely narrow bandwidth.

According to Eq. (10.3-30), the field of view of Lyot-1 filters is inversely proportional to the square root of the birefringence of the crystal. Thus, for the purpose of achieving a large field of view, it is desirable to use materials with small birefringence.

Wide fields of view can also be achieved by using the material dispersion of the birefringence. These filters will be the subject of the next two sections.

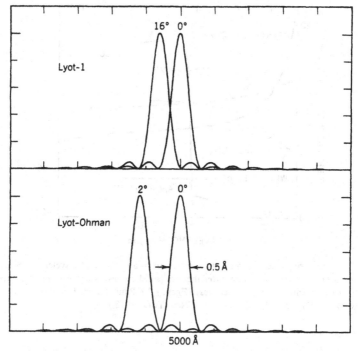

Figure 10.9   Comparison of field of view of Lyot–Ohman and Lyot-1 filters.

10.4   DISPERSIVE BIREFRINGENT FILTERS

In the derivation of the bandwidths for Lyot–Ohman and Solc filters, we assumed that the birefringence ($n_e - n_o$) is independent of the wavelength variation. In other words, the variation of the phase retardation is due to the $1/\lambda$ factor in Eq. (10.1-2). This assumption is no longer true when the material exhibits a dispersion of the birefringence.

In this section, we investigate new types of filter that use such material dispersion for the construction of narrow-band and wide-field filters. These filters are called *dispersive birefringent filters* [7]. Many crystals are found to exhibit anomalous dispersion of birefringence in the spectral regimes near the absorption band edges. These include cadmium sulfide (CdS) (5150–5400 Å) [8], silver thiogallate ($AgGaS_2$) (4900–5500 Å) [9], magnesium fluoride (MgF_2) (1150–1500 Å) [10], and sapphire (1430–1530 Å) [10]. In fact, there is even a change of sign of the birefringence (zero crossing) after passing through zero in the case of MgF_2, $AgGaS_2$, and CdS. Figure 10.10 shows the

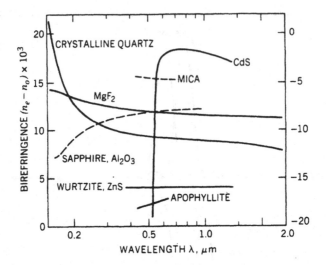

Figure 10.10 Birefringence of some optical materials as a function of wavelength. Scale at the left is for materials having a positive birefringence (solid curves); scale at right is for materials with a negative birefringence (dashed curves). (From J. M. Bennett and H. E. Bennett, *Handbook in Optics*, W. G. Driscoll ed., McGraw-Hill, New York, 1978.)

dispersion of birefringence for some materials. Table 10.3 lists the indices of refraction n_o, n_e for some selected materials at several wavelengths.

One of the most important concepts in dispersive birefringent filters is *differential phase retardation*. We start by rewriting phase retardation as

$$\Gamma = \frac{2\pi}{\lambda} \Delta n \, d, \qquad (10.4\text{-}1)$$

where $\Delta n = n_e - n_o$ is the birefringence of the crystal at wavelength λ. Here n_e and n_o are the refractive indices of the crystal. The crystals are cut with input and output faces parallel to the c axis. The transmission of polarized light through such a plate when sandwiched between parallel polarizers is given by

$$T = \cos^2 \frac{\Gamma}{2}. \qquad (10.4\text{-}2)$$

Maximum transmission occurs at the wavelengths when the phase retardation is an integral multiple of 2π (i.e., $\Gamma = 2m\pi$). The transmission characteristics are governed by the variation of the phase retardation Γ with respect to the wavelength λ. This variation is described by the following differential:

$$\frac{\partial \Gamma}{\partial \lambda} = \frac{2\pi d}{\lambda}\left(\alpha - \frac{\Delta n}{\lambda}\right), \qquad (10.4\text{-}3)$$

where α is the rate of dispersion of the birefringence and is defined as

$$\alpha = \frac{\partial \Delta n}{\partial \lambda}. \qquad (10.4\text{-}4)$$

The variation of phase retardation Γ with respect to wavelength λ [see Eq. (10.4-3)] consists of two parts. First, the variation of Γ with respect to λ can be due to the factor $1/\lambda$ in Eq. (10.4-1). This variation is presented by the second term, $-\Delta n/\lambda$, in the parentheses in Eq. (10.4-3). Second, the variation of Γ with respect to λ can also result from the dispersion of the birefringence. The latter contributed to differential phase retardation and is described by the first term, α, in the parentheses in Eq. (10.4-3). For a conventional wave plate such as quartz at 5000 Å, the variation of phase retardation with respect to wavelength is dominated by the term $\Delta n/\lambda$ ($\sim 1.8 \times 10^{-6}\,\text{Å}^{-1}$) because of the small dispersion (α of about $-1.6 \times 10^{-7}\,\text{Å}^{-1}$). For the case of a CdS plate at 5300 Å, $\Delta n/\lambda$ is only $1.1 \times 10^{-6}\,\text{Å}^{-1}$, whereas α is $0.7 \times 10^{-4}\,\text{Å}^{-1}$. Therefore, the variation of phase retardation Γ of a CdS plate with respect to wavelength λ is dominated by the rapid variation of Δn with respect to λ, while the factor $1/\lambda$ plays only a very minor role. Thus, the transmission of a relatively thin plate of those anomalous dispersion birefringent crystals will have a very narrow bandwidth. From Eqs. (10.4-1)–(10.4-3), the FWHM $\Delta\lambda_{1/2}$ for each peak ($\Gamma = 2m\pi$) is given by

$$\Delta\lambda_{1/2} = \frac{\pi}{|\partial\Gamma/\partial\lambda|} = \frac{\lambda}{2d|\alpha - \Delta n/\lambda|} \simeq \frac{\lambda}{2|\alpha|d}. \qquad (10.4\text{-}5)$$

The last equality is true provided $|\alpha| \gg |\Delta n/\lambda|$. We find that for birefringent crystals with a strong dispersion (large α), the bandwidth is inversely proportional to the product of the dispersion rate α and the plate thickness d. The separation between the peaks is given by

$$\Delta\lambda = \frac{2\pi}{|\partial\Gamma/\partial\lambda|} = \frac{\lambda}{d|\alpha - \Delta n/\lambda|} \simeq \frac{\lambda}{|\alpha d|}. \qquad (10.4\text{-}6)$$

This separation $\Delta\lambda$ is called the *free spectral range* (FSR).

Table 10.3. Refractive Indices of Selected Uniaxially Birefringent Crystals

Crystal	0.450 μm	0.488 μm	0.500 μm	0.515 μm	0.550 μm	0.600 μm	0.633 μm	1.000 μm
ADP								
n_e	1.487	1.484	1.483	1.483	1.481	1.478	1.477	1.469
n_o	1.534	1.531	1.529	1.529	1.526	1.523	1.522	1.508
BaTiO$_3$								
n_e	2.492	2.449	2.438	2.426	2.400	2.380	2.364	2.297
n_o	2.576	2.521	2.507	2.492	2.460	2.436	2.416	2.338
BeO								
n_e	1.744	1.741	1.740	1.739	1.736	1.734	1.732	—
n_o	1.728	1.725	1.724	1.723	1.721	1.718	1.717	—
Calcite								
n_e	1.493	1.491	1.490	1.489	1.488	1.486	1.485	1.480
n_o	1.673	1.668	1.666	1.665	1.661	1.658	1.656	1.644
CdS								
n_e	—	—	—	2.729	2.593	2.511	2.477	2.352
n_o	—	—	—	2.747	2.580	2.493	2.460	2.334
KDP								
n_e	1.4756	1.473	1.4721	1.4716	1.4698	1.4679	1.4669	1.4604
n_o	1.5188	1.5155	1.5145	1.5137	1.5115	1.509	1.5074	1.4954
LiNbO$_3$								
n_e	2.277	2.252	2.245	2.239	2.224	2.208	2.204	2.165
n_o	2.381	2.353	2.344	2.337	2.319	2.300	2.296	2.249
LiTaO$_3$								
n_e	2.247	2.227	2.221	2.216	2.204	2.188	2.182	2.143
n_o	2.242	2.222	2.216	2.211	2.200	2.183	2.177	2.139

	1	2	3	4	5	6	7	8
MgF₂								
n_e	1.382	1.380	1.380	1.379	1.379	1.378	1.377	—
n_o	1.393	1.392	1.392	1.391	1.390	1.389	1.389	—
PbMoO₄								
n_e	2.373	2.335	2.325	2.315	2.294	2.273	2.262	—
n_o	2.574	2.503	2.487	2.470	2.436	2.403	2.387	—
TcO₂								
n_e	2.540	2.497	2.486	2.474	2.450	2.424	2.412	2.352
n_o	2.368	2.332	2.322	2.312	2.291	2.269	2.260	2.208
TiO₂, Rutile								
n_e	3.174	3.058	3.034	3.011	2.953	2.904	2.879	2.747
n_o	2.820	2.731	2.712	2.694	2.648	2.609	2.590	2.484
Sapphire								
n_e	1.773	1.768	1.767	1.766	1.762	1.760	1.758	—
n_o	1.781	1.776	1.775	1.774	1.770	1.768	1.766	—
Quartz, SiO₂								
n_e	1.562	1.559	1.558	1.557	1.555	1.553	1.552	1.544
n_o	1.553	1.550	1.549	1.548	1.546	1.544	1.543	1.535
AgGaS₂								
n_e	—	2.722	2.681	2.650	2.580	2.525	2.505	—
n_o	—	2.706	2.686	2.664	2.615	2.569	2.551	—
ZnO								
n_e	2.123	2.081	2.068	2.060	2.041	2.015	2.007	1.959
n_o	2.106	2.064	2.051	2.044	2.025	1.999	1.990	1.944
ZnS								
n_e	2.477	2.435	2.425	2.415	2.392	2.368	2.356	2.303
n_o	2.473	2.430	2.421	2.410	2.386	2.363	2.352	2.301

Using the dispersive materials for the birefringent crytal plates in Lyot–Ohman filters, we obtain the following expression for bandwidth:

$$\Delta\lambda_{1/2} = 0.886 \frac{2\pi}{2^N |\partial\Gamma/\partial\lambda|} = 0.886 \frac{\lambda}{2^N |\alpha - \Delta n/\lambda|d}, \qquad (10.4\text{-}7)$$

where N is the number of stages. For anomalous dispersive materials with $|\Delta n/\lambda| \ll |\alpha|$, the bandwidth becomes

$$\Delta\lambda_{1/2} = 0.886 \frac{\lambda}{2^N |\alpha d|} = 0.886 \frac{\lambda}{2 |\alpha D|}, \qquad (10.4\text{-}8)$$

where D is the thickness of the thickest plate. The FSR of this filter is determined by the thickness of the thinnest plate and is given by

$$\Delta\lambda = \frac{2\pi}{|\partial\Gamma/\partial\lambda|} = \frac{\lambda}{|\alpha - \Delta n/\lambda|d} \simeq \frac{\lambda}{|\alpha d|}. \qquad (10.4\text{-}9)$$

The finesse of this filter, defined as the ratio of the FSR to the bandwidth, remains unchanged:

$$F = 1.13 \times 2^N. \qquad (10.4\text{-}10)$$

For the incidence of off-axis light, the spectral shift [see Eq. (10.3-15)] now becomes

$$\Delta\lambda = \frac{\Delta n}{2n_o^2 (\alpha - \Delta n/\lambda)} \sin^2\theta (1 - 2\cos^2\phi). \qquad (10.4\text{-}11)$$

In a conventional Lyot–Ohman filter made of crystals such as quartz, the dispersion α is small, and the spectral shift is

$$\Delta\lambda = -\frac{\lambda}{2n_o^2} \sin^2\theta (1 - 2\cos^2\phi) \quad \text{(conventional)}. \qquad (10.4\text{-}12)$$

In a dispersive Lyot–Ohman filter made of crystals such as CdS in the spectral regime where α is large (i.e., $\alpha \gg \Delta n/\lambda$), the spectral shift [Eq. (10.4-11)] becomes

$$\Delta\lambda = \frac{\Delta n}{2n_o^2 \alpha} \sin^2\theta (1 - 2\cos^2\phi) \quad \text{(dispersive)}. \qquad (10.4\text{-}13)$$

Notice that the spectral shift for the dispersive birefringent filter is small because of the small birefringence and strong dispersion (large α). As a result,

Table 10.4. Field of View and Plate Thickness of Birefringent CdS and Quartz Filters at 5300 Å

	CdS		Quartz	
	Simple Lyot–Ohman Filter	Wide-Angle Lyot-1 Filter	Simple Lyot–Ohman Filter	Wide-Angle Lyot-1 Filter
$\Delta\lambda_{1/2} = 1$ Å				
D (cm)	0.34	0.34	13.8	13.8
θ (degrees)	16	90	1.2	22.7
$\Delta\lambda_{1/2} = 0.1$ Å				
D (cm)	3.4	3.4	138	138
θ (degrees)	5.2	90	0.4	7.2

D = thickness of thickest plate; θ = field of view.

the angular field of view of a dispersive birefringent filter is expected to be very large compared with a conventional birefringent filter. The field of view of dispersive Lyot–Ohman birefringent filter is now given by

$$\theta = \pm n_o \left| \frac{(\alpha - \Delta n/\lambda)\,\Delta\lambda_{1/2}}{\Delta n} \right|^{1/2}, \qquad (10.4\text{-}14)$$

where $\Delta\lambda_{1/2}$ is the bandwidth. A conventional Lyot–Ohman filter would have a field of view of only $\theta = n_o\sqrt{\Delta\lambda_{1/2}/\lambda}$. We notice that the field of view is increased by a factor of $|(\alpha\lambda - \Delta n)/\Delta n|^{1/2}$. For example, a 1-Å dispersive Lyot–Ohman filter made of CdS at $\lambda = 5300$ Å has a field of view of approximately $\pm 16°$. A conventional Lyot–Ohman filter made of quartz crystal would have a field of view of only $\pm 2°$. Notice that, according to Eq. (10.4-14), the field of view is arbitrarily large for the passband at the *isotropic point* when $\Delta n = 0$.

A further increase in the field of view can be obtained by using wide-field elements. An increase by a factor of $(2n_e/\Delta n)^{1/2}$ in the field of view is obtained in both the conventional and the dispersive Lyot-1 filter. This field of view is thus given by

$$\theta = \pm n_o \left| \frac{2n_e}{\Delta n} \right|^{1/2} \left| \frac{\alpha\,\Delta\lambda_{1/2}}{\Delta n} \right|^{1/2} \quad (\text{for } \alpha \gg \Delta n/\lambda) \qquad (10.4\text{-}15)$$

for a Lyot-1 dispersive birefringent filter. Table 10.4 compares the field of view and plate thickness of a dispersive birefringent filter with those of a conventional birefringent filter.

As a result of the sharp birefringence dispersion and the smallness of the birefringence, the dispersive birefringent filter can offer simultaneously an extremely narrow bandwidth and an enormous field of view using crystals with moderate thicknesses. The sharp birefringence dispersion (i.e., $\alpha \gg \Delta n / \lambda$) allows the use of a thin crystal plate to achieve a narrow bandwidth. Such a small thickness and the smallness of birefringence lead to a phase retardation that is a small integral multiple of 2π. If we rewrite the field of view [both Eqs. (10.4-14) and (10.3-17)] as [according to Eqs. (10.4-7) and (10.1-13)]

$$\theta = \pm n_o \left| \frac{0.886\pi}{\Gamma_D} \right|^{1/2}, \qquad (10.4\text{-}16)$$

we find immediately that small phase retardation does lead to a wide field of view. The phase retardation Γ_D in Eq. (10.4-16) is that of the thickest plate.

10.5 ISO-INDEX FILTERS

A special case of the dispersive birefringent filter is the *iso-index filter* that operates at the *isotropic point*. The accidental crossing of the dispersion curves $n_o(\lambda)$ and $n_e(\lambda)$ of a uniaxial CdS crystal was first observed experimentally at the wavelength $\lambda = 5122 \text{ Å}$ at $T = 93 \text{ K}$ [8]. At this wavelength, which is often called the *iso-index point* or the *isotropic point*, the CdS crystal becomes optically isotropic. The zero crossing of the dispersion curves $n_e(\lambda) - n_o(\lambda)$ was subsequently reported for MgF_2 [10], Al_2O_3 [10], ZnO [11, 12], ZnS [13], CdS [11, 13–18], CdSe [17–26], SnO_2 [21], $AgGaS_2$ [9, 22–24], $CdGa_2S_4$ [25–27], and several other crystals with the chalcopyrite structure [28]. In addition, the zero crossing of the birefringence was also reported for several mixed crystals such as $CdS_{1-x}Se_x$ [29–32] and $Zn_xCd_{1-x}S$ [33]. These materials are called *iso-index crystals* because of their unique optical properties.

The operation of this new type of filter is based on the zero crossing of birefringence (see Fig. 10.11) [34, 35]. In the spectral regime around the isotropic point λ_c, birefringence can be approximately given as

$$n_e - n_o = \alpha(\lambda - \lambda_c) + \cdots, \qquad (10.5\text{-}1)$$

where α is the rate of dispersion of the birefringence at the crossing wavelength λ_c, that is,

$$\alpha = \frac{d}{d\lambda}(n_e - n_o). \qquad (10.5\text{-}2)$$

The phase retardation of a wave plate made of these crystals is certainly zero at the crossing wavelength λ_c. This plate is a *zero wave plate* for this

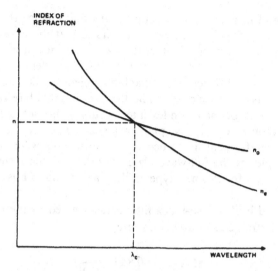

Figure 10.11 Dispersion curves for n_o and n_e of a uniaxial crystal. This material becomes isotropic at λ_c when $n_o = n_e$.

wavelength, λ_c, regardless of the angle of incidence. Thus, when a wave plate such as this is placed between a pair of parallel polarizers, the whole structure acts like a typical stage of a Lyot–Ohman filter.

Consider now a Lyot–Ohman filter that consists of iso-index crystal plates separated by parallel polarizers. In the special regimes around λ_c, the phase retardation of the thinnest plate is given approximately by

$$\Gamma = \frac{2\pi}{\lambda} \alpha(\lambda - \lambda_c)d. \tag{10.5-3}$$

The phase retardation is zero at the isotropic point $\lambda = \lambda_c$. In our earlier discussion on Lyot–Ohman filters, we have shown that peak transmission occurs at

$$\Gamma = 2m\pi, \qquad m = \text{integer}, \tag{10.5-4}$$

where the integer is never zero for ordinary material. Using iso-index crystals as the birefringent plates, $m = 0$ corresponds to a passband at the isotropic point λ_c.

The transmission of light at wavelength λ_c is obvious because no phase retardation is introduced by these crystal plates. The light therefore suffers no loss at the polarizers. For light with other wavelengths, the plates are birefringent and introduce phase retardations that change the polarization state of the light. This light therefore suffers loss at the polarizers. In this fashion, light with wavelengths other than λ_c are strongly discriminated, and the structures act similar to narrow-band filters. The sidelobes of these filters are similar to those of the conventional birefringent filters. The major difference between this new type of filter and the conventional birefringent filter is that the main peak of the iso-index filter occurs at the isotropic point when every plate becomes a zero wave plate, whereas the peak transmission of the conventional birefringent filters occurs at wavelengths where the plates are integral multiples of the full-wave plate. It is this difference that offers such large fields of view for this new type of filter, according to Eqs. (10.4-16) and (10.5-4).

The bandwidth of the iso-index filter is determined by the rate of dispersion of the birefringence α and is given by

$$\Delta\lambda_{1/2} = 0.886 \frac{\lambda}{2|\alpha D|}, \tag{10.5-5}$$

which is identical to Eq. (10.4-8) and can be considered as a special case of Eq. (10.4-7) when $\Delta n = 0$. The parameter D in Eq. (10.5-5) is the thickness of the thickest plate. Note that $\Delta\lambda_{1/2}$ is determined by the product of the thickness and the dispersion rate α. Many materials exhibit sharp birefringence crossing with a dispersion rate on the order of $10^{-4}\,\text{Å}^{-1}$ (e.g., CdS and AgGaS$_2$ with $\alpha \sim 2 \times 10^{-4}\,\text{Å}^{-1}$). Thus, crystals like CdS or AgGaS$_2$ with a total thickness of 2 mm would have a bandwidth of 1 Å according to Eq. (10.5-5). A conventional wide-field quartz Lyot-1 filter with the same bandwidth would require a thickest plate of about 14 cm and yet have a field of view of less than 30°.

Another important feature of the iso-index filter is its large field of view. This wide field of view arises from the operation of this filter at the isotropic point of the crystals. Since the transmission of light at the isotropic point is independent of the optical path in the crystals, this type of filter is expected to have a large field of view. The peak transmission at the isotropic point λ_c will be independent of the angle of incidence. The bandwidth $\Delta\lambda_{1/2}$, however, will depend slightly on the angle of incidence.

According to Eqs. (10.3-13) and (10.5-1), the phase retardation in the spectral regime around the isotropic point can be written as

$$\Gamma = \frac{2\pi}{\lambda} \alpha(\lambda - \lambda_c)d \left[1 - \frac{\sin^2\theta}{2n^2}(1 - 2\cos^2\phi) \right], \tag{10.5-6}$$

where n is the common value of n_o and n_e at the isotropic point λ_c. We note that $\Gamma = 0$ at $\lambda = \lambda_c$ regardless of the angle of incidence (θ, ϕ). Therefore,

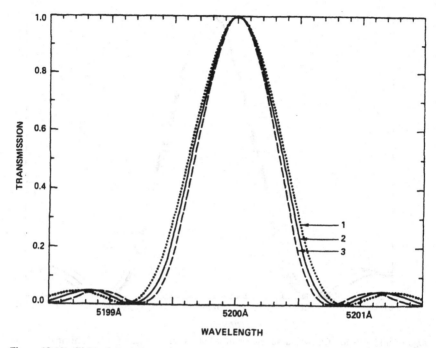

Figure 10.12 Field of view characteristics of a Lyot–Ohman iso-index filter. Curve 2, normal incident light; curve 1 is for extreme angle of incidence ($\theta = 90°$) with plane of incidence parallel to the c axis of plates ($\phi = 90°$); curve 3 is for the same extreme angle of incidence ($\theta = 90°$) with plane of incidence perpendicular to the c axis of plates ($\phi = 0°$).

the transmission maximum of the iso-index filter at λ_c is independent of the angle of incidence. The bandwidth, however, is expected to be different from that of the normal incidence. For the case of a Lyot– Ohman iso-index filter, the bandwidth $\Delta\lambda_{1/2}$ (FWHM) for light incident from the direction (θ, ϕ) is

$$\Delta\lambda_{1/2}(\theta, \phi) = \Delta\lambda_{1/2}\left[1 - \frac{\sin^2\theta}{2n^2}(1 - 2\cos^2\phi)\right]^{-1}, \quad (10.5\text{-}7)$$

where $\Delta\lambda_{1/2}$ is the bandwidth given by Eq. (10.5-5).

In the extreme angle of incidence $\theta = \frac{1}{2}\pi$ and $\phi = 0$ (or $\frac{1}{2}\pi$), the bandwidth is increased (or decreased) by a factor of $1/2n^2$. For the case of CdS with $\lambda_c = 5200$ Å and $n = 2.7$, the bandwidth is increased (or decreased) by 7%. In fact, the FSR is also scaled by the same factor. The calculated transmission spectrum for these extreme angles of incidence is shown in Fig. 10.12.

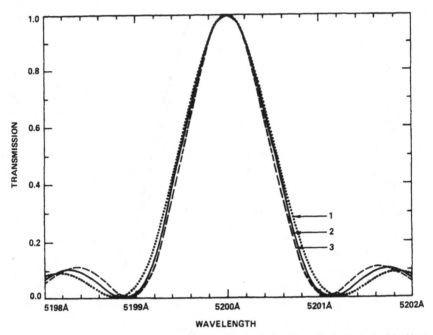

Figure 10.13 Field of view characteristics of a Solc iso-index filter. Curve 2, normal incident light; curve 1 is for extreme angle of incidence ($\theta = 90°$) with plane of incidence parallel to the averaged direction of the c axes of the plates; curve 3 for the same angle of incidence ($\theta = 90°$) with plane of incidence perpendicular to the averaged direction of the c axes of the plates.

For the case of a fan Solc filter, the c axis of each plate is oriented at a prescribed angle. Therefore, the phase retardation [Eq. (10.5-6)] for off-axis light is different from each plate in a fan Solc filter. The net result is that the broadening (or narrowing) in the spectral bandwidth is expected to be smaller than that of the Lyot–Ohman iso-index filter. Calculated results on the transmission characteristics for some cases of extreme angle of incidence are shown in Fig. 10.13. The FSR of this filter at nonnormal angles of incidence also increased (or decreased) by the same factor. Numerical results shown in Fig. 10.13 indicate that the broadening and narrowing are indeed smaller than that of the Lyot–Ohman iso-index filter shown in Fig. 10.12. No appreciable broadenings are shown in the results of calculations for the cases when the plane of incidence is parallel or perpendicular to the polarizer axis, which is 45° with respect to the average direction of the c axes of the plates. This corresponds to $\phi = 45°$ or $\phi = 135°$ for the case of a Lyot–Ohman filter when the phase retardation is independent of the angle of incidence θ.

10.5.1 Other Iso-Index Filters

Spectral filters can also be built by using the coupling of ordinary and extraordinary waves at the isotropic point.

Consider now the propagation of electromagnetic radiation in iso-index crystals. At the isotropic wavelength, ordinary and extraordinary modes in the crystals have exactly the same phase velocity. Therefore, a complete energy exchange between the two modes is possible, provided a coupling mechanism exists. The coupling may be induced by an external magnetic field, stress, or electric field. Since the energy exchange only happens at wavelengths near the isotropic point, it can be used to build a narrow-band filter with a passband at the iso-index wavelength. In 1966, C. H. Henry first proposed and demonstrated an optical filter based on the energy exchange at the iso-index point in CdS crystals with coupling induced by an appropriately applied magnetic field or stress [36]. Lotspeich [37] has proposed and demonstrated such a filter using electro-optic mode coupling in a $AgGaS_2$ crystal; the principle of operation and characteristics of this type of filter are described in Refs. 36–40.

Coupling of the orthogonally polarized ordinary and extraordinary modes of propagation may also be caused by the natural optical activity. The gyrotropic iso-index filter that uses the accidental isotropy and natural gyrotropy of certain crystals was first suggested by Hobden [23, 25] and demonstrated by Solovyov and Rudakov [20]. Recently, a narrow-band tunable optical filter made from a $CdGa_2S_4$ single crystal was demonstrated. A general analysis of the gyrotropic iso-index filter is found in Ref. 42.

10.6 LIGHT PROPAGATION IN TWISTED ANISOTROPIC MEDIA

In this section, the propagation of elecromagnetic radiation through a slowly twisting anisotropic medium is described by the Jones matrix method. The transmission of light through a twisted nematic liquid crystal is a typical example. This situation is similar to a fan-type Solc filter structure with the number of plates, N, tending to infinity and the plate thickness tending to zero as $1/N$. In fact, we will subdivide the twisted anisotropic medium into N plates and assume that each plate is a wave plate with a phase retardation and an azimuth angle. The overall Jones matrix can then be obtained by multiplying all the matrices associated with each plate.

We will limit ourselves to the case when the twisting is linear and the azimuth angle of the c axis is

$$\psi(z) \; = \; \alpha z, \qquad\qquad (10.6\text{-}1)$$

where z is the distance in the direction of propagation and α is a constant.

Let Γ be the phase retardation of the plate when it is untwisted. In particular, for the case of nematic liquid crystal with the c axis parallel to the plate surfaces, Γ is given by

$$\Gamma = \frac{2\pi}{\lambda}(n_e - n_o)d, \tag{10.6-2}$$

where d is the thickness of the plate. The total twist angle is

$$\phi \equiv \psi(d) = \alpha d. \tag{10.6-3}$$

To derive the Jones matrix for such a structure, we need to divide this plate into N equally thick plates. Each plate has a phase retardation of Γ/N. The plates are oriented at azimuth angles $\varrho, 2\varrho, 3\varrho, \ldots, (N - 1)\varrho, N\varrho$ with $\varrho = \phi/N$. The overall Jones matrix for these N plates is given by

$$M = \prod_{m=1}^{N} R(m\varrho)W_0 R(-m\varrho). \tag{10.6-4}$$

It is important to remember that the preceding matrix product, $m = 1$ appears on the right-hand side. Following procedure (10.2-25), this matrix can be written as

$$M = R(\phi)[W_0 R(-\phi/N)]^N, \tag{10.6-5}$$

where

$$W_0 = \begin{pmatrix} e^{-i\Gamma/2N} & 0 \\ 0 & e^{i\Gamma/2N} \end{pmatrix}. \tag{10.6-6}$$

Using Eqs. (9.3-12) and (10.6-6), we obtain

$$M = R(\phi)\begin{pmatrix} \cos\frac{\phi}{N}e^{-i\Gamma/2N} & \sin\frac{\phi}{N}e^{-i\Gamma/2N} \\ -\sin\frac{\phi}{N}e^{i\Gamma/2N} & \cos\frac{\phi}{N}e^{i\Gamma/2N} \end{pmatrix}^N. \tag{10.6-7}$$

Equation (10.6-7) can be further simplified by using the Chebyshev identity (10.2-4). In the limit when $N \to \infty$, the result is given by (see Problem 10.7)

$$M = R(\phi)\begin{pmatrix} \cos\chi - i\frac{\Gamma}{2}\frac{\sin\chi}{\chi} & \phi\frac{\sin\chi}{\chi} \\ -\phi\frac{\sin\chi}{\chi} & \cos\chi + i\frac{\Gamma}{2}\frac{\sin\chi}{\chi} \end{pmatrix}, \tag{10.6-8}$$

where

$$\chi = \sqrt{\phi^2 + (\Gamma/2)^2} \qquad (10.6\text{-}9)$$

Here, we have an exact expression for the Jones matrix of a linearly twisted anisotropic plate.

Let E be the initial polarization state. The polarization state E' after exiting the plate can be written as

$$E' = ME. \qquad (10.6\text{-}10)$$

10.6.1 Adiabatic Following

It often happens, especially in twisted nematic liquid crystals, that the phase retardation Γ is much larger than the twist angle ϕ. For example, consider a liquid crystal layer 25 μm thick with a twist angle of $\frac{1}{2}\pi$. The birefringence of the liquid crystal is typically $(n_e - n_o) = 0.1$. For wavelength $\lambda = 0.5\,\mu$m, $\Gamma/\phi = 20$. This number can be even bigger if the layer is thicker. If we assume $\Gamma \gg \phi$, the Jones matrix equation (10.6-8) becomes

$$M = R(\phi)\begin{pmatrix} e^{-i\Gamma 2} & 0 \\ 0 & e^{i\Gamma 2} \end{pmatrix}. \qquad (10.6\text{-}11)$$

If the incident light is linearly polarized either parallel or perpendicular to the c axis at the entrance plane, then according to Eq. (10.6-11), the light will remain linearly polarized either parallel or perpendicular to the "local" c axis. In a sense, the polarization vector follows the rotation of the local axes, provided the polarization vector is along one of the axes. The operation of the Jones matrix on any polarization vector can be divided into two parts. First, a phase retardation matrix operates on the Jones vector of the incident wave. For light linearly polarized along one of the principal axes, this phase retardation matrix only gives a phase shift to the light beam and leaves the polarization state unchanged. Second, the operation of $R(\phi)$ is to rotate the Jones vector by an angle ϕ. For linearly polarized light, this rotation makes the polarization vector parallel to the principal axes at the exit face of the plate. Thus, if the incident light is polarized along the direction of the normal modes at the input plane ($z = 0$), the polarization vector of the light wave will follow the rotation of the principal axes and remain parallel (or perpendicular) to the c axis provided the twist rate is small. This is called *adiabatic following*. This phenomenon has an important application in liquid crystal light valves. The principal of operation of these light valves is discussed in what follows.

We consider the case of a twisted nematic liquid crystal with a one-fourth turn ($\phi = \frac{1}{2}\pi$). If this liquid crystal layer is placed between a pair of parallel polarizers with their transmission axes (x) parallel to the c axis of the liquid crystal at the entrance plane ($z = 0$), the Jones vector representation of the wave immediately after transmitting through the first polarizer is

$$\begin{pmatrix} 1 \\ 0 \end{pmatrix}. \tag{10.6-12}$$

The rotation matrix $R(\phi)$ in Eq. (10.6-8) is

$$R(\tfrac{1}{2}\pi) = \begin{pmatrix} 0 & -1 \\ 1 & 0 \end{pmatrix}. \tag{10.6-13}$$

After transmitting through the liquid crystal, the polarization state of the beam [according to Eqs. (10.6-8), (10.6-12), and (10.6-13)] becomes

$$E' = \begin{pmatrix} \phi\, \dfrac{\sin\chi}{\chi} \\ \cos\chi - i\,\dfrac{\Gamma}{2}\dfrac{\sin\chi}{\chi} \end{pmatrix}. \tag{10.6-14}$$

The y component will be blocked by the second polarizer. The transmittance of the whole structure is thus given by

$$T = \frac{\sin^2[(\pi/2)\sqrt{1 + (\Gamma/\pi)^2}]}{1 + (\Gamma/\pi)^2}, \tag{10.6-15}$$

where we used Eq. (10.6-9) for χ and $\phi = \tfrac{1}{2}\pi$.

For a crystal layer with sufficient thickness, the phase retardation is much larger than π (i.e., $\Gamma \gg \pi$). The transmittance is virtually zero, according to Eq. (10.6-15). This is a result of adiabatic following. The polarization vector follows the rotation of the axes and therefore is rotated by an angle $\phi = \tfrac{1}{2}\pi$ equal to the twist angle. Since this direction is orthogonal to the transmission axis of the analyzer, the transmission is zero.

In many twisted liquid crystals, the c axis can be forced to align along a given direction by the application of an electric field (or stress). For these crystals, the application of an electric field along the z direction will destroy this twisted structure of liquid crystals. This leads to $\Gamma = 0$ and the total transmission of light, according to Eq. (10.6-14). When the electric field is removed, the liquid crystal restores its twisted structure and leads to the extinction of light. This is the basic principle of liquid crystal light valves.

REFERENCES

1. B. Lyot, Optical apparatus with wide field using interference of polarized light, *C. R. Acad. Sci., Paris* **197**, 1593 (1933).

2. B. Lyot, Filter monochromatique polarisane et ses applications en physique solaire, *Ann. Astrophys.* **7**, 31 (1944).

3. Y. Ohman, A new monochromator, *Nature* **41**, 157, 291 (1938).

4. Y. Ohman, On some new birefringent filter for solar research, *Ark. Astron.* **2**, 165 (1958).

5. I. Solc, *Czechslov. Cosopis pro Fysiku* **3**, 366 (1953); **4**, 607, 669 (1954); **5**, 114 (1955).

6. I. Solc, Birefringent chain filters, *J. Opt. Soc. Am.* **55**, 621 (1965).

7. P. Yeh, Dispersive birefringent filters, *Opt. Commun.* **37**, 153 (1981).

8. H. Gobrecht and A. Bartschat, *Z. fur Physik* **156**, 131 (1959).

9. M. V. Hodben, *Acta Cryst.* **A24**, 676 (1968).

10. V. Chandrasekharan and H. Damany, *Appl. Opt.* **8**, 671 (1969).

11. J. J. Hopfield and D. G. Thomas, *Phys. Rev. Lett.* **15**, 22 (1965).

12. Y. S. Park, *J. Appl. Phys.* **39**, 3049 (1968).

13. T. M. Bieniewski and S. J. Czyzak, *J. Opt. Soc. Am.* **53**, 496 (1963).

14. S. A. Abagyan, G. A. Ivanov, Y. A. Lomov, and Y. E. Shanurin, *Sov. Phys. Semicond.* **4**, 2049 (1971).

15. S. A. Abagyan, G. A. Ivanov, E. V. Markov, G. A. Koroleva, and N. N. Pogorelova, *Sov. Phys. Semicond.* **5**, 1751 (1972).

16. M. P. Lisitsa, V. N. Malinko, and S. F. Terekhova, *Sov. Phys. Semicond.* **6**, 798 (1972).

17. A. A. Reza, G. A. Babonas, and A. Y. Shileika, *Sov. Phys. Semicond.* **9**, 986 (1976).

18. I. V. Baranets, A. K. Zilbershtein, and L. E. Solovev, *Opt. Spectrosc.* **37**, 164 (1974).

19. R. B. Parsons, W. Wardzynski, and A. D. Yoffe, *Proc. Roy. Soc. London Ser. A* **262**, 120 (1961).

20. L. E. Solovyov and V. S. Rudakov, *Vestn. Leningr. Gos. Univ.* **4**, 23 (1968).

21. B. Thiel and R. Helbig, *J. Crystal Growth* **32**, 259 (1976).

22. D. S. Chemla, P. Kupecek, D. S. Robertson, and R. C. Smity, *Opt. Commun.* **3**, 29 (1971).

23. G. D. Boyd, H. Kasper, and J. H. McFee, *IEEE J. Quant. Electron.* **QE-7**, 563 (1971).

24. C. Schwartz, D. S. Chemla, and B. Ayrault, *Opt. Commun.* **5**, 244 (1972).

25. M. V. Hobden, *Acta Crystallogr. Sec. A* **25**, 633 (1969).

26. A. K. Zilbershtein and L. E. Solovyov, *Opt. Spectrosc.* **35**, 274 (1973).

27. L. M. Suslikov, Z. P. Gadmaski, I. F. Kopinets, E. Yu. Peresh, and V. Yu Slivka, *Opt. Spectrosc.* **49,** 51 (1981).

28. J. L. Shay and J. H. Wernick, *Ternary Chalcopyrite Semiconductors*, Pergamon, New York, 1975, Appendix.

29. M. P. Lisitsa, L. F. Gudymenko, V. N. Malinko, and S. F. Terekhova, *Phys. Stat. Sol.* **31,** 389b (1969).

30. M. P. Lisitsa, V. N. Malinko, and S. F. Terekhova, *Sov. Phys. Semicond.* **5,** 1318 (1972).

31. A. K. Zilbershtein, L. E. Solovev, I. G. Lyapichev, and A. G. Sysoev, *Sov. Phys. Semicond.* **12,** 236 (1977).

32. J. P. Laurnenti, K. C. Rustagi, and M. Rouzeyre, *Appl. Phys. Lett.* **28,** 212 (1976).

33. J. P. Laurenti, K. Rustagi, M. Rouzeyre, H. Ruber, and W. Ruppel, *J. Appl. Phys.* **48,** 203 (1977).

34. P. Yeh, Zero crossing birefringent filters, *Opt. Comm.* **35,** 15 (1980).

35. P. Yeh, Recent advances in polarization interference filters, *Proc. Int. Conf. Las.* 591 (STS Press, New Orleans, 1980).

36. C. H. Henry, *Phys. Rev.* **143,** 627 (1966).

37. J. F. Lotspeich, *IEEE J. Quant. Electron.* **QE-15,** 904 (1979).

38. J. P. Laurenti, P. Vagot, and M. Rouzeyre, *Rev. Phys. Appl.* **12** (11), 1755 (1977).

39. J. P. Laurenti and M. Rouzeyre, *J. Appl. Phys.* **52,** 6484 (1981).

40. J. F. Lotspeich and A. Yariv, Optical activity effects in $AgGaS_2$ in iso-index electro-optic filters, *J. Opt. Soc. Am.* **71,** 1629 (1981).

41. V. V. Badikov, I. N. Matveev, S. M. Pshenichchnikov, O. V. Rychik, N. K. Trotsenko, and N. D. Ustinov, *Sov. J. Quant. Electron.* **11,** 548 (1981).

42. P. Yeh, Gyrotropic iso-index filters, *Appl. Opt.* **21,** 4054 (1982).

43. P. Yeh, Extended Jones matrix method, *J. Opt. Soc. Am.* **72,** 507 (1982).

PROBLEMS

10.1   Use the following integral form:

$$\int_0^{\pi/2} \left(\frac{\sin ax}{\sin x} \right)^2 dx \;=\; \tfrac{1}{2}\pi a.$$

(a) Show that the area of the whole curve in Fig. 10.3, including the main peak and all the sidelobes, is $2^{-N}\pi$, that is,

$$\int_0^{\pi} T(\Gamma) \, d\Gamma \;=\; \pi 2^{-N}.$$

(b) Show that, according to Eq. (10.1-7), the transmittance drops to half maximum at

$$\Gamma = 2m\pi + 2^{-N} \times 0.886\pi.$$

(c) According to (b), we may assume that the area under the main peak is approximately $2^{-N}0.886\pi$. Show that the total area of the sidelobes is only 11.4% of the total transmission.

10.2 *Sinc function.* The transmission of light through a Solc filter is described by Eq. (10.2-18), which is known as *the sinc function.* Such a function also describes the transmission of light through acousto-optic tunable filters. Use the following integral formula:

$$\int_0^\infty \frac{\sin^2[(\alpha\pi/2)\sqrt{1 + x^2}]}{1 + x^2}\, dx = \frac{\pi}{4} \int_0^{\alpha\pi} J_0(\xi)\, d\xi,$$

where J_0 is the zero-order Bessel function.

(a) Show that the total area under the transmission curve of a Solc filter ($\alpha = 1$), as shown in Fig. 10.5, is

$$\frac{\pi}{4} \int_0^\pi J_0(\xi)\, d\xi = 1.059.$$

(b) Show that, for $\alpha = 1$, half-maximum transmission occurs at $x = 0.80$.

(c) Show that the total area of the sidelobes is approximately 24% of the total area.

(d) The transmittance of the maxima is given by

$$T = \frac{1}{(2l + 1)^2}, \quad l = 0, 1, 2, \ldots .$$

Using the formula

$$\sum_{l=0}^\infty \frac{1}{(2l + 1)^2} = \tfrac{1}{8}\pi^2 = 1.234,$$

show that the area under all the sidelobes is approximately 19% of the total area. The discrepancy between (c) and (d) lies in the poor approximation of the area of the main peak.

10.3 *Achromatic half-wave plate.* An achromatic half-wave plate, by definition, is a crystal plate whose phase retardation is π for all wavelengths in the spectral regime of interest.

(a) Show that, for an achromatic half-wave plate,

$$\frac{\partial \Gamma}{\partial \lambda} = 0.$$

(b) Show that, for an achromatic half-wave plate,

$$\frac{\Delta n}{\lambda} = \frac{\partial \, \Delta n}{\partial \lambda},$$

where $\Delta n = n_e - n_o$.

(c) Show that a half-wave plate that is achromatic in the visible spectrum from 390 to 780 nm must have a birefringence that increases by a factor of 2 from red to violet. Find the best candidate material from Fig. 10.10.

10.4 *Hybrid wide-field element.* Without using an achromatic half-wave plate, try to design a wide-field element by using two materials with different birefringences. Let d_1, d_2 be the thickness of the crystal plates and let Δn_1, Δn_2 be the corresponding birefringences.

(a) Show that, for normal incidence, the total phase retardation is

$$\Gamma = \frac{2\pi}{\lambda} \Delta n_1 \, d_1 \pm \frac{2\pi}{\lambda} \Delta n_2 \, d_2,$$

where the plus sign is for parallel c axes and the minus is for crossed c axes configuration.

(b) Assuming that the birefringences are small so that Eq. (10.3-14) can be used for the angular dependence of a plate, show that a wide-field element is possible provided that the following condition is satisfied:

$$\frac{\Delta n_1}{n_1^2} d_1 + \frac{\Delta n_2}{n_2^2} d_2 = 0,$$

where n_1 and n_2 are the average indices of refraction of the materials.

(c) According to (b), a wide-field element is possible only when the two materials have opposite sign of birefringence. Using Table 10.3, find a few candidate material pairs.

(d) Ignoring dispersion, show that the bandwidth of such a hybrid wave plate is

$$\Delta\lambda_{1/2} = \frac{\lambda^2}{2|d_1 \, \Delta n_1 \pm d_2 \, \Delta n_2|}.$$

(e) Design a wide-field element by using quartz and sapphire crystals, which provide a bandwidth of 1 Å at $\lambda = 0.5 \, \mu m$. Find the plate thickness.

10.5 *A LiNbO$_3$ Lyot–Ohman filter.* Consider a simple Lyot–Ohman filter that consists of four LiNbO$_3$ crystal plates with thicknesses 1, 2, 4, and 8 cm. Use $n_e = 2.245$ and $n_o = 2.344$.

(a) Find the bandwidth of such a filter at $\lambda = 0.5 \, \mu m$. *Answer:* 0.14 Å.

(b) Find the free spectral range (FSR) of such a filter at $\lambda = 0.5 \, \mu m$. *Answer:* 2.5 Å.

(c) Find the field of view of such a filter. *Answer:* $\pm 0.2°$.

10.6 *Sapphire Lyot–Ohman filter.* Design a Lyot–Ohman filter that has a passband at $\lambda = 0.5 \, \mu m$ with a bandwidth of $\Delta\lambda_{1/2} = 1$ Å and a FSR of at least 100 Å.

(a) Find the number of stages needed. *Answer:* 7.

(b) Using sapphire as the crystal plates, find the plate thicknesses of all the plates. Also find the field of view of such a filter (use Table 10.3 for the refractive indices).

(c) Using the wide-field element described in Fig. 10.8, find the field of view of such a filter.

(d) Repeat (b) and (c) by using ZnS as the crystal plates.

11

Guided Waves in Layered Media

So far, we have discussed various properties of layered media, including reflection and transmission properties. In addition to these properties, layered media can also support confined electromagnetic propagation. These modes of propagation are the so-called guided waves (or guided modes), and the structures that support guided waves are called *waveguides*. In this chapter, we discuss the propagation of guided waves in layered media. As we know, any light beam with a finite transverse dimension will diverge as it propagates in a homogeneous medium. This divergence disappears in guiding dielectric structures under appropriate conditions. The transverse dimension of these modes of propagation is determined by the dielectric waveguide.

We shall derive first the properties of guided modes in a slab dielectric structure. Optical modes are presented as the solution of the eigenvalue equation, which is derived from Maxwell's equations subject to the boundary conditions imposed by waveguide geometry. Both TE and TM modes of propagation are derived. The physics of confined propagation is explained in terms of the total internal reflection of plane waves from the dielectric interfaces. In addition, modes of propagation in multilayer waveguides are also derived by using the matrix formulation.

We then discuss surface plasmons and Bloch guided modes. In the last part of the chapter, we present a perturbation theory that is very useful for the study of mode couplings. The theory is then applied to coupling between two dielectric waveguides. We shall also discuss the simple theory of effective index, which is useful in understanding waveguiding in two-dimensional structures.

11.1 SYMMETRIC SLAB WAVEGUIDES

Dielectric slabs are the simplest optical waveguides. Figure 11.1 shows a typical example of a slab waveguide. It consists of a thin dielectric layer (called the *guiding layer*, or simply the *core*) sandwiched between two semi-infinite bounding media. Generally speaking, the index of refraction of the guiding layer must be greater than those of the surrounding media. In

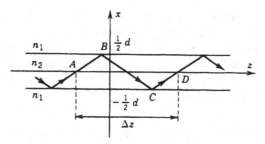

Figure 11.1 Schematic drawing of a slab waveguide.

addition, the thickness of the guiding layer is typically on the order of a wavelength. In symmetric slab waveguides, the two bounding media are identical. The simplest example will be a thin glass film (or layer) immersed in air or another fluid of a low index of refraction.

The following equation describes the index profile of a symmetric dielectric waveguide:

$$n(x) = \begin{cases} n_2, & |x| < d/2, \\ n_1, & \text{otherwise,} \end{cases} \qquad (11.1\text{-}1)$$

where d is the thickness of the guiding layer (core), n_2 is the index of refraction of the core, and n_1 is the index of refraction of the bounding media. To support guided modes, n_2 must be greater than n_1. The problem at hand is to find these guided modes.

The electromagnetic treatment of such a problem is relatively easy because the medium is homogeneous in each segment of the dielectric structure. In addition, the solutions of Maxwell's equations in homogeneous media are simply plane waves. Thus, all we need to do is to write the plane-wave solutions for each segment and then match the boundary conditions at the interfaces. We now consider the propagation of monochromatic radiation along the z axis. Maxwell's equation can be written in the form

$$\nabla \times \mathbf{H} = i\omega\varepsilon_0 n^2 \mathbf{E}, \qquad \nabla \times \mathbf{E} = -i\omega\mu\mathbf{H}, \qquad (11.1\text{-}2)$$

where n is the refractive index profile given in Eq. (11.1-1). Since the whole structure is homogeneous along the z axis, solutions to the wave equations (11.1-2) can be taken as

$$\mathbf{E}(x, t) = \mathbf{E}_m(x)\exp[i(\omega t - \beta z)],$$
$$\mathbf{H}(x, t) = \mathbf{H}_m(x)\exp[i(\omega t - \beta z)], \qquad (11.1\text{-}3)$$

where β is the z component of the wave vectors and is known as the *propagation constant* to be determined from Maxwell's equations and $\mathbf{E}_m(x)$ and $\mathbf{H}_m(x)$ are wavefunctions of the guided modes, the subscript m being an integer called the *mode number*. For layered dielectric structures that consist of homogeneous and isotropic materials, the wave equation can be obtained by eliminating $\mathbf{H}$ from Eq. (11.1-2):

$$\left[\frac{d^2}{dx^2} + \left(\frac{\omega}{c}n\right)^2 - \beta^2\right]\mathbf{E}_m(x) = 0. \tag{11.1-4}$$

Note that the preceding equation is not valid at the dielectric interfaces for TM waves (the $\mathbf{H}$ vector is transverse to the xz plane). We will solve for this equation separately in each segment of the structure and then match the tangential components of the field vectors at each interface. In addition to the continuity conditions at the interfaces, another important boundary condition for guided modes is that the field amplitudes be zero at infinity. A discussion of the general properties of the solutions of Eq. (11.1-4) will be given in Section 11.6.

We now proceed with the solution of Eq. (11.1-4) in each segment of the dielectric structure. The propagation constant is a very important parameter because it determines whether the field varies sinusoidally or exponentially. For confined modes, the field amplitude must fall off exponentially outside the guide structure. Consequently, the quantity $(n\omega/c)^2 - \beta^2$ must be negative for $|x| > \frac{1}{2}d$. In other words, the propagation constant β of a confined mode must be such that

$$\beta > \frac{n_1\omega}{c}, \tag{11.1-5}$$

where we recall that n_1 is the index of refraction of the bounding media. On the other hand, the continuity of the field requires that the magnitude of the field $E_m(x)$ attain a maximum value. The existence of a maximum requires that the Laplacian of the field be negative. In other words, the propagation constant of a confined mode must be such that

$$\beta < \frac{n_2\omega}{c}. \tag{11.1-6}$$

Thus, we will find confined modes whose propagation constant satisfies these conditions, Eqs. (11.1-5) and (11.1-6).

The modes can also be classified as either TE or TM modes. The TE modes have their electric field perpendicular to the xz plane (plane of incidence) and thus have only the field components E_y, H_x, and H_z. The TM modes have the field components H_y, E_x, and E_z.

11.1.1 Guided TE Modes

The electric field amplitude of the guided TE modes can be written in the form

$$E_y(x, z, t) = E_m(x) \exp[i(\omega t - \beta z)]. \qquad (11.1\text{-}7)$$

In a manner very similar to the wave function of a particle in a square-well potential, the mode function $E_m(x)$ is taken as

$$E_m(x) = \begin{cases} A \sin hx + B \cos hx, & |x| < \tfrac{1}{2}d, \\ C \exp(-qx), & \tfrac{1}{2}d < x, \\ D \exp(qx), & x < -\tfrac{1}{2}d, \end{cases} \qquad (11.1\text{-}8)$$

where A, B, C, and D are constants, and the parameters h and q are related to the propagation constant by

$$h = \left[\left(\frac{n_2\omega}{c}\right)^2 - \beta^2\right]^{1/2},$$

$$q = \left[\beta^2 - \left(\frac{n_1\omega}{c}\right)^2\right]^{1/2}. \qquad (11.1\text{-}9)$$

The parameter h may be considered as the transverse component of the wave vector in the guide layer. To be acceptable solutions, the tangential component of the electric and magnetic fields E_y, H_z must be continuous at the interfaces. Since $H_z = (i/\omega\mu)(\partial E_y/\partial x)$, we must match the magnitude as well as the slope of the mode functions $E_m(x)$ at the interfaces. This leads to

$$A \sin\left(\tfrac{1}{2}hd\right) + B \cos\left(\tfrac{1}{2}hd\right) = C \exp\left(-\tfrac{1}{2}qd\right),$$

$$hA \cos\left(\tfrac{1}{2}hd\right) - hB \sin\left(\tfrac{1}{2}hd\right) = -qC \exp\left(-\tfrac{1}{2}qd\right),$$

$$-A \sin\left(\tfrac{1}{2}hd\right) + B \cos\left(\tfrac{1}{2}hd\right) = D \exp\left(-\tfrac{1}{2}qd\right),$$

$$hA \cos\left(\tfrac{1}{2}hd\right) + hB \sin\left(\tfrac{1}{2}hd\right) = qD \exp\left(-\tfrac{1}{2}qd\right),$$

from which we obtain

$$2A \sin\left(\tfrac{1}{2}hd\right) = (C - D) \exp\left(-\tfrac{1}{2}qd\right), \qquad (11.1\text{-}10)$$

$$2hA \cos \left(\tfrac{1}{2}hd\right) = -q(C - D) \exp \left(-\tfrac{1}{2}qd\right), \qquad (11.1\text{-}11)$$

$$2B \cos \left(\tfrac{1}{2}hd\right) = (C + D) \exp \left(-\tfrac{1}{2}qd\right), \qquad (11.1\text{-}12)$$

$$2hB \sin \left(\tfrac{1}{2}hd\right) = q(C + D) \exp \left(-\tfrac{1}{2}qd\right). \qquad (11.1\text{-}13)$$

Unless $A = 0$ and $C = D$, Eqs. (11.1-10) and (11.1-11) yield

$$h \cot \left(\tfrac{1}{2}hd\right) = -q. \qquad (11.1\text{-}14)$$

Similarly, unless $B = 0$ and $C = -D$, Eqs. (11.1-12) and (11.1-13) give

$$h \tan \left(\tfrac{1}{2}hd\right) = q. \qquad (11.1\text{-}15)$$

Notice that both Eqs. (11.1-14) and (11.1-15) cannot be satisfied simultaneously since the elimination of q would lead to a pure imaginary h and a negative q. However, these two equations can be combined into a single equation:

$$\tan (hd) = \frac{2hq}{h^2 - q^2} . \qquad (11.1\text{-}16)$$

The solutions of TE modes may thus be divided into two classes: for the first class

$$A = 0, \qquad C = D, \qquad h \tan \left(\tfrac{1}{2}hd\right) = q, \qquad (11.1\text{-}17)$$

and for the second class

$$B = 0, \qquad C = -D, \qquad h \cot \left(\tfrac{1}{2}hd\right) = -q. \qquad (11.1\text{-}18)$$

Notice that the solutions in the first class have symmetric wavefunctions, whereas those of the second class have anitsymmetric wavefunctions.

The propagation constants of the TE modes are found from a numerical or graphical solution of Eqs. (11.1-17) and (11.1-18), with the definition of h and q given by Eq. (11.1-9). A very simple and well-known graphic solution is described here, since it clearly shows the way in which the number of TE modes depend on both the thickness d and the difference of indices of refraction. By putting $u = \tfrac{1}{2}hd$ and $v = \tfrac{1}{2}qd$, Eq. (11.1-17) becomes $u \tan u = v$, with

$$u^2 + v^2 = (n_2^2 - n_1^2) \left(\frac{\omega d}{2c}\right)^2$$

$$= (n_2^2 - n_1^2) \left(\frac{\pi d}{\lambda}\right)^2 \equiv V^2. \qquad (11.1\text{-}19)$$

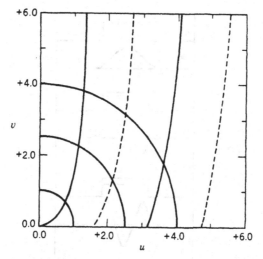

Figure 11.2 Graphic solution of Eqs. (11.1-17) and (11.1-18) for three values of V; solid curves are $v = u \tan u$, and the dashed curves are $v = -u \cot u$.

Since u and v are restricted to positive values, the propagation constants may be found in this case from the intersection of both the curve $v = u \tan u$ and a circle of known radius $V = (n_2^2 - n_1^2)^{1/2} (\pi d/\lambda)$ in the first quadrant of the uv plane. A similar graphic construction for the solution of Eq. (11.1-18) can be obtained by plotting $v = -u \cot u$ and the circle on the uv plane. Figure 11.2 shows such a graphic method for three values of V. For $V = 1$, there is only one solution. There are two solutions when V is 2.5 and three solutions when V is 4. Notice that the number of solutions depends on the value of V.

From Fig. 11.2, it is clear that the number of confined TE modes depends on the magnitude of the parameter V. For V between zero and $\frac{1}{2}\pi$, there is just one TE mode of the first class. The first mode of the second class appears when the parameter V is greater than $\frac{1}{2}\pi$. As this parameter V increases, confined modes appear successively, first of one class and then of the other. Figure 11.3 plots the wavefunctions of a symmetrical slab waveguide with $n_2 = 2.5$, $n_1 = 1.5$, and $d = \lambda$. According to Eq. (11.1-19), the parameter V is 2π. This waveguide supports four TE modes. It is not difficult to see from Fig. 11.3 that, when ordered according to the propagation constant β, the mth wavefunction has $m - 1$ nodes. We also notice that the wavefunctions are either symmetric or antisymmetric with respect to the origin $x = 0$. It follows from the discussion earlier that the wavefunctions are divided into

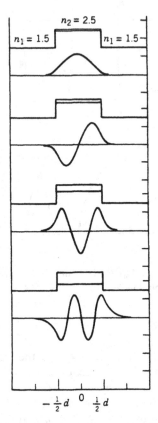

Figure 11.3 Wavefunctions of a symmetrical slab waveguide with $n_2 = 2.5$ and $n_1 = 1.5$. The thickness of the core is equal to the wavelength. The normalized propagation constants are also shown as the straight line under the index profile.

two classes [see Eqs. (11.1-17) and (11.1-18)]. This division is a direct consequence of the fact that the index profile $n(x)$ is symmetric about $x = 0$.

Knowledge that the solution possesses a definite symmetry sometimes simplifies the determination of the propagation constant, since we need only find the solution for positive x. Even solutions have zero slope and odd solutions have zero value at the origin $x = 0$. Thus, the wavefunction of the even solutions can be written as $\cos hx$, whereas those of the odd solutions can be written as $\sin hx$. Both types of solutions decay exponentially in the region $\frac{1}{2}d < |x|$. The solutions are then obtained by matching the value and the slope at $|x| = \frac{1}{2}d$.

For the purpose of describing and comparing the confined modes, it is convenient to define the normalized propagation constant as

$$\bar{\beta} = \frac{\beta}{\omega/c}.$$ (11.1-20)

Such a normalized propagation is often called *the effective index of refraction of the mode* n_{eff} and is related to the phase velocity of the mode:

$$\bar{\beta} = n_{\text{eff}} = \frac{c}{v_p},$$ (11.1-21)

where v_p is the phase velocity of the mode $v_p = \omega/\beta$. Thus, for confined modes the normalized propagation constant $\bar{\beta}$ or the effective index n_{eff} is between n_2 and n_1.

11.1.2 Guided TM Modes

The derivation of the confined TM modes is similar in principle to that of the TE modes. The field amplitudes are written

$$H_y(x, z, t) = H_m(x) \exp [i(\omega t - \beta z)],$$

$$E_x(x, z, t) = \frac{i}{\omega \varepsilon} \frac{\partial}{\partial z} H_y,$$ (11.1-22)

$$E_z(x, z, t) = -\frac{i}{\omega \varepsilon} \frac{\partial}{\partial x} H_y.$$

The wavefunction $H_m(x)$ is

$$H_m(x) = \begin{cases} A \sin hx + B \cos hx, & |x| < \tfrac{1}{2}d, \\ C \exp (-qx), & \tfrac{1}{2}d < x, \\ D \exp (qx), & x < -\tfrac{1}{2}d, \end{cases}$$ (11.1-23)

where A, B, C, and D are constants, and the parameters h and q are given by Eq. (11.1-9).

The continuity of H_y and E_z at the two interfaces $x = \pm \tfrac{1}{2}d$ leads, in a manner similar to Eqs. (11.1-14) and (11.1-15), to the following eigenvalue equation:

$$h \tan (\tfrac{1}{2}hd) = \frac{n_2^2}{n_1^2} q \quad \text{for even solutions,}$$

$$h \cot (\tfrac{1}{2}hd) = -\frac{n_2^2}{n_1^2} q \quad \text{for odd solutions.}$$ (11.1-24)

These two equations can also be combined into a single equation,

$$\tan(hd) = \frac{2h\bar{q}}{h^2 - \bar{q}^2}, \qquad (11.1\text{-}25)$$

where

$$\bar{q} = \frac{n_2^2}{n_1^2} q. \qquad (11.1\text{-}26)$$

Equation (11.1-24) can also be solved by using the graphic method described earlier.

11.1.3   Geometric Optics Treatment

In the preceding, we solved the wave equation for the TE and TM modes of a symmetric slab waveguide. These modes can also be derived by using geometric optics. This is possible because the waveguide consists of layers of homogeneous dielectric materials. Wave propagation in each of the homogeneous regions can be represented by a superposition of two plane waves. One of the plane waves may be considered as the incident wave, whereas the other may be viewed as the reflected one. According to Eqs. (11.1-5) and (11.1-6), the plane waves that represent the confined modes experience total internal reflection at both interfaces $|x| = \frac{1}{2}d$.

It seems that the condition of total internal reflection at both interfaces were enough to assure the confinement of energy in the guiding layer. However, it turns out that the total internal reflection is only a necessary condition. In other words, not all rays trapped by internal reflection constitute a mode. A mode, by definition, must have a unique propagation constant and a well-defined field amplitude at each point in space and time. Consider the representation of a mode $E_m(x)\exp[i(\omega t - \beta z)]$ by a zigzagging plane wave of the form

$$E(x, z, t) = E_0 \exp[i(\omega t - \beta z - hx)], \qquad (11.1\text{-}27)$$

where E_0 is a constant, and h is the transverse wave number. Let the plane wave travel a distance Δz in a time Δt in one zigzag $ABCD$ (See Fig. 11.1). If we follow the ray path $ABCD$ and add the phase shifts due to propagation and total reflections at points B and C, we obtain a total phase shift of

$$\omega \, \Delta t - \beta \, \Delta z - 2hd + 2\phi, \qquad (11.1\text{-}28)$$

where ϕ is the phase shift upon reflection at one of the interfaces. A mode of the form $E_m(x)\exp[i(\omega t - \beta z)]$ propagating from point A to point D will gain a phase shift of $\omega \Delta t - \beta \, \Delta z$. Therefore, a total reflecting zigzag ray will become a mode only when the extra transverse phase shift is an integral multiple of 2π, that is,

$$-2hd + 2\phi = -2m\pi, \qquad (11.1\text{-}29)$$

where m is an integer. The minus sign preceding $2m\pi$ is chosen such that m corresponds to the TE_m and TM_m modes of the waveguide. This phase shift, 2ϕ, for TE or TM waves, according to Eqs. (3.2-5) and (3.2-6), can be expressed conveniently in terms of h and q as

$$\phi = \begin{cases} 2 \tan^{-1} \dfrac{q}{h}, & \text{TE}, \\[2ex] 2 \tan^{-1} \dfrac{n_2^2 q}{n_1^2 h}, & \text{TM}. \end{cases} \qquad (11.1\text{-}30)$$

In Eq. (11.1-30), we drop the term $-\pi$ for TM waves [see Eq. (3.2-6)] because a round trip would yield -2π, which does not affect the field amplitude. These phase shifts are limited between 0 and π. Therefore, the fundamental mode corresponds to the situation when $2hd = 2\phi$ (i.e., $m = 0$). Higher order modes involve larger hd. The integer m in (11.1-27) therefore assumes only nonnegative values, that is, $m = 0, 1, 2, 3, \ldots$. Condition (11.1-29) is equivalent to the eigenvalue equations (11.1-16) and (11.1-25).

Consider the example shown in Fig. 11.3. This waveguide has $n_1 = 1.5$, $n_2 = 2.5$, and $d = \lambda$. There are four confined TE modes in this slab waveguide. The normalized propagation constant $\beta/(\omega/c)$ for these modes are 2.463, 2.348, 2.148, and 1.850. These normalized propagation constants correspond to the ray incident angles of 80.13°, 69.92°, 59.23°, and 47.73°, in that order. Note that all these angles are greater than the critical angle 36.87°. The fundamental mode ($m = 0$) always has the largest propagation constant, and the angle of incidence with the normalized propagation constant approaches n_2. The highest order mode has the least normalized propagation constant approaching n_1 and the smallest angle of incidence approaching the critical angle. Figure 11.4 shows the zigzag rays corresponding to the modes previously discussed.

Ray optics offers a convenient way of obtaining the mode condition of a slab waveguide. In a more complicated waveguide structure that involves multilayers or inhomogeneous layers, it is very difficult to use the ray optics approach. In addition, the ray optics approach only yields the mode condition. Nothing about the field distribution and the mode orthogonality is obtained.

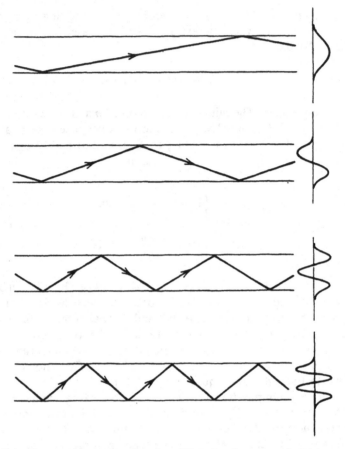

Figure 11.4   Zigzag representation of the four TE modes of a symmetrical slab waveguide.

11.2   ASYMMETRIC SLAB WAVEGUIDES

The symmetric slab waveguides described in the previous section are useful for introducing the concept of confined propagation because of its simplicity in mathematics. In practice, however, most slab waveguides are not symmetric. In fact, because most of the guiding layers are so thin that a substrate is necessary for the support of the structure. In this section, we treat the confined modes in asymmetric slab waveguides. The symmetric slab waveguides described in the last section may be viewed as a special case of the asymmetric ones.

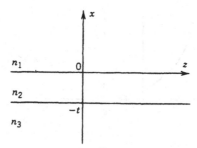

Figure 11.5 Schematic drawing of an asymmetric slab waveguide.

Referring to Fig. 11.5, we now consider the propagation of confined modes in an asymmetric slab waveguide whose index profile is given by

$$n(x) = \begin{cases} n_1, & 0 < x, \\ n_2, & -t < x < 0, \\ n_3, & x < -t, \end{cases} \qquad (11.2\text{-}1)$$

where t is the thickness of the guiding layer and the index of refraction of the guiding media, n_2, is greater than those of the bounding media, n_1 and n_3. Without loss of generality, we assume that $n_1 < n_3 < n_2$.

To be a satisfactory wavefunction, a solution to the Maxwell wave equation must be continuous, single valued, and finite throughout the space. We will first examine the physical nature of the solution as a function of the propagation constant β at a fixed frequency ω. For $\beta > n_2\omega/c$, it follows directly from Eq. (11.1-4) that $(1/E)(\partial^2 E/\partial x^2) > 0$ everywhere, and $E(x)$ is exponential at all three regions of the waveguides. If we take $E(x) = \exp(-qx)$, which decays to zero at $x = +\infty$, because of the need to match both $E(x)$ and its derivatives at the two interfaces, the resulting field distribution is infinite at $x = -\infty$, as shown in Fig. 11.6(a). Such a solution would correspond to a field of infinite energy. It is not *physically realizable*, and thus does not correspond to a real wave.

For $n_3(\omega/c) < \beta < n_2(\omega/c)$, as in Fig. 11.6(b), it follows from Eq. (11.1-4) that the solution is sinusoidal in the core ($-t < x < 0$), since $(1/E)(\partial^2 E/\partial x^2) < 0$, but is exponential in the bounding media. This makes it possible to have a solution $E(x)$ that satisfies the boundary conditions while *decaying* exponentially in the regions $x < -t$ and $0 < x$. One such solution is shown in Fig. 11.6(b). The energy carried by these modes, as represented by the Poynting vector, is confined to the vicinity of the guiding layer, and consequently, we will refer to them as *confined* or *guided modes*. In fact, only

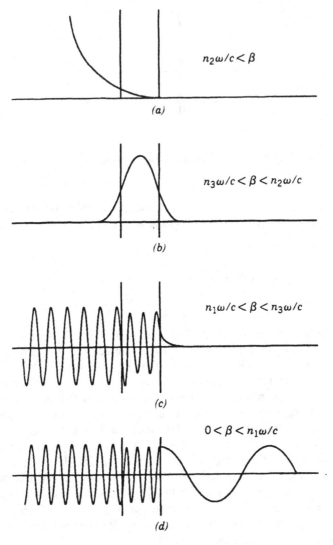

Figure 11.6 Typical field distributions corresponding to different values of β.

a small fraction of the energy is flowing outside the guiding layer. From the preceding discussion, it follows that a necessary condition for their existence is that $n_1(\omega/c), n_3(\omega/c) < \beta < n_2(\omega/c)$, so that confined modes are possible only when $n_2 > n_1, n_3$; that is, the inner layer possesses the highest index of refraction. In some basic sense, the confined modes in this regime are reminiscent of quantized states of an electron in a potential well, in which the electron is trapped by the potential well.

Mode solutions for $n_1(\omega/c) < \beta < n_3(\omega/c)$ [regime (c) in Fig. 11.6] correspond, according to Eq. (11.1-4), to exponential behavior in the region $0 < x$ and to sinusoidal behavior in the regions $x < 0$, as illustrated in Fig. 11.6(c). In this regime, almost all the energy is flowing in the substrate. We will refer to these modes as substrate radiation modes. For $0 < \beta < n_1(\omega/c)$, as in Fig. 11.6(d), the solution for $E(x)$ becomes sinusoidal in all three regions. These are the so-called radiation modes of the waveguides.

A solution of Eq. (11.1-4), subject to the boundary conditions at the interfaces given in what follows, shows that while in regimes (c) and (d) of Fig. 11.6, β is a continuous variable, the values of allowed β in the propagation regime $n_3(\omega/c) < \beta < n_2(\omega/c)$ are *discrete*. The number of confined modes depends on the thickness t, the frequency, and the indices of refraction n_1, n_2, n_3. At a given wavelength, the number of confined modes increases from zero with increasing t. At some cutoff thickness t, the mode TE_0 becomes confined. Further increases in the layer thickness t will allow TE_1 to exist as well, and so on.

We now turn our attention to the solution of the wave equation (11.1-4) for the dielectric waveguide sketched in Fig. 11.5. The derivation will be limited to the guided modes, which, according to Fig. 11.6, have a propagation constant β such that

$$n_3 \frac{\omega}{c} < \beta < n_2 \frac{\omega}{c},$$

where $n_1 < n_3$. This guide can, in the general case, support a finite number of confined TE modes with field components E_y, H_x, and H_z and TM modes with components H_y, E_x, and E_z. We will derive the mode wavefunction and the corresponding propagation constants.

11.2.1 Guided TE modes

For TE modes, the electric field vector is perpendicular to the plane of incidence (xz plane). Thus, E_y is the only component of the electric field vector. The field component E_y of the TE mode can be taken in the form

$$E_y(x, y, z, t) = E_m(x) \, e^{i(\omega t - \beta z)}, \tag{11.2-2}$$

where β is the propagation constant and $E_m(x)$ is the wavefunction of the mth mode.

The function $E_m(x)$ assumes the following forms in each of the three regions:

$$E_m(x) = \begin{cases} C \exp(-qx), & 0 \leqslant x, \\[2mm] C\left[\cos(hx) - \dfrac{q}{h}\sin(hx)\right], & -t \leqslant x \leqslant 0, \\[2mm] C\left[\cos(ht) + \dfrac{q}{h}\sin(ht)\right]\exp[p(x+t)], & x \leqslant -t, \end{cases}$$

$$(11.2\text{-}3)$$

where C is a normalization constant and h, q, and p are given by

$$h = \left[\left(\frac{n_2\omega}{c}\right)^2 - \beta^2\right]^{1/2},$$

$$q = \left[\beta^2 - \left(\frac{n_1\omega}{c}\right)^2\right]^{1/2}, \qquad (11.2\text{-}4)$$

$$p = \left[\beta^2 - \left(\frac{n_3\omega}{c}\right)^2\right]^{1/2}.$$

These relationships are obtained by substituting Eq. (11.2-3) into Eq. (11.1-4).

The boundary conditions require that E_y and $H_z = (i/\omega\mu)(\partial E_y/\partial x)$ be continuous at both $x = 0$ and $x = -t$. The wavefunction in Eq. (11.2-3) has been chosen such that E_y is continuous at both interfaces as well as $\partial E_y/\partial x$ at $x = 0$. By imposing the continuity requirements on $\partial E_y/\partial x$ at $x = -t$, we get, from Eq. (11.2-3),

$$h \sin(ht) - q\cos(ht) = p\left[\cos(ht) + \frac{q}{h}\sin(ht)\right]$$

or

$$\tan(ht) = \frac{p+q}{h(1 - pq/h^2)}. \qquad (11.2\text{-}5)$$

This is the so-called *mode condition*. The propagation constant β of a guided TE mode must satisfy this condition. Given a set of refractive indices n_1, n_2, and n_3 of a slab waveguide, Eq. (11.2-5) in general yields a finite number of solutions for β provided the thickness t is large enough. These modes are mutually orthogonal.

The normalization constant C is chosen so that the field $E_m(x)$ in Eq. (11.2-3) corresponds to a power flow of 1 W (per unit width in the y direction) along the z axis in the mode. A mode for which $E_y = AE_m(x)$ will thus correspond to a power flow of $|A|^2$ W/m. The normalization condition is given by

$$S_z = \tfrac{1}{2} \int \operatorname{Re}[\mathbf{E} \times \mathbf{H}^*]_z \, dx = 1$$

or, equivalently,

$$-\frac{1}{2} \int_{-\infty}^{\infty} E_y H_x^* \, dx = \frac{\beta_m}{2\omega\mu} \int_{-\infty}^{\infty} [E_m(x)]^2 \, dx = 1, \qquad (11.2\text{-}6)$$

where m denotes the mth confined TE mode [corresponding to the mth eigenvalue of Eq. (11.2-5)] and $H_x = -i(\omega\mu)^{-1}\, \partial E_y/\partial z$.

Substitution of Eq. (11.2-3) for the wavefunction in Eq. (11.2-6) and carrying out the integration lead to, after a few steps of algebraic manipulation,

$$C_m = 2h_m \left(\frac{\omega\mu}{|\beta_m|[t + (1/q_m) + (1/p_m)](h_m^2 + q_m^2)} \right)^{1/2}. \qquad (11.2\text{-}7)$$

The orthonormalization of the modes can be written as

$$\int_{-\infty}^{\infty} E_m E_l \, dx = \frac{2\omega\mu}{|\beta_m|} \delta_{l,m}. \qquad (11.2\text{-}8)$$

where $\delta_{l,m}$ is a Kronecker delta.

11.2.2 Guided TM Modes

For TM modes, the magnetic field vector is perpendicular to the plane of incidence (xz plane). The derivation of the confined TM modes is similar in principle to that of the TE modes. The field components are

$$H_y(x, z, t) = H_m(x)e^{i(\omega t - \beta z)},$$

$$E_x(x, z, t) = \frac{i}{\omega\varepsilon} \frac{\partial H_y}{\partial z} = \frac{\beta}{\omega\varepsilon} H_m(x)\, e^{i(\omega t - \beta z)}, \qquad (11.2\text{-}9)$$

$$E_z(x, z, t) = -\frac{i}{\omega\varepsilon} \frac{\partial H_y}{\partial x}.$$

The function $H_m(x)$ assumes the following forms in each of the three regions:

$$H_m(x) = \begin{cases} -C\left[\dfrac{h}{\bar{q}}\cos(ht) + \sin(ht)\right]e^{p(x+t)}, & x < -t, \\[2ex] C\left[-\dfrac{h}{\bar{q}}\cos(hx) + \sin(hx)\right], & -t < x < 0, \\[2ex] -\dfrac{h}{\bar{q}}\,Ce^{-qx}, & 0 < x, \end{cases}$$

$$\text{(11.2-10)}$$

where C is a normalization constant and h, q, and p are given by Eq. (11.2-4), and $\bar{q}$ is defined as follows.

The continuity of H_y and E_z at the two interfaces leads, in a manner similar to Eq. (11.2-5), to the eigenvalue equation

$$\tan(ht) = \frac{h(\bar{p} + \bar{q})}{h^2 - \bar{p}\bar{q}}, \qquad \text{(11.2-11)}$$

where

$$\bar{p} = \frac{n_2^2}{n_3^2}\,p, \qquad \bar{q} = \frac{n_2^2}{n_1^2}\,q.$$

The normalization constant C is again chosen so that the field represented by Eqs. (11.2-9) and (11.2-10) carries 1 W of power flow along the z axis per unit width in the y direction. Thus, we have

$$\frac{1}{2}\int_{-\infty}^{\infty} H_y E_x^* \, dx = \frac{\beta}{2\omega}\int_{-\infty}^{\infty} \frac{H_m^2(x)}{\varepsilon(x)} \, dx = 1,$$

or, using $n^2(x) = \varepsilon(x)/\varepsilon_0$,

$$\int_{-\infty}^{\infty} \frac{[H_m(x)]^2}{n^2(x)} \, dx = \frac{2\omega\varepsilon_0}{|\beta_m|}. \qquad \text{(11.2-12)}$$

Carrying out the integration using Eq. (11.2-10) gives

$$C_m = 2\sqrt{\frac{\omega\varepsilon_0}{|\beta_m|t_{\text{eff}}}},$$

$$t_{\text{eff}} \equiv \frac{\bar{q}^2 + h^2}{\bar{q}^2}\left(\frac{t}{n_2^2} + \frac{\bar{q}^2 + h^2}{\bar{q}^2 + h^2}\frac{1}{n_1^2 q} + \frac{\bar{p}^2 + h^2}{\bar{p}^2 + h^2}\frac{1}{n_3^2 p}\right). \qquad \text{(11.2-13)}$$

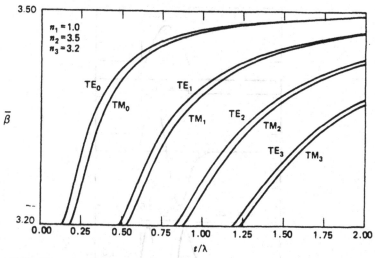

Figure 11.7 Dispersion curves for the confined modes of GaAs film on AlGaAs substrate: $n_1 = 1.0, n_2 = 3.50, n_3 = 3.20$. (Adapted from A. Yariv and P. Yeh, *Optical Waves in Crystals,* Wiley, New York, 1984, p. 420. Copyright © 1984. By permission of John Wiley & Sons, Inc.)

The TM modes are also orthogonal. In fact, all TE and TM modes are mutually orthogonal. The general orthonormality property will be discussed further in Section 11.6.

Figure 11.7 illustrates the dependence of propagation constants β on the waveguide thickness (t/λ) for a waveguide with $n_2 = 3.50$, $n_3 = 3.20$, and $n_1 = 1.0$. According to this figure, such a waveguide can support three TE and three TM modes when the core thickness is equal to the wavelength (i.e., $t = \lambda$). In fact, at $t/\lambda = 1$, the normalized propagation constants (or effective index) for the TE modes are 3.4734, 3.3938, and 3.2640. Wavefunctions of these three modes are plotted in Fig. 11.8.

Generally speaking, a mode becomes confined above a certain (cutoff) value of t/λ. At the cutoff value, $p = 0$ and $\beta = n_3\omega/c$, and the mode extends to $x = -\infty$. According to the mode conditions (11.2-5), (11.2-11), and (11.2-4), cutoff values of t/λ for TE and TM modes are given, respectively, by

$$\left(\frac{t}{\lambda}\right)_{TE} = \frac{1}{2\pi\sqrt{n_2^2 - n_3^2}}\left[m\pi + \tan^{-1}\left(\frac{n_3^2 - n_1^2}{n_2^2 - n_3^2}\right)^{1/2}\right]$$

and   (11.2-14)

$$\left(\frac{t}{\lambda}\right)_{TM} = \frac{1}{2\pi\sqrt{n_2^2 - n_3^2}}\left[m\pi + \tan^{-1}\frac{n_2^2}{n_1^2}\left(\frac{n_3^2 - n_1^2}{n_2^2 - n_3^2}\right)^{1/2}\right],$$

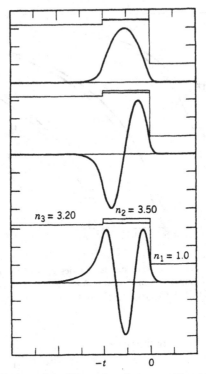

Figure 11.8 Field distribution of the TE modes of a waveguide with $n_2 = 3.50$, $n_3 = 3.20$, $n_1 = 1.0$, and $t/\lambda = 1.0$.

where m is an integer ($m = 0, 1, 2, 3, \ldots$) that denotes the mth confined TE (or TM) mode. Note that the cutoff thickness of the TM_m mode is always larger than that of the TE_m mode because $n_1 < n_2$. For values of (t/λ) slightly above the cutoff value, $p \geqslant 0$, and the mode is poorly confined. As the values of t/λ increase, so does the value of p, and the mode becomes increasingly confined to layer 2. This is reflected in the normalized propagation constant or the effective index, $\beta\lambda/2\pi$, which at cutoff is equal to n_3 and for large t/λ approaches n_2. In a symmetric waveguide ($n_1 = n_3$), the lowest order modes TE_0 and TM_0 have no cutoff and are confined for all values of t/λ. The confinement is, however, poor when t/λ becomes small.

The total number of confined modes that can be supported by a waveguide depends on the value of t/λ. To study the number of confined modes, we defined the parameter

$$V \equiv \frac{2\pi t}{\lambda} \sqrt{n_2^2 - n_3^2}. \qquad (11.2\text{-}15)$$

Let us now consider what happens about the TE modes in a given waveguide (i.e., fixed n_1, n_2, n_3, and t) as the wavelength of the light decreases gradually, assuming that the medium remains transparent and the indices of refraction n_1, n_2, and n_3 do not vary significantly. Since $\omega/c = 2\pi/\lambda$, the effect of decreasing the wavelength is to increase the value of ω/c. At long wavelengths (low frequencies), such that

$$0 < V < \tan^{-1}\left(\frac{n_3^2 - n_1^2}{n_2^2 - n_3^2}\right)^{1/2}, \qquad (11.2\text{-}16)$$

the value of t/λ is below the cutoff value, and no confined mode exists in the waveguide. As the wavelength decreases such that

$$\tan^{-1}\left(\frac{n_3^2 - n_1^2}{n_2^2 - n_3^2}\right)^{1/2} < V < \pi + \tan^{-1}\left(\frac{n_3^2 - n_1^2}{n_2^2 - n_3^2}\right)^{1/2}, \qquad (11.2\text{-}17)$$

one solution exists to the mode condition (11.2-5). The mode is designated as TE_0 and has a transverse h parameter falling within the range

$$0 < ht < \pi$$

so that it has no zero crossings in the interior of the guiding layer $(-t < x < 0)$. When the wavelength decreases further such that the parameter falls within the range

$$\pi + \tan^{-1}\left(\frac{n_3^2 - n_1^2}{n_2^2 - n_3^2}\right)^{1/2} < V < 2\pi + \tan^{-1}\left(\frac{n_3^2 - n_1^2}{n_2^2 - n_3^2}\right)^{1/2}, \qquad (11.2\text{-}18)$$

the mode condition (11.2-5) yields two solutions. One corresponds to a value of $ht < \pi$ and is thus that of the lowest order TE_0 mode. In the second mode,

$$\pi < ht < 2\pi, \qquad (11.2\text{-}19)$$

and consequently, it has one zero crossing (i.e., the point where $E_y = 0$) in the guiding region $(-t < x < 0)$. This is the so-called TE_1 mode. Both of these modes correspond to the same frequency and can thus be excited simultaneously by the same input field. We notice, however, that the TE_0 mode has a larger value of p (i.e., $p_0 > p_1$) and is therefore more tightly confined to the guiding slab. It follows from Eq. (11.2-4) that $\beta_0 > \beta_1$, so that the phase velocity $v_0 = \omega/\beta_0$ of the TE_0 mode is smaller than that of the TE_1 mode. We can now generalize and state that the mth mode (TE or TM) satisfies

$$(m - 1)\pi < ht < m\pi \qquad (11.2\text{-}20)$$

and has $m - 1$ zero crossings in the guiding layer $(-t < x < 0)$. The general features of TM modes are similar to those of TE modes, except that the corresponding values of p are somewhat smaller, indicating a lesser degree of confinement. A larger fraction of the total TM mode power thus propagates in the outer medium compared to a TE mode of the same order. This point is taken up in Problem 11.5.

11.2.3 Geometric Optics Treatment

In a manner similar to that leading to Eq. (11.1-29), geometric optics can also be used to obtain the mode condition. Following the procedure described in Section 11.1, we obtain

$$-2ht + \phi_{21} + \phi_{23} = -2m\pi, \qquad (11.2-21)$$

where m is an integer, ϕ_{21} is the phase shift upon total reflection from the interface $x = 0$, and ϕ_{23} is the phase shift upon total reflection from the interface $x = -t$. According to Eqs. (3.2-5) and (3.2-6), these phases are

$$\phi_{21} = \begin{cases} 2 \tan^{-1} \dfrac{q}{h}, & \text{TE}, \\[3mm] 2 \tan^{-1} \dfrac{n_2^2 q}{n_1^2 h}, & \text{TM}, \end{cases} \qquad (11.2-22)$$

$$\phi_{23} = \begin{cases} 2 \tan^{-1} \dfrac{p}{h}, & \text{TE}, \\[3mm] 2 \tan^{-1} \dfrac{n_2^2 p}{n_3^2 h}, & \text{TM}. \end{cases} \qquad (11.2-23)$$

Here, again, we drop the term $-\pi$ for TM waves because a round trip would yield -2π, which does not affect the field amplitude. The phase shifts in the preceding equations are again limited between 0 and π. Therefore, the fundamental modes correspond to the situation when $m = 0$.

Substitution of Eqs. (11.2-22) and (11.2-23) into Eq. (11.2-21) will lead to the mode condition in Eqs. (11.2-5) and (11.2-11).

11.3 MULTILAYER WAVEGUIDES

We have seen that the dielectric layer sandwiched between media of lower indices of refraction (i.e., slab waveguides) can support propagation of confined modes. The slab waveguides are not the only dielectric structures that support confined propagation. Waveguides can also be made of multiple dielectric layers depending on the specific applications. For example, the two-channel waveguide has been studied extensively in the theory of branching waveguides, which can be used as mode selectors, switches, and directional couplers in integrated optics. In this section, we consider a more general type of waveguide that consists of an arbitrary layered dielectric medium sandwiched between two semi-infinite bounding media. We will first present a general formulation using the 2×2 matrix method. We then discuss the properties of two-channel waveguides and periodic multichannel waveguides.

11.3.1 General Formulation

Consider a multilayer dielectric waveguide with an index profile described by

$$
n(x) = \begin{cases}
n_0, & x < 0, \\
n_1, & 0 < x < x_1, \\
n_2, & x_1 < x < x_2, \\
\vdots & \\
n_N, & x_{N-1} < x < x_N, \\
n_s, & x_N < x,
\end{cases} \tag{11.3-1}
$$

where n_0 and n_s are the indices of refraction of the bounding media. Notice that this structure is formally identical to that given by Eq. (5.1-16). Wave propagation in such a structure has been discussed in Section 5.1 and can be described by Eq. (5.1-19). To study the propagation of confined modes, we rewrite the same equation here:

$$
E(x) = \begin{cases}
B_0 e^{q_0 x}, & x < 0, \\
A_l e^{-ik_{lx}(x - x_l)} + B_l e^{ik_{lx}(x - x_l)}, & x_{l-1} < x < x_l, \\
A'_s e^{-q_s(x - x_N)}, & x_N < x,
\end{cases} \tag{11.3-2}
$$

with $x_0 = 0$, $x_s = x_N$, and

$$k_{lx} = \left[\left(\frac{n_l \omega}{c} \right)^2 - \beta^2 \right]^{1/2},$$

$$q_0 = ik_{0x} = \left[\beta^2 - \left(\frac{n_0 \omega}{c} \right)^2 \right]^{1/2},$$

(11.3-3)

$$q_s = ik_{sx} = \left[\beta^2 - \left(\frac{n_s \omega}{c} \right)^2 \right]^{1/2},$$

where β is the propagation constant along the z axis and k_{lx} is the x component of the wave vector in the layer whose index of refraction is n_l. For confined modes, the two parameters q_0 and q_s must remain positive.

Following the formulation in Section 5.1, we obtain a linear relation between the field amplitudes on both sides of the layered medium:

$$\begin{pmatrix} 0 \\ B_0 \end{pmatrix} = \begin{pmatrix} m_{11} & m_{12} \\ m_{21} & m_{22} \end{pmatrix} \begin{pmatrix} A'_s \\ 0 \end{pmatrix},$$

(11.3-4)

where we recall that we have set $A_0 = B'_s = 0$ for confined modes whose field amplitudes must vanish at infinity. The matrix m_{ij} is obtained by multiplying all the matrices associated with the layers in sequence according to Eq. (5.1-27). Here B_0 is the amplitude of the plane wave on the left side ($x < 0$) of the structure, and A'_s is that of the plane wave on the right (or substrate) side of the structure.

The mode condition, according to Eq. (11.3-4), is simply

$$m_{11} = 0,$$

(11.3-5)

where m_{11} is a function of β, ω, indices n_l, and thickness t_l. Given a structure (n_l, t_l) and a frequency ω, Eq. (11.3-5) can be used to solve for the propagation constant β of all the confined modes. To illustrate the use of Eq. (11.3-5), we revisit the slab waveguide in the following example.

11.3.2 Example: Matrix Treatment of Slab Waveguides

The matrix elements that link the amplitudes of TE wave on both sides of the guiding layer are given in Eq. (5.1-29). The matrix element m_{11} is rewritten in terms of p, q, and h as

$$m_{11} = \frac{1}{2} \left(1 + \frac{p}{q} \right) \cos ht + \frac{1}{2} i \sin ht \left(\frac{h}{-iq} + \frac{-ip}{h} \right),$$

where we have used

$$n_2 \cos \theta_2 = \frac{h}{\omega/c},$$

$$n_1 \cos \theta_1 = -\frac{iq}{\omega/c},$$

$$n_3 \cos \theta_3 = -\frac{ip}{\omega/c},$$

$$\phi = \frac{\omega}{c} n_2 \cos \theta_2 t = ht.$$

The mode condition in Eq. (11.3-5) thus becomes

$$\frac{1}{2}\left(1 + \frac{p}{q}\right) \cos ht - \frac{1}{2} \sin ht \left(\frac{h}{q} - \frac{p}{h}\right) = 0,$$

which is identical to the mode condition of TE waves derived earlier [Eq. (11.2-5)].

As illustrated in the preceding example, the mode condition for the waveguide mode of a layered structure can be easily obtained by using the matrix method. The matrix method becomes almost essential when the number of layers, N, is large because the conventional approach, similar to the one used in the previous section, involves the solution of $2(N + 1)$ linear equations. We will now use this matrix method to study the confined modes of periodic multilayer (or multichannel) dielectric waveguides. The study of such waveguides is important in understanding the optical properties of supermodes in laser diode arrays.

The periodic multilayer waveguides consist of a periodic layered medium (see Fig. 11.9). Analytic expressions for the mode dispersion relations and field distributions can be obtained by the matrix method.

We are looking for guided waves propagating in the positive z direction. Two important periodic multichannel waveguides will be considered in the following.

(a) Symmetric Type. Consider the simplest kind of symmetric periodic multilayer waveguide with the index of refraction given by

$$n(x) = \begin{cases} n_2, & m\Lambda \leqslant x \leqslant m\Lambda + b \ (m = 0, 1, 2, \ldots, N - 1), \\ n_1, & \text{otherwise,} \end{cases}$$

$$(11.3\text{-}6)$$

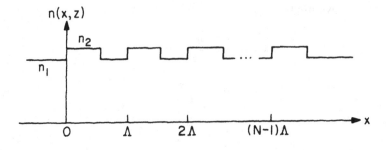

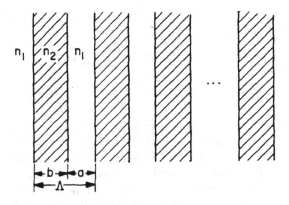

Figure 11.9 Section view of a typical N-channel symmetric waveguide.

with

$$n_1 < n_2. \tag{11.3-7}$$

The geometry of the waveguide is sketched in Fig. 11.9. We will limit our analysis to TE waves only. It was shown in Section 6.1 that the translation matrix T, which relates the field vector in one period to that of the next one, is given by

$$T = \begin{pmatrix} A & B \\ C & D \end{pmatrix}, \tag{11.3-8}$$

where, after defining $ik_{1x} = q$ and $k_{2x} = p$,

$$A = e^{qa}\left[\cos pb - \frac{1}{2}\left(\frac{p}{q} - \frac{q}{p}\right)\sin pb\right], \qquad (11.3\text{-}9)$$

$$B = e^{-qa}\left[-\frac{1}{2}\left(\frac{p}{q} + \frac{q}{p}\right)\sin pb\right], \qquad (11.3\text{-}10)$$

$$C = e^{qa}\left[\frac{1}{2}\left(\frac{p}{q} + \frac{q}{p}\right)\sin pb\right], \qquad (11.3\text{-}11)$$

$$D = e^{-qa}\left[\cos pb + \frac{1}{2}\left(\frac{p}{q} - \frac{q}{p}\right)\sin pb\right], \qquad (11.3\text{-}12)$$

with

$$q = \sqrt{\beta^2 - [(\omega/c)n_1)]^2} = ik_{1x}, \qquad (11.3\text{-}13)$$

$$p = \sqrt{[(\omega/c)n_2]^2 - \beta^2} = k_{2x}. \qquad (11.3\text{-}14)$$

Since we are interested in guided waves only, the fields must be evanescent in the n_1 layer. The matrix equation (6.1-16) for this case can be written as

$$\begin{pmatrix} A_0 \\ B_0 \end{pmatrix} = \begin{pmatrix} A & B \\ C & D \end{pmatrix}^N \begin{pmatrix} A_N \\ B_N \end{pmatrix}. \qquad (11.3\text{-}15)$$

For confined modes, we must set $A_0 = B_N = 0$ [see Eq. (11.3-4)].

Using Eq. (11.3-5) and the Chebyshev identity equation (6.3-6), the mode dispersion relation is immediately given by

$$A \frac{\sin NK\Lambda}{\sin K\Lambda} - \frac{\sin(N-1)K\Lambda}{\sin K\Lambda} = 0, \qquad (11.3\text{-}16)$$

where A is given by Eq. (11.3-9), N is the number of guiding layers (or channels), and $K\Lambda$ is given by

$$\cos K\Lambda = \tfrac{1}{2}(A + D)$$
$$= \cosh qa \cos pb - \frac{1}{2}\left(\frac{p}{q} - \frac{q}{p}\right)\sinh qa \sin pb. \qquad (11.3\text{-}17)$$

Given a structure with n_1, n_2, a, b, and N fixed, parameters such as A, D, and $K\Lambda$ can be considered as functions of β.

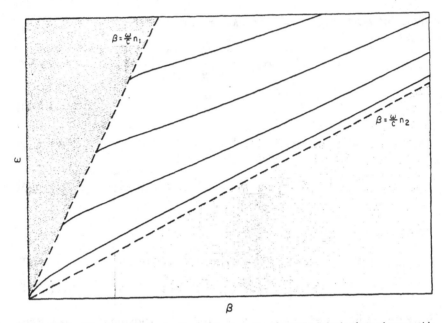

Figure 11.10 Dispersion curves for the confined modes of a typical single-channel waveguide ($N = 1$).

If the left side of Eq. (11.3-16) is plotted using Eq. (11.3-17) as a function of β for a given frequency ω, the zeros are the mode propagation constants (β's). It can be shown mathematically that there are exactly N zeros in each allowed band where $K\Lambda$ varies from $m\pi$ to $(m + 1)\pi$ and none elsewhere (see Appendix A). Physically, the waveguide can be considered as a system of N interacting symmetric slab waveguides. The N modes are simply due to the splitting of an N-fold degenerate band as the separations between the N identical slab waveguides are reduced from infinity. Each confined mode of the single slab waveguide thus gives rise to a band with N nondegenerate modes. The dispersion relation (ω vs. β) is shown in Figs. 11.10 and 11.11.

(b) Asymmetric Type. Consider a simple asymmetric N-channel waveguide with the following index of refraction:

$$n(x, z) = \begin{cases} n_a, & x < 0 \\ n_2, & m\Lambda \leqslant x < m\Lambda + b \ (m = 0, 1, 2, \ldots, N - 1), \\ n_1, & \text{otherwise,} \end{cases}$$

$$(11.3\text{-}18)$$

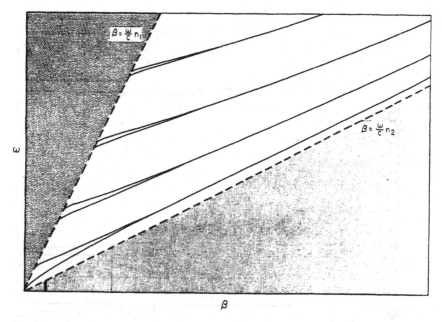

Figure 11.11 Dispersion curves for the confined modes of a typical two-channel waveguide ($N = 2$). Notice the splitting in comparison with Fig. 10.10.

where n_a is different from n_1. The matrix in this case is easily obtained from the formulation in Eq. (5.1-27) and is given by

$$M = \begin{pmatrix} \frac{1}{2}\left(1 + \frac{q}{q_a}\right) & \frac{1}{2}\left(1 - \frac{q}{q_a}\right) \\ \frac{1}{2}\left(1 - \frac{q}{q_a}\right) & \frac{1}{2}\left(1 + \frac{q}{q_a}\right) \end{pmatrix} \begin{pmatrix} A & B \\ C & D \end{pmatrix}^N, \quad (11.3\text{-}19)$$

where

$$q_a = \left[\beta^2 - \left(\frac{n_a \omega}{c}\right)^2\right]^{1/2}. \quad (11.3\text{-}20)$$

Again, by using the Chebyshev identity and Eq. (11.3-5), the mode dispersion relation is given by

$$\left(A + \frac{q_a - q}{q_a + q}\, C\right)\frac{\sin NK\Lambda}{\sin K\Lambda} - \frac{\sin (N-1)K\Lambda}{\sin K\Lambda} = 0, \quad (11.3\text{-}21)$$

where A and C are given by Eqs. (11.3-9) and (11.3-11).

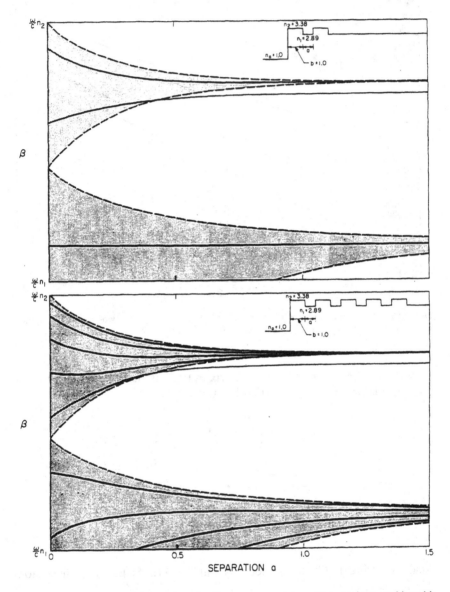

Figure 11.12 The β versus separation of a for two asymmetric multichannel waveguides with $N = 2$ (upper diagram) and $N = 5$ (lower diagram) at $b(\omega/c) = 3\pi/4$. The dark zones are the allowed bands. Dashed curves are band edges. The inset shows the refractive index profile.

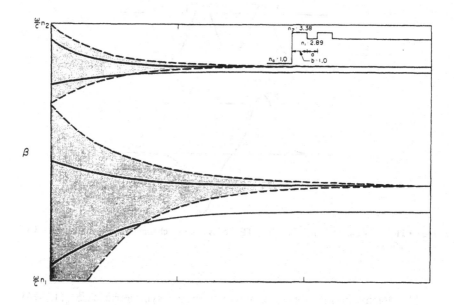

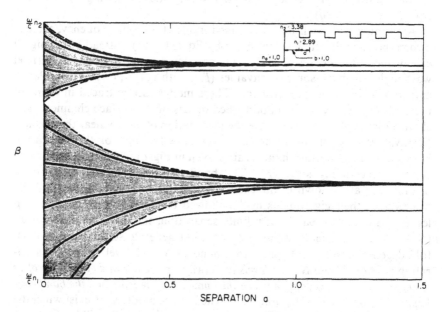

Figure 11.13 The β versus separation of a for two asymmetric multichannel waveguides with $N = 2$ (upper diagram) and $N = 5$ (lower diagram) at $h(\omega/c) = \pi$. The dark zones are the allowed bands. Dashed curves are band edges. The inset shows the refractive index profile.

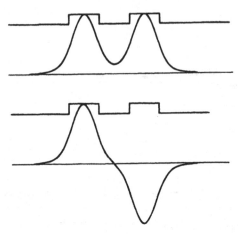

Figure 11.14 Wavefunctions of confined TE modes of a two-channel waveguide with $n_2 = 3.38$, $n_1 = 2.89$, $a = b = 0.5\lambda$.

The eigenvalues β are determined as in the symmetric case (11.3-16). Equation (11.3-21) can be reduced to Eq. (11.3-16), which is the mode condition for a symmetric N-channel waveguide by setting $n_a = n_1$ (or equivalently, $q_a = q$).

Associated with each β of a confined mode at a given frequency, there is a corresponding Bloch K vector given by Eq. (11.3-17). Instead of having all eigenvalues (β's) in the allowed band, an asymmetric periodic N-channel waveguide can have some eigenvalues (β's) with corresponding complex K and thus be in the forbidden band. These modes can be traced in terms of perturbation theory to the unperturbed modes of the surface channels. The characteristics of those modes are the localization of energy near the surface. Eigenvalues (β's) of the confined modes as a function of the separation between the neighboring channels are shown in Figs. 11.12 and 11.13 for two typical waveguiding structures. The band edges of the infinite periodic medium are also shown in the same figures. For small separations, all the modes have their eigenvalues in the allowed bands. There are exactly N β levels in a complete band. At infinite separation, the β levels consist of an $(N - 1)$-fold degenerate state and one nondegenerate state. The $(N - 1)$-fold degenerate state will split into a band of $N - 1$ levels when the separation is finite. *Those $N - 1$ levels are always in the allowed band regardless of separation. The crossing between the nondegenerate state and the band edge happens at some critical separation a_c.* The surface modes only exist when the separation is larger than a_c. The properties of surface mode will be discussed more thoroughly in Section 11.5.

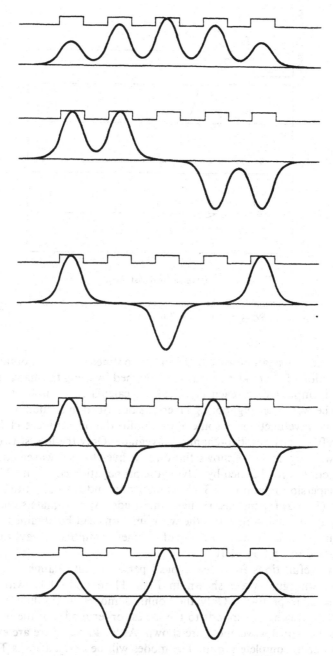

Figure 11.15 Wavefunctions of confined TE modes of a five-channel waveguide with $n_2 = 3.38$, $n_1 = 2.89$, $a = b = 0.5\lambda$.

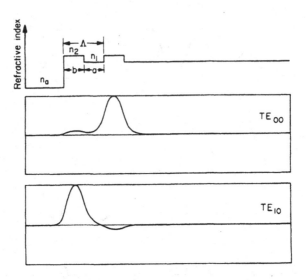

Transverse field distribution

Figure 11.16 Wavefunctions for the confined modes in the first band of a two-channel waveguide, with $n_1 = 2.89$, $n_2 = 3.38$, $n_a = 1.0$, $a = b = 0.5\lambda$.

Once the propagation constants β of the confined modes are obtained, the wavefunction of the modes can also be obtained by using the matrix method. The field amplitudes of each layer can be calculated by using the matrix relation between the neighboring layers, as described in Section 5.1. We will now show wavefunctions of some of the confined modes. Figure 11.14 shows the wavefunctions of a two-channel waveguide. Only the lowest two modes are shown. Figure 11.15 shows those for a five-channel waveguide. These wavefunctions are obtained by solving the propagation constants β from the mode dispersion relation (11.3-21) numerically and then substituting them into Eq. (11.3-2) for the electric field amplitude. Approximate solutions for the propagation constants and the wavefunctions can be obtained by using the perturbation approach. The case of symmetric multilayer waveguides will be discussed in Section 11.10.

The wavefunctions for a few typical periodic multichannel asymmetric dielectric waveguides are shown in Figs. 11.16 and 11.17 with $N = 2$ and $N = 5$, respectively. Only the confined modes in the first band (see Fig. 11.12), which corresponds to the lowest order modes of the uncoupled individual channel waveguide, are shown. As we know, there are exactly N modes in each complete group. The modes will be designated as TE_{mn} and TM_{mn}, with n the band index ($n = 0, 1, 2, \ldots$) and m the mode index

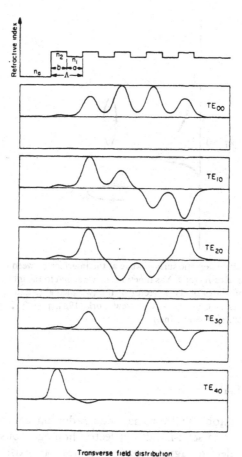

Figure 11.17 Wavefunctions for the confined modes in the first band of a five-channel waveguide, with $n_1 = 2.89$, $n_2 = 3.38$, $n_a = 1.0$, $a = b = 0.5\lambda$.

$(m = 0, 1, 2, \ldots, N - 1)$. The wavefunction of TE_{mn} (or TM_{mn}) has exactly $m + nN$ zero crossings. There are n zero crossings in each guiding channel and m zero crossings in the $N - 1$ separation layers. The field can have at most one zero crossing in each separation layer where the wave is evanescent.

We notice that the wavefunction of TE_{40} in Fig. 11.17 is very different from the rest. Its energy is mostly confined near the surface $x = 0$. This mode has a propagation constant β that corresponds to the one in the forbidden band (see Fig. 11.13). Such a mode also exists in a semi-infinite structure and will be discussed further in Section 11.5.

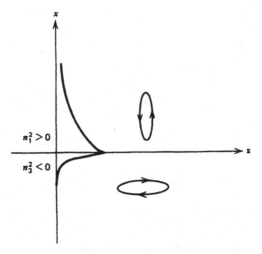

Figure 11.18 An electromagnetic surface wave at the interface between two media. The graph shows the field amplitudes H_y (or E_z) as functions of x normal to the interface. The two ellipses show the polarization states of the E vector in the two media. (Adapted from A. Yariv and P. Yeh, *Optical Waves in Crystals*, Wiley, New York, 1984, p. 490. Copyright © 1984. By permission of John Wiley & Sons, Inc.)

11.4 SURFACE PLASMONS

Confined propagation of electromagnetic radiation can also exist at the interface between two semi-infinite dielectric homogeneous media. We will show that such electromagnetic surface waves can exist at the interface between two media, provided the dielectric constants of the media are opposite in sign (e.g., air and silver). Only a single TM mode exists at a given frequency. The amplitude of the wave decreases exponentially in the two directions normal to the interface. These modes are also called surface plasmon waves because of the electron plasma contribution to the negative dielectric constant of metals [see Eq. (2.3-13)] when the optical frequency is lower than the plasma frequency (i.e., $\omega < \omega_p$). In what follows, we derive the propagation characteristics of such electromagnetic surface waves.

Referring to Fig. 11.18, we consider the wave propagation along the interface between a medium with a positive dielectric constant ($n_1^2 > 0$) and a medium with a negative dielectric constant ($n_3^2 < 0$). A typical example is the interface between air and silver where $n_1^2 = 1$ and $n_3^2 = -16.40 - i0.54$ at $\lambda = 6328\,\text{Å}$. If we neglect the small imaginary part ($-i0.54$), we may consider silver as a material with a negative dielectric constant. The presence of the imaginary part will cause attenuation in propagation along the interface. We will first solve for the surface modes by neglecting the imaginary part of the dielectric constant. The propagation attenuation coefficient will then be obtained by using a perturbation theory or directly from the complex root of the mode condition.

The structure shown in Fig. 11.18 may be viewed as a special case of the waveguide structure shown in Fig. 11.5 with $t = 0$. Therefore, the field expressions (11.2-3) and (11.2-10) and the mode conditions (11.2-5) and (11.2-11) for TE and TM waves, respectively, can also be applied to the guide structure shown in Fig. 11.18. By putting $t = 0$ in Eq. (11.2-5), we obtain the mode condition for the TE surface waves,

$$p + q = 0, \tag{11.4-1}$$

where p and q are the exponential attenuation constants in media 3 and 1, respectively. Since a confined mode requires that $p > 0$ and $q > 0$, Eq. (11.4-1) can never be satisfied. This proves that a TE surface mode cannot exist at the interface of two homogeneous media. For TM waves, the mode function $H_y(x)$ can be written as

$$H_y(x) = \begin{cases} C \exp(-qx), & x \geq 0, \\ C \exp(px), & x \leq 0, \end{cases} \tag{11.4-2}$$

where C is the constant of normalization. The mode condition can be obtained by insisting on the continuity of E_z at the interface $x = 0$ or directly from Eq. (11.2-11) by setting $t = 0$. The result is

$$\frac{p}{n_3^2} + \frac{q}{n_1^2} = 0. \tag{11.4-3}$$

Since $n_1^2 n_3^2 < 0$, the mode condition (11.4-3) for TM surface waves can be satisfied for confined propagation with $p > 0$ and $q > 0$. Using Eq. (11.2-4) for p and q, the mode condition (11.4-3) can be solved, and the propagation constant β is given by

$$\beta = \left(\frac{n_1^2 n_3^2}{n_1^2 + n_3^2} \right)^{1/2} \frac{\omega}{c}. \tag{11.4-4}$$

A confined propagating mode must have a real propagation constant. According to Eq. (11.4-4) and $n_1^2 n_3^2 < 0$, this requires that

$$n_1^2 + n_3^2 < 0, \tag{11.4-5}$$

that is, the sum of the dielectric constant of the media must also be negative. The attenuation constants p and q are given, according to Eqs. (11.2-4) and (11.4-4), by

$$p^2 = \frac{-n_3^4}{n_1^2 + n_3^2} \left(\frac{\omega}{c}\right)^2,$$

$$q^2 = \frac{-n_1^4}{n_1^2 + n_3^2} \left(\frac{\omega}{c}\right)^2,$$

(11.4-6)

which are all positive according to Eq. (11.4-5). The electric field components are given, according to Eqs. (11.2-9) and (11.2-2), by

$$E_x = \begin{cases} \dfrac{\beta}{\omega\varepsilon_0} \dfrac{C}{n_1^2} \exp(-qx)\exp[i(\omega t - \beta z)], & x \geqslant 0, \\[2mm] \dfrac{\beta}{\omega\varepsilon_0} \dfrac{C}{n_3^2} \exp(px)\exp[i(\omega t - \beta z)], & x \leqslant 0, \end{cases}$$

(11.4-7)

and

$$E_z = \begin{cases} \dfrac{iq}{\omega\varepsilon_0} \dfrac{C}{n_1^2} \exp(-qx)\exp[i(\omega t - \beta z)], & x \geqslant 0, \\[2mm] -\dfrac{ip}{\omega\varepsilon_0} \dfrac{C}{n_3^2} \exp(px)\exp[i(\omega t - \beta z)], & x \leqslant 0. \end{cases}$$

(11.4-8)

We note that the electric field vector $\mathbf{E}$ is elliptically polarized in the xz plane with the principal axes of the ellipses parallel to the coordinate axes. For propagation in the $+z$ direction ($\beta > 0$), the $\dot{E}$ vector is right-handed elliptically polarized in the upper half-space $x \geqslant 0$ and left-handed elliptically polarized for the upper and lower half-space, respectively (see Fig. 11.18).

The normalization constant C is chosen so that the surface mode represented by Eqs. (11.4-2), (11.4-7), and (11.4-8) carried 1 W per unit width in the y direction. Using Eqs. (11.2-12) and (11.2-2), we obtain

$$C^2 = \frac{4\omega\varepsilon_0}{\beta} \left(\frac{1}{qn_1^2} + \frac{1}{pn_3^2}\right)^{-1}.$$

(11.4-9)

11.4.1 Attenuation Coefficient

In the preceding derivation, we assumed that both n_1^2 and n_3^2 are real numbers. In the case of the air–silver interface, for example, $n_1^2 = 1$ for air and n_3^2 has a small imaginary part for silver. In fact, for most metals the dielectric constants $\varepsilon_0 n^2$ are complex numbers in the optical frequency regions (see Table 2.1). Surface wave propagation at the interface between a metal and a dielectric medium suffers ohmic losses. The propagation therefore attenuates in the z direction. This corresponds to a complex propagation constant β,

$$\beta = \beta^{(0)} - \tfrac{1}{2}i\alpha, \qquad (11.4\text{-}10)$$

where α is the power attenuation coefficient and $\beta^{(0)}$ is the real part of the propagation constant. The expression for propagation constant β derived in Eq. (11.4-4) is, in fact, valid also when n_1^2 and n_3^2 are complex. We can thus obtain the attenuation coefficient α directly from Eq. (11.4-4), which is complex if n_1^2 and n_3^2 are complex. In the case of a dielectric–metal interface, n_1^2 is a real positive number and n_3^2 is a complex number $(n - i\kappa)^2$, and the propagation constant of the surface wave is thus given, according to Eq. (11.4-4), by

$$\beta = \left[\frac{n_1^2(n - i\kappa)^2}{n_1^2 + (n - i\kappa)^2}\right]\frac{\omega}{c}, \qquad (11.4\text{-}11)$$

where κ is the extinction coefficient of the metal (see Section 2.3). In the spectral regimes of interest, n is much smaller than κ (i.e., $n \ll \kappa$). This corresponds to a small imaginary part of n_3^2, since $n_3^2 = (n^2 - \kappa^2) - 2in\kappa$. In this case, the propagation constant β can be written approximately as

$$\beta = \beta^{(0)}\left[1 - \frac{in\kappa n_1^2}{(n^2 - \kappa^2)(n_1^2 + n^2 - \kappa^2)}\right], \qquad (11.4\text{-}12)$$

where

$$\beta^{(0)} = \left[\frac{n_1^2(n^2 - \kappa^2)}{n_1^2 + (n^2 - \kappa^2)}\right]^{1/2}\frac{\omega}{c}. \qquad (11.4\text{-}13)$$

Thus, the attenuation coefficient α can be written, according to Eqs. (11.4-10), (11.4-12), and (11.4-13), as

$$\alpha = \frac{2n\kappa n_1^3}{[(n^2 - \kappa^2)(n_1^2 + n^2 - \kappa^2)^3]^{1/2}}\frac{\omega}{c}. \qquad (11.4\text{-}14)$$

In expressions (11.4-12)–(11.4-14), we keep the $n^2 - \kappa^2$ term because it is proportional to the real part of the dielectric constant of the metal. Thus, in terms of the dielectric constants,

$$\varepsilon_1 = \varepsilon_0 n_1^2,$$
$$\varepsilon_R - i\varepsilon_I = (n - k)^2 - i2n\kappa, \qquad (11.4\text{-}15)$$

and the propagation and attenuation constants can be written, respectively, as

$$\beta^{(0)} = \left(\frac{\varepsilon_1 \varepsilon_R}{\varepsilon_1 + \varepsilon_R}\right)^{1/2} \frac{\omega}{c} \qquad (11.4\text{-}16)$$

and

$$\alpha = \frac{\varepsilon_I \varepsilon_1^{3/2}}{[\varepsilon_R(\varepsilon_1 + \varepsilon_R)^3]^{1/2}} \frac{\omega}{c}, \qquad (11.4\text{-}17)$$

where ε_1 is the dielectric constant of the dielectric medium and $\varepsilon_R - i\varepsilon_I$ is the complex dielectric constant of the metal. Note that $\varepsilon_1 > 0$, $\varepsilon_R < 0$, and $\varepsilon_1 + \varepsilon_R < 0$. The value of ε_I is always greater than zero.

The attenuation coefficient can also be obtained from the ohmic loss calculation and can be written as

$$\alpha = \tfrac{1}{2} \int_{-\infty}^{\infty} \sigma |E|^2 \, dx, \qquad (11.4\text{-}18)$$

where σ is the conductivity and is related to the complex dielectric constant by

$$\sigma = 2\omega\varepsilon_0 n\kappa = \omega\varepsilon_0 \varepsilon_I. \qquad (11.4\text{-}19)$$

Using expressions (11.4-7) and (11.4-8) for the field E and (11.4-4) and (11.4-6) and carrying out the simple integration (11.4-18), we arrive at the same expression [(11.4-14)] for the attenuation coefficient α.

11.4.2 Example: Surface Waves at Air–Silver Interface

At $\lambda = 6328 \text{ Å}$, the complex refractive index of silver is $n - i\kappa = 0.067 - i4.05$. The surface mode at the interface between air and silver has a propagation constant, according to Eq. (11.4-13), of

$$\beta = 1.032 \frac{\omega}{c} = 10.25 \, \mu m^{-1}$$

and an attenuation coefficient, according to Eq. (11.4-14), of

$$\alpha = 0.0022 \frac{\omega}{c} = 220 \, cm^{-1}.$$

The field attenuation constants p and q are

$$p \;=\; 4.18\,\frac{\omega}{c} \;=\; \frac{26.26}{\lambda}\,, \qquad q \;=\; 0.25\,\frac{\omega}{c} \;=\; \frac{1.60}{\lambda}\,.$$

Note that the field amplitude decreases to $\exp(-26.26) = 4 \times 10^{-12}$ at a depth of $\lambda = 6328\,\text{Å}$ in silver. Thus, the wave is essentially propagating in the air while guided by the surface of the silver. This is also reflected from the phase velocity of the surface wave, which is $c/1.032$, very close to the velocity of light in air.

At $\lambda = 10\,\mu\text{m}$, $n - i\kappa = 10.8 - i60.7$ for silver, and the attenuation coefficient α becomes

$$\alpha \;=\; 0.6\,\text{cm}^{-1}.$$

If we put $n_1^2 = 1$ and neglect n^2 in the denominator in Eq. (11.4-14) since $n^2 \ll \kappa^2$, the attenuation coefficient α can be written as

$$\alpha \;=\; \frac{4\pi n}{\kappa^3 \lambda}\,. \tag{11.4-20}$$

We note that α decreases as κ^{-3}. For metals such as silver, gold, aluminum, and copper, the extinction coefficient κ becomes very large. Thus, according to Eq. (11.4-20), the attenuation coefficient α for the surface waves is expected to be smaller for these metals in the infrared spectral regimes.

11.5 ELECTROMAGNETIC BLOCH SURFACE WAVES

Electromagnetic surface waves can also be guided by the boundary of a semi-infinite periodic multilayer dielectric medium. In this section, the band theory described in Section 6.2 of periodic layered media is used to study the surface wave with an eigenvalue (or Bloch wave number) in the forbidden band.

The existence of a surface state can be explained as follows. In Section 6.2, we have shown that, at a given frequency, there are regions of β for which K is complex and $K = m\pi/\Lambda \pm iK_i$. For an infinite periodic medium, the exponential intensity variation cannot exist, and we refer to these regions as forbidden. If the periodic medium is semi-infinite, the exponentially damped solution is a legitimate solution near the interface, and the field envelope decays as $\exp(-K_i x)$, where x is the distance from the interface.

The existence of surface states can also be argued using perturbation theory. According to perturbation theory, the periodic multilayer dielectric medium, which consists of alternating layers of different indices of refraction, can be considered as a system of interacting waveguides. These waveguides are identical to each other except for the one near the surface. The interaction strength between the waveguides depends on the separation between the neighboring waveguides due to overlap of the evanescent field distributions. When the separation is infinite, there is no interaction, and the guides can be considered as independent of each other. The eigenvalues (β's) thus fall into two groups: one is an infinitely degenerate state; the other is a nondegenerate state that corresponds to the extreme guide near the surface. As the waveguides are brought together, the interaction between the waveguides causes the eigenvalues to split. The splitting for the cases of $N = 2$ and $N = 5$ is shown in Figs. 11.12 and 11.13. As the eigenvalues split, the allowed energy band for the infinite structure is fully occupied by the levels originating in the infinitely degenerate level. As a result, the nondegenerate level corresponding to the waveguide near the surface will be "expelled" out of the allowed energy band. The only place where this state can be accommodated is in the forbidden gap. The wavefunction for this state is localized near the surface because the corresponding eigenvalue is in the forbidden band, that is, $K = m\pi/\Lambda + iK_i$.

To investigate the properties and the Bloch surface modes, we consider a semi-infinite periodic mutilayer dielectric medium that consists of alternating layers of different indices of refraction. The index profile is given by

$$
n(x, z) = \begin{cases}
n_a, & x \leqslant 0, \\
n_2, & m\Lambda \leqslant x < m\Lambda + b, \\
n_1, & m\Lambda + b \leqslant x < (m + 1)\Lambda \quad (m = 0, 1, 2, \ldots),
\end{cases}
\tag{11.5-1}
$$

where n_1, n_2 are the indices of refraction of the layers of the periodic layered medium and n_a is the index of refraction of the semi-infinite medium.

The geometry of the structure is sketched in Fig. 11.19. We look for the possibility of waves propagating in the positive z direction. Since the structure is semi-infinite, we are only interested in the surface wave as far as guiding is concerned. Although both TE and TM modes exist, here we consider only the case of TE surface modes where the electric field is polarized in the y direction. The wavefunction of the electric field (TE) obeys the wave equation

$$
\frac{\partial^2}{\partial x^2} E(x) + \frac{\omega^2}{c^2} n^2(x) E(x) = \beta^2 E(x),
\tag{11.5-2}
$$

where β is a propagation constant.

Figure 11.19 A semi-infinite periodic layered medium. (From A. Yariv and P. Yeh, *Optical Waves in Crystals*, Wiley, New York, 1984, p. 210. Copyright © 1984. Reprinted by permission of John Wiley & Sons, Inc.)

We take the solution in the following form:

$$E(x) = \begin{cases} \alpha e^{q_a x}, & x \leqslant 0, \\ E_K(x)e^{-iKx}, & x \geqslant 0, \end{cases} \qquad (11.5\text{-}3)$$

where α is a constant, q_a is given by

$$q_a = \sqrt{\beta^2 - [(\omega/c)n_a]^2}, \qquad (11.5\text{-}4)$$

and $E_K(x)e^{-iKx}$ is the Bloch wavefunction given by Eq. (6.2-9).

To be a guided wave, the constant K in Eq. (11.5-3) must be complex so that the field decays as x goes to infinity. This is possible only when the propagating conditions (i.e., β) in the periodic medium correspond to a forbidden band. Another condition is that $E(x)$ and its x derivative be continuous at the interface with medium a. By using Eq. (11.5-3) and the explicit form of the Bloch wavefunction equation (6.2-9) at $n = 0$, the boundary conditions at $z = 0$ become

$$\alpha = a_0 + b_0,$$
$$q_a \alpha = -ik_{1x}(a_0 - b_0), \qquad (11.5\text{-}5)$$

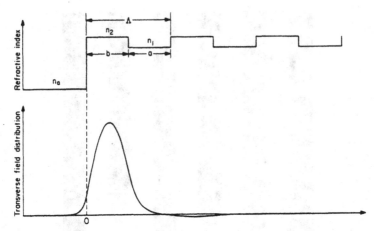

Figure 11.20 Wavefunction of a fundamental Bloch surface mode guided by surface of a semi-infinite periodic layered medium. (From A. Yariv and P. Yeh, *Optical Waves in Crystals*, Wiley, New York, 1984, p. 212. Copyright © 1984. Reprinted by permission of John Wiley & Sons, Inc.)

where a_0 and b_0 are given by Eq. (6.2-7). By eliminating α from Eq. (11.5-5) and using Eq. (6.2-7), we obtain

$$q_a = ik_{1x} \frac{e^{iK\Lambda} - A - B}{e^{iK\Lambda} - A + B}, \qquad (11.5-6)$$

where A and B are given by Eq. (6.1-13) for TE waves and $e^{iK\Lambda}$ is given by Eq. (6.2-6). The sign of K is chosen to make its imaginary part negative so that the field amplitude decays exponentially as x increases.

The wavefunctions of some typical surface waves are shown in Figs. 11.20 and 11.21. It is evident that the energy is more or less concentrated in the first few periods of the semi-infinite periodic medium. It can easily be shown that

$$\frac{\text{Energy in first period}}{\text{Energy in whole semi-infinite periodic structure}} = 1 - e^{-2K_i\Lambda}, \qquad (11.5-7)$$

where K_i is the imaginary part of $-K$. Generally speaking, the fundamental surface wave has the highest K_i and hence the highest degree of localization. The fundamental Bloch surface wave may happen to be in the zeroth or the first forbidden gap. It depends on the magnitude of the index of refraction n_a. For $n_a < n_1$, which is the practical case of interest (n_a is the index of refraction of air), the fundamental surface wave has a Bloch wave vector in the first forbidden gap. This is due to the fact that when the waveguides are separated

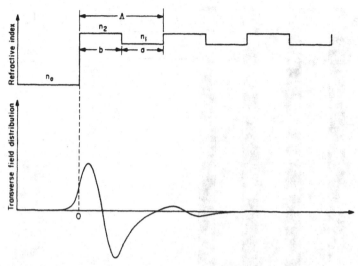

Figure 11.21 Wavefunction of a higher order Bloch surface mode guided by surface of a semi-infinite periodic layered medium. (From A. Yariv and P. Yeh, *Optical Waves in Crystals*, Wiley, New York, p. 212. Copyright © 1984. Reprinted by permission of John Wiley & Sons, Inc.)

infinitely from each other, the singlet state has an eigenvalue β lower than that of the infinitely degenerate state. The field amplitude in each period is similar to that of the distribution in the preceding period, except that the amplitude is reduced by a factor of $(-1)^m e^{-K_i \Lambda}$, where m is the integer corresponding to the mth forbidden gap.

We have derived the mode condition for the surface wave by matching the boundary condition between an evanescent wave and a decaying Bloch wave. This electromagnetic Bloch surface wave is almost completely analogous to the surface state in solid-state physics. The existence of the surface mode in a semi-infinite structure is independent of the separation between waveguides because the allowed band is always fully occupied. However, in a finite system, the allowed band is not fully occupied. As a result, the surface wave appears only when the separation is *large* enough so that one of the eigenvalues falls within the forbidden gap (see Figs. 11.12 and 11.13). The number of surface modes equals the number of modes that can be guided by the waveguide near the surface. This is consistent with the perturbation theory.

Electromagnetic Bloch surface waves can also exist at the interface between two periodic layered media. The analysis is similar to that of the surface modes, except that the air is now replaced by another layered

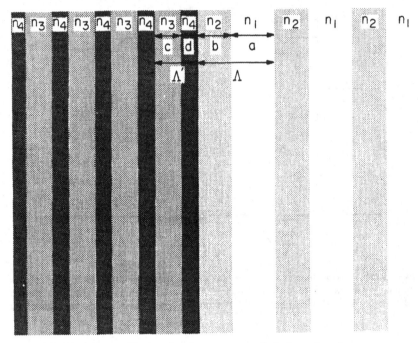

Figure 11.22 Interface between two periodic layered media.

medium. The index profile is now given by ($m = 0, 1, 2, 3, \ldots$)

$$
n(x, z) = \begin{cases}
n_1, & m\Lambda + b \leqslant x < (m + 1)\Lambda, \\
n_2, & m\Lambda \leqslant x < m\Lambda + b, \\
n_3, & -(m + 1)\Lambda' \leqslant x < -m\Lambda' - c, \\
n_4, & -m\Lambda' - c \leqslant x < -m\Lambda',
\end{cases} \tag{11.5-8}
$$

where n_1, n_2, n_3, and n_4 are the indices of refraction of the layers, Λ and Λ' are the periods, and b and c are the thicknesses of the layers. The geometry of the structure is sketched in Fig. 11.22.

We again look for the possibility of guided propagation in the positive z direction. Since the structure is infinite, we are interested only in the inter-facial waves as far as guiding is concerned. We will again analyze this problem for TE waves only. The analysis for TM waves is similar. We take the solution of the wave equation in the following form:

$$E(x) = \begin{cases} E_K(x)e^{-iKx}, & x \geqslant 0, \\ E_{K'}(x)e^{-iK'x}, & x \leqslant 0. \end{cases} \qquad (11.5\text{-}9)$$

To be a guided wave, the Bloch wave numbers K and K' must be complex so that the field decays as x goes to infinity, that is,

$$K = \frac{l\pi}{\Lambda} + iK_i, \qquad K' = \frac{l'\pi}{\Lambda'} + iK'_i, \qquad (11.5\text{-}10)$$

where l and l' are integers.

In addition, the sign must be properly chosen such that $K_i < 0$ and $K'_i > 0$. This is possible only when the forbidden bands of both layered media have some overlap and the propagation condition (i.e., β) has to be in these overlap regions. Another condition is that $E(x)$ and its x derivative be continuous at the interface. This gives us the dispersion relation

$$ik_{4x}\left(\frac{e^{iK'\Lambda'} - A' - B'}{e^{iK'\Lambda'} - A' + B'}\right) = ik_{1x}\left(\frac{e^{iK\Lambda} - A - B}{e^{iK\Lambda} - A + B}\right), \qquad (11.5\text{-}11)$$

where A and B are given by Eq. (6.1-13) and A', B' are those of the other periodic layered medium. The A' and B' may be obtained formally from Eq. (6.1-13) by replacing n_1 and n_2 with n_4 and n_3, respectively and also by replacing a and b with d and c, respectively.

The optical energy of these interface modes is also localized near the interface. A special case of particular interest is the symmetric case when $n_3 = n_1, n_4 = n_2, c = a$, and $d = b$ (see Fig. 11.22). Under these conditions, Eq. (11.5-11) becomes

$$(e^{iK\Lambda} - A + B)(e^{iK\Lambda} - A - B) = 0.$$

Physically, this means that either $E(x)$ or its derivative must vanish at $x = 0$. The modes can be divided into two categories: (1) even modes with their maxima right at the interface and (2) odd modes with their node at the interface. A typical wavefunction of this structure is shown in Fig. 11.23.

This special case is useful to illustrate the existence of the interface mode. Referring to Fig. 11.23, we consider the case of an odd mode whose wavefunction is an odd function of x and vanishes at $x = 0$. Since the Bloch

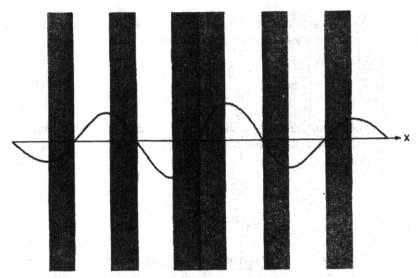

Figure 11.23 An odd interface Bloch mode.

wavefunction must be periodic, it must vanish at every other interface, as shown in Fig. 11.23. Thus, the electric field in the first period is

$$E(x) = \begin{cases} C_2 \sin h_2 x, & 0 < x < b, \\ C_1 \sin h_1(x - \Lambda), & b < x < \Lambda, \end{cases} \tag{11.5-12}$$

where C_1 and C_2 are two constants. By imposing the continuity of $E(x)$ and $dE(x)/dx$ at the interface $x = b$, we obtain the condition

$$h_1 \cot h_1 a + h_2 \cot h_2 b = 0, \tag{11.5-13}$$

where we recall that h_1 and h_2 are given by

$$h_1 = \left[\left(\frac{n_1 \omega}{c} \right)^2 - \beta^2 \right]^{1/2}, \qquad h_2 = \left[\left(\frac{n_2 \omega}{c} \right)^2 - \beta^2 \right]^{1/2}. \tag{11.5-14}$$

Solutions of Eq. (11.5-13) can always exist by choosing proper values of the layer thicknesses a and b. For example, an immediate solution is

$$h_1 a = h_2 b = \tfrac{1}{2}\pi, \tag{11.5-15}$$

which corresponds to quarter-wave layers.

The electric field amplitude decays like $\exp[-K_i|x|]$ away from the interface $x = 0$. The decay constant K_i is the imaginary part of the Bloch wave number K given by

$$\cos K\Lambda = \cos h_1 a \cos h_2 b - \frac{1}{2}\left(\frac{h_2}{h_1} + \frac{h_1}{h_2}\right)\sin h_1 a \sin h_2 b, \qquad (11.5\text{-}16)$$

which, for the special case described by Eq. (11.5-15), becomes

$$\cos K\Lambda = \cosh K_i\Lambda = -\frac{1}{2}\left(\frac{h_2}{h_1} + \frac{h_1}{h_2}\right). \qquad (11.5\text{-}17)$$

For $n_2 > n_1$ and consequently $h_2 > h_1$, the field amplitude decreases by a factor of

$$e^{-iK\Lambda} = -e^{-K_i\Lambda} = -\frac{h_1}{h_2} \qquad (11.5\text{-}18)$$

per period as $|x|$ increases.

This special example shows the existence of the interface mode. It is possible that there is no solution of Eq. (11.5-11) for some particular cases.

11.6 GENERAL PROPERTIES OF DIELECTRIC WAVEGUIDES

All cylindrical waveguides have many properties in common. The general requirement for a guide of electromagnetic radiation is that there be a flow of energy only along the guiding structure and not perpendicular to it. This means that the fields will be appreciable only in the immediate neighborhood of the guiding structure. A dielectric cylinder of arbitrary cross section, such as shown in Fig. 11.24, can serve as a waveguide provided its dielectric constant is large enough. For optical waves, this means that the core of the guide must have a higher refractive index compared to its surroundings. Generally speaking, a beam propagating in a transversely inhomogeneous medium tends to bend toward the high-refractive-index region according to the ray equation (8.6-6). Thus, the higher refractive index in the guiding region (core) has an effect similar to that of a converging lens. Under the appropriate conditions, this converging effect due to a higher core index may cancel out exactly the spreading due to diffraction. When this happens, a guided mode is supported by the dielectric structure.

We begin by considering a guide with an arbitrary cross section, as illustrated in Fig. 11.24. The axis of the guide will be taken as the z axis, and

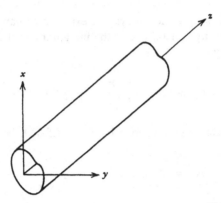

Figure 11.24 A section of a general cylindrical waveguide. (From A. Yariv and P. Yeh, *Optical Waves in Crystals*, Wiley, 1984, p. 406. Copyright © 1984. Reprinted by permission of John Wiley & Sons, Inc.)

the time variation of the modes is of the form $\exp(i\omega t)$. Since the whole dielectric structure is homogeneous along the z axis, solutions to the wave equations can be taken as

$$\mathbf{E} = \mathbf{E}_m(x, y)\exp[i(\omega t - \beta z)], \qquad (11.6\text{-}1)$$

where β is the propagation constant and $\mathbf{E}_m(x, y)$ is the wavefunction. We shall limit ourselves to dielectric structures that consist of piecewise homogeneous and isotropic materials, or materials with a small gradient in the distribution of the refractive index so that the wave equations reduce to (1.4-9). Substitution of Eq. (11.6-1) for $\mathbf{E}$ in Eq. (1.4-9) yields

$$\left\{\nabla_t^2 + \left[\frac{\omega^2}{c^2}n^2(x, y) - \beta^2\right]\right\}\mathbf{E}_m(x, y) = 0, \qquad (11.6\text{-}2)$$

where $\nabla_t^2 = \nabla^2 - \partial^2/\partial z^2$ is the transverse Laplacian operator. Equation (11.6-2) governs the transverse behavior of the field. In the case of piecewise homogeneous dielectric structures (e.g., layered structures), Eq. (11.6-2) holds separately in each homogeneous region. Therefore, the field must be solved separately in each region, and then the tangential components of the field must be matched at each interface. Another important boundary condition for guided modes is that the field amplitudes be zero at infinity. The axial propagation constant β must be the same throughout the guide structure to satisfy the boundary conditions at all points on the interfaces of those homogeneous media.

The basic prolem is that of finding the solution to the eigenvalue equation (11.6-2) subject to the continuity conditions of the tangential component of the fields at the dielectric interfaces and the boundary conditions at infinity. Given a refractive index profile $n^2(x, y)$, there are, in general, an infinite number of eigenvalues β^2 corresponding to an infinite number of modes. However, normally, only a finite number of these modes are confined near the core and will propagate freely along the guide. One of the necessary conditions for a guided mode is that there be no transverse flow of energy. This requires that the fields fall off exponentially outside the guide structure. Consequently, the quantity $(\omega^2/c^2)n^2 - \beta^2$ must be negative in the region far from the guiding region (core). In other words, the propagation constant β of a confined mode must be such that

$$\beta^2 > \frac{\omega^2}{c^2} n^2(\infty), \tag{11.6-3}$$

where $n^2(\infty)$ is the index of refraction at infinity ($\sqrt{x^2 + y^2} \to \infty$). On the other hand, if the fields vanish at infinity (i.e., $|E(\infty)|^2 = 0$), the continuity of the fields requires that the magnitude of the field, $|E(x, y)|$, attain at least a maximum value at some point in the xy plane. Normally, $E(x, y)$ is a smooth function of space. The existence of a maximum requires that the Laplacian of the field be negative. In other words, the propagation constant β of a confined mode must be such that

$$\beta^2 < \frac{\omega^2}{c^2} n^2(x, y) \tag{11.6-4}$$

in some region of the xy plane (usually the core). In particular, if n_c is the maximum value of the refractive index profile $n(x, y)$, the propagation constant of a confined mode must always satisfy the condition

$$\beta^2 < \frac{\omega^2}{c^2} n_c^2. \tag{11.6-5}$$

In the region where condition (11.6-4) is satisfied, the solutions to the wave equation (11.6-2) are oscillatory. These oscillatory solutions must be matched to the exponential solutions at the boundary of the dielectric interfaces. Therefore, not all the β's satisfying conditions (11.6-5) and (11.6-3) are legitimate eigenvalues of the confined modes. This was illustrated in Sections 11.1 and 11.2 when we studied the modes of slab waveguides. In what follows, we investigate some fundamental properties of waveguide modes.

11.6.1 Orthogonality of Modes

The waveguide modes supported by an arbitrary dielectric structure have an important and useful orthogonal property. Let $\mathbf{E}_1$ and $\mathbf{E}_2$ be the electric field for two linearly independent solutions to the wave equation (11.6-2). In other words,

$$\left[\nabla_t^2 + \frac{\omega^2}{c^2} n^2(x, y) \right] \mathbf{E}_1 = \beta_1^2 \mathbf{E}_1, \qquad (11.6\text{-}6)$$

$$\left[\nabla_t^2 + \frac{\omega^2}{c^2} n^2(x, y) \right] \mathbf{E}_2 = \beta_2^2 \mathbf{E}_2, \qquad (11.6\text{-}7)$$

where β_1 and β_2 are two different eigenvalues (propagation constants).

We now scalar multiply both sides of Eq. (11.6-6) by $\mathbf{E}_2$ and similarly scalar multiply both sides of Eq. (11.6-7) by $\mathbf{E}_1$. Subtracting the results [i.e., $\mathbf{E}_1 \cdot (11.6\text{-}7) - \mathbf{E}_2 \cdot (11.6\text{-}6)$], we obtain

$$(\mathbf{E}_1 \cdot \nabla_t^2 \mathbf{E}_2 - \mathbf{E}_2 \cdot \nabla_t^2 \mathbf{E}_1) = (\beta_2^2 - \beta_1^2)\mathbf{E}_2 \cdot \mathbf{E}_1. \qquad (11.6\text{-}8)$$

We now integrate both sides of this equation over the entire xy plane. By using integration by parts (or, equivalently, Green's identity), we obtain

$$\lim_{C \to \infty} \int_C [(\mathbf{E}_1 \cdot \nabla_t)\mathbf{E}_2 - (\mathbf{E}_2 \cdot \nabla_t)\mathbf{E}_1] \, dl = (\beta_2^2 - \beta_1^2) \iint \mathbf{E}_2 \cdot \mathbf{E}_1 \, dx \, dy,$$

$$(11.6\text{-}9)$$

where C denotes a closed path of integration at infinity and $\nabla_t = \mathbf{x} \, \partial/\partial x + \mathbf{y} \, \partial/\partial y$. For confined modes, $\mathbf{E}_1$ and $\mathbf{E}_2$ vanish at infinity. Therefore, the contour integration vanishes. Thus, we obtain

$$\iint \mathbf{E}_2 \cdot \mathbf{E}_1 \, dx \, dy = 0, \qquad (11.6\text{-}10)$$

provided $\beta_1 \neq \beta^2$.

The preceding proof breaks down when these two modes are degenerate such that $\beta_1 = \beta_2$. However, for degenerate modes, it is always possible to choose a suitable linear combination of the degenerate modes such that this subset of modes forms an orthogonal set.

If the wavefunctions of the modes are normalized to a power of 1 W, the orthonormalization of the modes can be written as

$$\iint \mathbf{E}_l \cdot \mathbf{E}_m \, dx \, dy = \frac{2\omega\mu}{|\beta_m|} \delta_{lm}, \qquad (11.6\text{-}11)$$

where δ_{lm} is the Kronecker delta and l, m are two arbitrary mode indices.

The orthonormality relation can also be written as

$$\tfrac{1}{2} \iint (\mathbf{E}_l \times \mathbf{H}_m) \cdot \mathbf{z} \, dx \, dy \; = \; \delta_{lm}, \qquad (11.6\text{-}12)$$

where $\mathbf{H}_m$ is the magnetic field vector and $\mathbf{z}$ is the unit vector along the positive z axis. In fact, Eq. (11.6-12) is the most general form of orthonormality. It is valid even when the approximation Eq. (1.4-9) breaks down (see Problems 11.1 and 11.2). When there are no losses present, Eq. (11.6-12) becomes

$$\tfrac{1}{2} \iint (\mathbf{E}_l \times \mathbf{H}_m^*) \cdot \mathbf{z} \, dx \, dy \; = \; \delta_{lm}. \qquad (11.6\text{-}13)$$

According to this latter relation, the power flow in a lossless dielectric waveguide is the sum of the power carried by each mode individually.

For TE or TM modes, the orthonormality [Eq. (11.6-13)] reduces to (see Problems 11.1 and 11.2)

$$\iint \mathbf{E}_l \cdot \mathbf{E}_m^* \, da \; = \; \frac{2\omega\mu}{|\beta_m|} \, \delta_{lm}, \quad \text{TE,}$$

$$\iint \mathbf{H}_l \cdot \frac{1}{\varepsilon} \, \mathbf{H}_m^* \, da \; = \; \frac{2\omega}{|\beta_m|} \, \delta_{lm}, \quad \text{TM,} \qquad (11.6\text{-}14)$$

respectively, where $da = dx \, dy$ is the area element of integration.

11.6.2   Energy Transport

In a lossless dielectric waveguide, each mode carries power and propagates along the guide independent of the presence of other modes. This is apparent from the orthogonal properties of the modes expressed by Eq. (11.6-13). The transport of power is given by the real part of the integral of the complex Poynting vector over the entire xy plane. Because of the orthogonal property of the modes, we are able to treat the power transport of one mode at a time. For a given mode of propagation, the time-averaged power flow is given by

$$P \; = \; \tfrac{1}{2} \operatorname{Re} \iint \mathbf{E} \times \mathbf{H}^* \cdot da. \qquad (11.6\text{-}15)$$

The field energy per unit length of the guide of the mode is given by

$$U \; = \; \tfrac{1}{4} \iint (\mathbf{E} \cdot \varepsilon\mathbf{E}^* + \mathbf{H} \cdot \mu\mathbf{H}^*) \, da, \qquad (11.6\text{-}16)$$

where we assume that ε and μ are real so that the integrand is also real. Since Maxwell's equations are linear, the power flow P and the energy density U

are proportional. The constant of proportionality has the dimension of velocity and is called *the velocity of energy propagation*:

$$v_e = \frac{P}{U}.$$ (11.6-17)

This energy velocity can be shown to equal the group velocity, which is defined as

$$v_g = \frac{\partial \omega}{\partial \beta}.$$ (11.6-18)

In a dielectric waveguide, the propagation constant β for each mode is a function of ω. When a light pulse with a finite spread in frequency, $\delta\omega$, is propagating in a guide, it is possible that the power of each spectral component is carried by only one mode. If $\delta\beta$ is the corresponding spread in the propagation constant of the mode, the velocity of the pulse is given by Eq. (11.6-18).

To prove that the energy velocity v_e and the group velocity v_g are equal, we start from Maxwell's equation by substituting $\nabla_t + \partial/\partial z$ for ∇ and obtain

$$\nabla_t \times \mathbf{H} + \mathbf{z} \times \frac{\partial}{\partial z}\mathbf{H} = i\omega\varepsilon\mathbf{E},$$ (11.6-19)

$$\nabla_t \times \mathbf{E} + \mathbf{z} \times \frac{\partial}{\partial z}\mathbf{E} = -i\omega\mu\mathbf{H},$$ (11.6-20)

where we recall that $\mathbf{z}$ is a unit vector along the guide (z axis). Since the z dependence of the mode is in the form $\exp(-i\beta z)$, Eqs. (11.6-19) and (11.6-20) can be written as

$$\nabla_t \times \mathbf{H} - i\beta\mathbf{z} \times \mathbf{H} = i\omega\varepsilon\mathbf{E},$$ (11.6-21)

$$\nabla_t \times \mathbf{E} - i\beta\mathbf{z} \times \mathbf{E} = -i\omega\mu\mathbf{H}.$$ (11.6-22)

Suppose now that β is changed by an infinitesimal amount, $\delta\beta$. If $\delta\omega$, $\delta\mathbf{E}$, and $\delta\mathbf{H}$ are the corresponding changes in ω, $\mathbf{E}$, and $\mathbf{H}$, respectively, we obtain, after a few steps of algebra,

$$\nabla_t \cdot F - 4i\,\delta\beta\,\mathrm{Re}\,[(\mathbf{E} \times \mathbf{H}^*) \cdot \mathbf{z}] = -2i\,\delta\omega[\mathbf{E} \cdot \varepsilon\mathbf{E}^* + \mathbf{H} \cdot \mu\mathbf{H}^*],$$ (11.6-23)

where F is given by

$$F = \delta\mathbf{E} \times \mathbf{H}^* + \delta\mathbf{H}^* \times \mathbf{E} + \mathbf{H} \times \delta\mathbf{E}^* + \mathbf{E}^* \times \delta\mathbf{H}.$$ (11.6-24)

If we perform an integration over the entire xy plane on Eq. (11.6-23) and use the two-dimensional divergence theorem,

$$\iint_S \nabla_t \cdot F \, da = \int_C F \cdot n \, dl, \qquad (11.6\text{-}25)$$

we obtain

$$\int_C F \cdot n \, dl - 8i \, \delta\beta \, P = -8i \, \delta\omega \, U, \qquad (11.6\text{-}26)$$

where C is a contour at infinity and P, U are given by Eqs. (11.6-15) and (11.6-16), respectively. The contour integral in Eq. (11.6-26) vanishes because the field amplitudes of the confined modes are zero at infinity. This leads to

$$\delta\beta \, P = \delta\omega \, U. \qquad (11.6\text{-}27)$$

By using the definition of group velocity and energy velocity, Eq. (11.6-27) can be written as

$$v_e \, \delta\beta = \delta\omega = \frac{\partial \omega}{\partial \beta} \delta\beta = v_g \, \delta\beta. \qquad (11.6\text{-}28)$$

Since $\delta\beta$ is an arbitrary infinitesimal number, we conclude that

$$v_e = v_g. \qquad (11.6\text{-}29)$$

11.7 PERTURBATION THEORY AND MODE COUPLING

In the preceding sections, we derived some general properties of dielectric waveguide modes and specifically obtained solutions for the confined modes supported by a slab waveguide. The guided modes can actually be excited and propagate along the axis (z) of the dielectric waveguide independently of each other, provided the dielectric function $\varepsilon(x, y) = \varepsilon_0 n^2(x, y)$ remains independent of z. In the case when there is a dielectric perturbation $\Delta\varepsilon(x, y, z)$ due to waveguide imperfections, bending, or surface corrugations, and so on, the eigenmodes are coupled among each other. In other words, if a pure mode is excited at the beginning of the waveguide, some of its power may be transferred to other modes. An increasingly large number of experiments and devices involve intentional coupling between such modes. Two typical examples involve TM $\leftrightarrow$ TE mode conversion by the electro-optic or acousto-optic effect or coupling of forward-to-backward modes by means of corrugation in one of the waveguide interfaces. In this section, we will derive the coupled-mode theory which may be used to describe such mode coupling. Dielectric perturbation is represented by a small perturbing term, $\Delta\varepsilon(x, y, z)$, such that the dielectric tensor as a function of space can be written as

$$\varepsilon(x, y, z) = \varepsilon_0(x, y) + \Delta\varepsilon(x, y, z), \qquad (11.7\text{-}1)$$

where $\varepsilon_0(x, y)$ is the unperturbed dielectric function describing the waveguide structure.

If an arbitrary field of frequency ω is excited at $z = 0$, the propagation of this field in such a perturbed waveguide can always be expressed in terms of a linear combination of the unperturbed eigenmodes, that is,

$$\mathbf{E} = \sum A_m(z)\mathbf{E}_m(x, y)\exp[i(\omega t - \beta_m z)], \qquad (11.7\text{-}2)$$

where $A_m(z)$ are the mode amplitudes and are assumed to be functions of z. The wavefunctions $\mathbf{E}_m(x, y)\exp[i(\omega t - \beta_m z)]$ satisfy the wave equation

$$\left[\frac{\partial^2}{\partial x^2} + \frac{\partial^2}{\partial y^2} + \omega^2 \mu \varepsilon_0(x, y) - \beta_m^2\right]\mathbf{E}_m(x, y) = 0, \qquad (11.7\text{-}3)$$

where we have neglected the term $\nabla(\nabla \cdot \mathbf{E})$, which is legitimate, provided the variation of $\varepsilon_0(x, y)$ over a wavelength is small. For the case of slab waveguides, $\varepsilon_0(x, y)$ is discontinuous at the dielectric interfaces, and $\nabla \cdot \mathbf{E}$ is zero for TE modes, since $\mathbf{E} \cdot \nabla\varepsilon = 0$. Therefore, Eq. (11.7-3) holds. For TM modes of the slab waveguides, Eq. (11.7-3) is not valid at the interfaces.

These modes are mutually orthogonal and satisfy the following orthonormality condition:

$$\langle m \,|\, k \rangle \equiv \int \mathbf{E}_m^* \cdot \mathbf{E}_k \, dx \, dy = \frac{2\omega\mu}{|\beta_m|} \delta_{km}. \qquad (11.7\text{-}4)$$

This condition is always valid provided Eq. (11.7-3) is true.

Substituting Eq. (11.7-2) into the wave equation

$$[\nabla^2 + \omega^2 \mu \varepsilon_0(x, y) + \omega^2 \mu \, \Delta\varepsilon(x, y, z)]\mathbf{E} = 0 \qquad (11.7\text{-}5)$$

and using Eq. (11.7-3) lead to

$$\sum_m \left(\frac{d^2}{dz^2} A_m - 2i\beta_m \frac{d}{dz} A_m\right)\mathbf{E}_m(x, y)e^{-i\beta_m z}$$

$$= -\omega^2 \mu \sum_m \Delta\varepsilon(x, y, z)A_m\mathbf{E}_m(x, y)e^{-i\beta_m z}. \qquad (11.7\text{-}6)$$

We now assume further that the dielectric perturbation is "weak" so that the variation of the mode amplitudes are "slow" and satisfies the condition

$$\left|\frac{d^2}{dz^2} A_k\right| \ll \left|\beta_k \frac{d}{dz} A_k\right|. \qquad (11.7\text{-}7)$$

This is the parabolic approximation used earlier in Eq. (8.5-8). Thus, neglecting the second-order derivatives and carrying out the scalar product of Eq. (11.7-6) with $E_k(x, y)$ and integrating over the entire xy plane, we obtain

$$\frac{d}{dz} A_k(z) = -\frac{i\omega}{4} \frac{\beta_k}{|\beta_k|} \sum_m \langle k|\Delta\varepsilon|m \rangle A_m(z) e^{i(\beta_k - \beta_m)z}, \qquad (11.7\text{-}8)$$

where we have used Eq. (11.7-4), and $\langle k|\Delta\varepsilon|m \rangle$ is defined as

$$\langle k|\Delta\varepsilon|m \rangle = \int E_k^* \cdot \Delta\varepsilon \, E_m \, dx \, dy. \qquad (11.7\text{-}9)$$

Equation (11.7-8) describes the evolution of the mode amplitudes $A_k(z)$ as the wave propagates along the waveguide. It is a set of coupled differential equations and can be solved for a large variety of mode interactions. Two important examples are considered here. These are the uniform dielectric perturbation in which $\Delta\varepsilon = \Delta\varepsilon(x, y)$ is a function of x and y only (i.e., $\partial\Delta\varepsilon/\partial z = 0$) and the periodic perturbation in which $\Delta\varepsilon$ is a periodic function of z.

11.7.1   Uniform Dielectric Perturbation

The coupled-mode approach discussed in the preceding provides a way of investigating the coupling between the unperturbed modes due to the presence of the dielectric perturbation, especially when the perturbation is z dependent. There are many practical situations where it is desirable or necessary to obtain the propagation constants and the modes of the whole waveguide structure, including the dielectric perturbation $\Delta\varepsilon(x, y)$. In the event that the wave equation is difficult to solve (or cannot be directly solved), the perturbation theory provides a method of treating such problems. Often, the perturbation $\Delta\varepsilon(x, y)$ is chosen such that the unperturbed waveguide structure can be easily solved or has known solutions.

Let the unperturbed modes of propagation be

$$E = E_m(x, y) \exp[i(\omega t - \beta_m z)], \qquad m = 0, 1, 2, \ldots, \qquad (11.7\text{-}10)$$

whose transverse mode functions E_m's satisfy the perturbed wave equation

$$[\nabla_t^2 + \omega^2 \mu\varepsilon(x, y)] E_m(x, y) = \beta_m^2 E_m(x, y). \qquad (11.7\text{-}11)$$

These modes form a complete orthogonal set and obey the orthonormalization relation Eq. (11.7-4).

We now consider the effect of a dielectric perturbation $\Delta\varepsilon(x, y)$ that is small compared with $\varepsilon(x, y)$. We assume that the application of such a small perturbation will only cause small changes in the mode functions and propagation constants. Let δE_m and $\delta\beta_m^2$ be the changes in the mode functions and propagation constants, respectively. The true wave equation now takes the form

$$[\nabla_t^2 + \omega^2\mu\varepsilon + \omega^2\mu\,\Delta\varepsilon](E_m + \delta E_m) = (\beta_m^2 + \delta\beta_m^2)(E_m + \delta E_m).$$

$$(11.7\text{-}12)$$

If we neglect the second-order terms $\Delta\varepsilon\,\delta E_m$ and $\delta\beta^2\,\delta E_m$ and use Eq. (11.7-11), Eq. (11.7-12) is simplifed to

$$[\nabla_t^2 + \omega^2\mu\varepsilon]\delta E_m + \omega^2\mu\,\Delta\varepsilon\,E_m = \beta_m^2\,\delta E_m + \delta\beta_m^2\,E_m. \qquad (11.7\text{-}13)$$

To solve this, we expand δE_m in terms of the unperturbed mode functions

$$\delta E_m(x, y) = \sum_l a_{ml}E_l(x, y), \qquad (11.7\text{-}14)$$

where the a_{ml}'s are constants. Substituting Eq. (11.7-14) for δE_m in Eq. (11.7-13) and using Eq. (11.7-11), we obtain

$$\sum_l a_{ml}(\beta_l^2 - \beta_m^2)E_l = (\delta\beta_m^2 - \omega^2\mu\,\Delta\varepsilon)E_m. \qquad (11.7\text{-}15)$$

If we now scalar multiply by E_m^* and integrate over the whole xy plane, we observe that the expression on the left vanishes because of the orthogonal property [Eq. (11.7-4)]. Thus, we obtain the equation

$$\int E_m^* \cdot (\delta\beta_m^2 - \omega^2\mu\,\Delta\varepsilon)E_m\,dx\,dy = 0. \qquad (11.7\text{-}16)$$

Since $\delta\beta_m^2$ is a constant, Eq. (11.7-16) can be written as

$$\delta\beta_m^2 = \frac{\int E_m^* \cdot \omega^2\mu\,\Delta\varepsilon E_m\,dx\,dy}{\int E_m^* \cdot E_m\,dx\,dy}. \qquad (11.7\text{-}17)$$

This gives the first-order correction to the propagation constant β_m^2. Using Eq. (11.7-4) and $\delta\beta_m^2 = 2\beta_m\,\delta\beta_m$, Eq. (11.7-17) can also be written as

$$\delta\beta_m = \frac{\omega}{4}\int E_m^* \cdot \Delta\varepsilon(x, y)\,E_m\,dx\,dy. \qquad (11.7\text{-}18)$$

To obtain the correction δE_m for the mode function, we scalar multiply each side of Eq. (11.7-15) by E_l^* ($l \neq m$) and integrate over x and y. This leads to

$$a_{ml}(\beta_l^2 - \beta_m^2)\frac{2\omega\mu}{\beta_l} = -\int E_l^* \cdot \omega^2\mu\,\Delta\varepsilon(x, y)E_m\,dx\,dy, \qquad (11.7\text{-}19)$$

where we have used the orthonormalization properties [Eq. (11.7-4)] of the unperturbed modes. The coefficients in the expansion [Eq. (11.7-14)] of δE_m in terms of the set E_l are thus given by

$$a_{ml} = \frac{\omega \beta_l}{2(\beta_m^2 - \beta_l^2)} \int E_l^* \cdot \Delta\varepsilon(x, y)E_m \, dx \, dy, \qquad l \neq m. \qquad (11.7\text{-}20)$$

The value of a_{mm} is not given by this process; it is to be chosen so as to normalize the resultant mode function according to Eq. (11.7-4) and is given by (see Problem 11.7)

$$a_{mm} = -\frac{1}{4}\frac{\delta\beta_m^2}{\beta_m^2} = -\frac{1}{2}\frac{\delta\beta_m}{\beta_m}. \qquad (11.7\text{-}21)$$

Using Eq. (11.7-18) for $\delta\beta_m$, the coefficient a_{mm} can thus be written as

$$a_{mm} = -\frac{\omega}{8\beta_m} \int E_m^* \cdot \Delta\varepsilon(x, y)E_m \, dx \, dy. \qquad (11.7\text{-}22)$$

It is convenient to use the "coupling coefficients"

$$\kappa_{lm} = \frac{\omega}{4} \int E_l^* \cdot \Delta\varepsilon(x, y)E_m \, dx \, dy \qquad (11.7\text{-}23)$$

so that the expression for the first-order correction to the mode function can be written, according to Eqs. (11.7-20) and (11.7-14), as

$$\delta E_m = \sum_{l \neq m} \frac{2\beta_l}{\beta_m^2 - \beta_l^2} \kappa_{lm} E_l - \frac{\kappa_{mm}}{2\beta_m} E_m. \qquad (11.7\text{-}24)$$

The correction to the propagation constant, $\delta\beta_m$ [Eq. (11.7-18)], is thus

$$\delta\beta_m = \kappa_{mm}. \qquad (11.7\text{-}25)$$

The results [Eqs. (11.7-23) and (11.7-17)] can be used to evaluate the attenuation coefficients of the modes in a waveguide, where loss due to bulk absorption is small and can be considered as a small perturbation.

11.7.2 Periodic Dielectric Perturbation

Consider now a special case of dielectric perturbation that is periodic in z. For the sake of clarity in describing the perturbation theory, we assume that such a perturbation is sinusoidal and is given by

$$\Delta\varepsilon = \varepsilon_1 e^{-iKz} + \varepsilon_1^* e^{iKz}, \qquad (11.7\text{-}26)$$

where ε_1 is the amplitude of the periodic perturbation and can be a function of x and y, ε_1^* is the complex conjugate of ε_1, and K is given by

$$K = \frac{2\pi}{\Lambda}, \tag{11.7-27}$$

with Λ the period of the perturbation.

Substituting Eq. (11.7-26) for $\Delta\varepsilon$ into the coupled-mode equation (11.7-8), we notice that coupling is significant only when

$$\beta_k - \beta_m \pm K = 0. \tag{11.7-28}$$

When condition (11.7-28) is not satisfied, the fast-varying phase on the right side leads to vanishing contributions to the integration. Thus, Eq. (11.7-28) is also called a phase-matching condition.

Let the two coupled modes be designated as 1 and 2. Neglecting interaction with any of the other modes, Eq. (11.7-8) for $k = 1, 2$ can be written as

$$\frac{d}{dz} A_1 = -i \frac{\beta_1}{|\beta_1|} \kappa_{12} A_2 e^{i\Delta\beta z},$$

$$\frac{d}{dz} A_2 = -i \frac{\beta_2}{|\beta_2|} \kappa_{12}^* A_1 e^{-i\Delta\beta z}, \tag{11.7-29}$$

where

$$\Delta\beta = \beta_1 - \beta_2 - \frac{2\pi}{\Lambda}, \tag{11.7-30}$$

and κ_{12} is the coupling coefficient [according to Eq. (11.7-9)]

$$\kappa_{12} = \frac{\omega}{4} \iint \mathbf{E}_1^* \cdot \varepsilon_1 \mathbf{E}_2 \, dx \, dy, \tag{11.7-31}$$

where we recall that ε_1 is the amplitude of the periodic perturbation.

Notice that the coupling coefficients [Eq. (11.7-31)] depend on the polarization states of the coupled modes as well as on the tensor property of ε_1. We also note that Eq. (11.7-31) is formally identical to Eq. (11.7-23).

The signs of the factors $\beta_1/|\beta_1|$ and $\beta_2/|\beta_2|$ in the coupled equations (11.7-29) are very important and will determine the behavior of the coupling. These signs, of course, depend on the direction of propagation of the coupled modes. The coupling is therefore divided into two categories: codirectional and contradirectional coupling. Equation (11.7-29) can be used to study a wide variety of mode couplings. Interested readers are referred to Ref. 3 for details.

11.8 COUPLING OF TWO WAVEGUIDES

In the previous section, we treated the coupling between the modes of an arbitrary waveguide due to dielectric perturbation. Confined modes of two different waveguides can also couple with each other. Coupling between the modes of two different waveguides is significant only when these two guides are close to each other such that there is some overlap of the mode wavefunctions. Wave propagation in such a composite waveguide structure can be treated by two approaches. In the coupled-mode analysis (or perturbation theory), the electric field is represented by a linear combination of the unperturbed normal modes with varying amplitudes. Variation of the amplitudes with distance indicates the transfer of energy between the unperturbed modes. In the other approach, the electric field is represented by the actual normal modes of the composite structure. In Section 11.3, when we studied electromagnetic propagation in multichannel waveguides, especially in the two-channel, such actual normal modes are derived as solutions of the wave equation. There are many situations when solutions of the wave equation for a composite waveguide structure, which may consist of two individual channels, are not available. Under such circumstances, wave propagation in two-channel waveguides can be treated by using the perturbation theory developed in the last section.

We now consider coupling between the modes of two parallel waveguides separated by a finite distance from each other. Exchange of power between guided modes of adjacent waveguides is known as *directional coupling*. This phenomenon is similar to electron motion in a two-atom molecule. The coupling of the two waveguides plays a key role in a number of useful functions in thin-film devices, including power division, modulation, switching, frequency selection, and polarization selection.

Waveguide coupling can be treated by the coupled-mode theory. Consider the case of two cylindrical waveguides illustrated in Fig. 11.25. Let $\mathbf{E}_a(x, y)$ $\exp[i(\omega t - \beta_a z)]$ and $\mathbf{E}_b(x, y) \exp[i(\omega t - \beta_b z)]$ be the modes of propagation of the individual waveguides when they are far apart. The electric field in the coupled-guide structure can be approximated by

$$\mathbf{E}(x, y, z, t) = A(z)\mathbf{E}_a(x, y)e^{i(\omega t - \beta_a z)} + B(z)\mathbf{E}_b(x, y)e^{i(\omega t - \beta_b z)} \qquad (11.8\text{-}1)$$

provided the two waveguides are not too close to each other. In the absence of coupling—that is, if the distance between guides a and b is infinite—$A(z)$

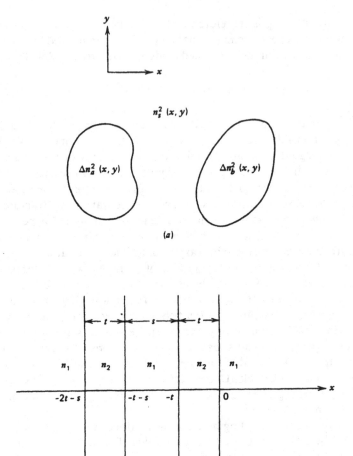

Figure 11.25 (a) A general waveguide structure consisting of two parallel cylindrical guides; (b) two identical slab waveguides separated by a distance t: $\Delta n_b^2(x) = (n_2^2 - n_1^2)$ for $-t < x < 0$; $\Delta n_a^2(x) = (n_2^2 - n_1^2)$ for $-2t - s < x < -t - s$. (Adapted from A. Yariv and P. Yeh, *Optical Waves in Crystals*, Wiley, New York, p. 460. Copyright © 1984. By permission of John Wiley & Sons, Inc.)

and $B(z)$ do not depend on z and will be independent of each other, since each of the two terms on the right side of Eq. (11.8-1) satisfies the wave equation (11.7-3) separately.

Let $n^2(x, y)$ be the refractive index distribution of the composite waveguide structure. To study the mode coupling, we decompose the refractive index profile into three parts, and we get

$$n^2(x, y) = n_s^2(x, y) + \Delta n_a^2(x, y) + \Delta_b^2(x, y), \qquad (11.8\text{-}2)$$

where $n_s^2(x, y)$ represents the refractive index distribution of the supporting medium (cladding), $\Delta n_a^2(x, y)$ represents the presence of waveguide a, and $\Delta n_b^2(x, y)$ represents the presence of waveguide b. It is thus obvious that the individual waveguide modes $E_a(x, y)$ satisfy the equation

$$\left\{ \frac{\partial^2}{\partial x^2} + \frac{\partial^2}{\partial y^2} + \frac{\omega^2}{c^2} [n_s^2(x, y) + \Delta n_a^2(x, y)] \right\} E_a(x, y) = \beta_a^2 E_a(x, y),$$

$$\alpha = a, b. \qquad (11.8\text{-}3)$$

The presence of waveguide b imposes a dielectric perturbation $\varepsilon_0 \, \Delta n_b^2(x, y)$ to the propagation of the modes $E_a(x, y) \exp[i(\omega t - \beta_a z)]$ and vice versa. The total electric field [Eq. (11.8-1)] must obey the wave equation

$$\left\{ \frac{\partial^2}{\partial x^2} + \frac{\partial^2}{\partial y^2} + \frac{\partial^2}{\partial z^2} + \frac{\omega^2}{c^2} [n_s^2(x, y) + \Delta n_a^2(x, y) + \Delta n_b^2(x, y)] \right\} E = 0.$$

$$(11.8\text{-}4)$$

To obtain the coupled equations for the mode amplitudes $A(z)$ and $B(z)$, we substitute Eqs. (11.8-1) into (11.8-4) and use Eq. (11.8-3) and the assumption of slow variation of mode amplitudes over z [Eq. (11.7-7)]. These lead to

$$-2i\beta_a \frac{dA}{dz} E_a e^{i(\omega t - \beta_a z)} - 2i\beta_b \frac{dB}{dz} E_b e^{i(\omega t - \beta_b z)}$$

$$= -\frac{\omega^2}{c^2} \Delta n_b^2(x, y) A E_a e^{i(\omega t - \beta_a z)} - \frac{\omega^2}{c^2} \Delta n_a^2(x, y) B E_b e^{i(\omega t - \beta_b z)}. \quad (11.8\text{-}5)$$

We now take the scalar product of Eq. (11.8-5) with $E_a^*(x, y)$ and $E_b^*(x, y)$, respectively, and integrate over all x and y. The result, using the normalization condition (11.7-4), is

$$\frac{dA}{dz} = -i\kappa_{ab} B e^{i(\beta_a - \beta_b)z} - i\kappa_{aa} A,$$

$$(11.8\text{-}6)$$

$$\frac{dB}{dz} = -i\kappa_{ba} A e^{-i(\beta_a - \beta_b)z} - i\kappa_{bb} B,$$

where

$$\kappa_{ab} = \frac{\omega\varepsilon_0}{4} \int \mathbf{E}_a^* \cdot \Delta n_a^2(x, y)\mathbf{E}_b \, dx \, dy, \tag{11.8-7}$$

$$\kappa_{ba} = \frac{\omega\varepsilon_0}{4} \int \mathbf{E}_b^* \cdot \Delta n_b^2(x, y)\mathbf{E}_a \, dx \, dy,$$

$$\kappa_{aa} = \frac{\omega\varepsilon_0}{4} \int \mathbf{E}_a^* \cdot \Delta n_b^2(x, y)\mathbf{E}_a \, dx \, dy, \tag{11.8-8}$$

$$\kappa_{bb} = \frac{\omega\varepsilon_0}{4} \int \mathbf{E}_b^* \cdot \Delta n_a^2(x, y)\mathbf{E}_b \, dx \, dy.$$

In arriving at Eq. (11.8-6), we use the assumption that the waveguides are not too close, so that the overlap integral of the mode functions is small, that is,

$$\int \mathbf{E}_a^* \cdot \mathbf{E}_b \, dx \, dy \ll \int \mathbf{E}_a^* \cdot \mathbf{E}_a \, dx \, dy. \tag{11.8-9}$$

The terms κ_{aa} and κ_{bb} result from the dielectric perturbations to one of the waveguides due to the presence of the other waveguide and represent only a small correction to the propagation constants β_a and β_b, respectively [see Eqs. (11.7-18) and (11-7-25)]. The terms κ_{ab} and κ_{ba} represent the exchange coupling between the two guides. So, if we take the total field as

$$\mathbf{E} = A(z)\mathbf{E}_a e^{i[\omega t - (\beta_a + \kappa_{aa})z]} + B(z)\mathbf{E}_b e^{i[\omega t - (\beta_b + \kappa_{bb})z]} \tag{11.8-10}$$

instead of Eq. (11.8-1), Eq. (11.8-6) becomes

$$\frac{dA}{dz} = -i\kappa_{ab} B e^{i\Delta\beta z}, \qquad \frac{dB}{dz} = -i\kappa_{ba} A e^{-i\Delta\beta z}, \tag{11.8-11}$$

where

$$\Delta\beta = (\beta_a + \kappa_{aa}) - (\beta_a + \kappa_{bb}). \tag{11.8-12}$$

The solution of Eq. (11.8-11) subject to a single input at guide a [i.e., $A(0) = A_0$, $B(0) = 0$] can be solved easily (see Problem 11.9). In terms of powers $P_a = A^*A$ and $P_b = B^*B$ in the two guides, the solution in the case $\kappa_{ab} = \kappa_{ba} = \kappa$ becomes

$$P_a(z) = P_0 - P_b(z),$$

$$P_b(z) = P_0 \frac{\kappa^2}{\kappa^2 + (\Delta\beta/2)^2} \sin^2\left\{ \left[\kappa^2 + \left(\frac{\Delta\beta}{2}\right)^2 \right]^{1/2} z \right\}, \tag{11.8-13}$$

where $P_0 = |A(0)|^2$ is the input power to guide a. Complete power transfer occurs in a distance $L = \pi/2\kappa$ provided $\Delta\beta = 0$; that is, equal phase velocities in both modes. For $\Delta\beta \neq 0$, the maximum fraction of power that can be transferred is, from Eq. (11.8-13),

$$\frac{\kappa^2}{\kappa^2 + (\Delta\beta/2)^2} \tag{11.8-14}$$

The coupling constant $\kappa = \kappa_{ab} = \kappa_{ba}$ is given by Eq. (11.8-7). It can be evaluated straghtforwardly using the field expressions (11.2-3) and (11.2-10) for the case of two parallel slab dielectric waveguides. In the special case of two identical slab waveguides, the coupling constant κ for the same TE modes of propagation is

$$\kappa = \frac{2h^2 p e^{-ps}}{\beta[t + (2/p)](h^2 + p^2)^2} \left(\frac{2\pi}{\lambda}\right)^2 (n_2^2 - n_1^2), \tag{11.8-15}$$

where t is the guide width, s is the separation, and $p = q$ and h are given by Eq. (11.2-4). In the well-confined case, $t \gg 2/p$, and Eq. (11.8-15) simplifies to

$$\kappa = \frac{2h^2 p e^{-ps}}{\beta t (h^2 + p^2)^2} \left(\frac{2\pi}{\lambda}\right)^2 (n_2^2 - n_1^2). \tag{11.8-16}$$

A typical value of κ obtained at $\lambda \sim 1\,\mu m$ with t, $s \sim 3\,\mu m$ and $\Delta n \sim 5 \times 10^{-3}$ is $\kappa \sim 5\,cm^{-1}$, so that coupling distances are on the order of magnitude of $\kappa^{-1} \simeq 2\,mm$.

11.8.1 Approximate Wavefunctions and Propagation Constants

In addition to the coupled-mode analysis, perturbation theory can also be used to find approximate solutions to the wave equation that governs the propagation of electromagnetic radiation in the composite structure. In other words, the approximation of the actual normal modes can be obtained by carrying out a perturbation analysis. This technique is particularly useful when the wavefunctions as well as the eigenvalues (propagation constants) of the individual guides are known and the closed-form solutions of the wavefunctions and eigenvalues of the composite waveguide structure are not available.

Again, let $E_a(x, y)\exp[i(\omega t - \beta_a z)]$ and $E_b(x, y)\exp[i(\omega t - \beta_b z)]$ be the fundamental modes of the individual waveguides when they are apart. The waveguide modes of the composite structure can be approximated by

$$E(x, y)e^{-i\beta z} = [C_a E_a(x, y) + C_b E_b(x, y)]e^{-i\beta z} \tag{11.8-17}$$

provided that the two guides are not too close to each other.

To calculate the propagation constant β and evaluate the constants C_a and C_b, we decompose the refractive index profile into three parts, according to Eq. (11.8-2). Substituting Eq. (11.8-17) into the wave equation (11.8-4) and using Eq. (11.8-8), we obtain

$$
C_a \left[\beta_a^2 + \frac{\omega^2}{c^2} \Delta n_b^2(x, y) - \beta^2 \right] \mathbf{E}_a + C_b \left[\beta_b^2 + \frac{\omega^2}{c^2} \Delta n_a^2(x, y) - \beta^2 \right] \mathbf{E}_b = 0.
$$

$$
(11.8\text{-}18)
$$

We scalar multiply Eq. (11.8-18) with $\mathbf{E}_a^*$ and $\mathbf{E}_b^*$ one at a time and then integrate over all x and y. These lead to the following linear equation for C_a and C_b:

$$
\begin{pmatrix} \beta_a^2 - \beta^2 + K_a & J_a + I^*(\beta_b^2 - \beta^2) \\ J_b + I(\beta_a^2 - \beta^2) & \beta_b^2 - \beta^2 + K_b \end{pmatrix} \begin{pmatrix} C_a \\ C_b \end{pmatrix} = 0, \quad (11.8\text{-}19)
$$

with

$$
I = \int \mathbf{E}_b^* \cdot \mathbf{E}_a \, dx \, dy,
$$

$$
J_a = \left(\frac{\omega}{c} \right)^2 \int \mathbf{E}_a^* \cdot \Delta n_a^2 \, \mathbf{E}_b \, dx \, dy,
$$

$$
J_b = \left(\frac{\omega}{c} \right)^2 \int \mathbf{E}_b^* \cdot \Delta n_b^2 \, \mathbf{E}_a \, dx \, dy, \qquad (11.8\text{-}20)
$$

$$
K_b = \left(\frac{\omega}{c} \right)^2 \int \mathbf{E}_b^* \cdot \Delta n_a^2 \, \mathbf{E}_b \, dx \, dy,
$$

$$
K_a = \left(\frac{\omega}{c} \right)^2 \int \mathbf{E}_a^* \cdot \Delta n_b^2 \, \mathbf{E}_a \, dx \, dy,
$$

and

$$
\int |\mathbf{E}_a|^2 \, dx \, dy = \int |\mathbf{E}_b|^2 \, dx \, dy = 1. \qquad (11.8\text{-}21)
$$

These integrals are carried out over the entire xy plane. Here, $\Delta n_a^2(x, y)$ and $\Delta n_b^2(x, y)$ vanish except at the cores a and b, respectively. Physically, I is the overlap integral of the two individual wave functions that are not orthogonal to each other, K_a and K_b are the dielectric perturbations to one of the waveguides due to the presence of the other waveguide, and J_a and J_b represent the exchange coupling between the two waveguides.

The eigenvalues β_1 and β_2 are the solutions of the secular equation

$$\begin{vmatrix} \beta_a^2 - \beta^2 + K_a & J_a + I^*(\beta_b^2 - \beta^2) \\ J_b + I(\beta_a^2 - \beta^2) & \beta_b^2 - \beta^2 + K_b \end{vmatrix} = 0. \qquad (11.8\text{-}22)$$

This is a quadratic equation in β^2 and, in general, will yield two desired propagation constants β_1, β_2. The corresponding wavefunctions [Eq. (11.8-17)] can be obtained from Eq. (11.8-19) by solving for C_a and C_b.

The expressions are much simpler in the degenerate case when the two waveguides are identical, that is, $\beta_a^2 = \beta_b^2 \equiv \beta_0^2$, $K_a = K_b \equiv K$, and $J_a = J_b \equiv J$. In this case, the propagation constants and the corresponding wavefunctions are given by

$$\beta^2 = \beta_0^2 + \frac{K \pm J}{1 \pm I}, \qquad (11.8\text{-}23)$$

$$\mathbf{E} = \frac{1}{[2(1 \pm I)]^{1/2}} (\mathbf{E}_a \pm \mathbf{E}_b). \qquad (11.8\text{-}24)$$

The plus sign is for the symmetric one and the minus sign is for the anti-symmetric one. The factor $1/[2(1 \pm I)]^{1/2}$ is the normalization coefficient. We also note that these two solutions are already mutually orthogonal.

If we take $\mathbf{E}_a$ and $\mathbf{E}_b$ as the fundamental mode shown in Fig. 11.3 and then compare the result in Eq. (11.8-24) with the exact solution in Fig. 11.14, we find that the approximation is very good.

11.9 EFFECTIVE INDEX THEORY

Although simple layered structures such as dielectric slabs can be used for waveguiding purposes, the confinement of energy is only limited to one dimension. In practice, more complicated waveguide structures are used. For example, the waveguides used in integrated optics or guided-wave optics are usually two-dimensional waveguides and ridge waveguides. Figure 11.26 illustrates the geometry of such two-dimensional waveguides. Exact analytical treatment of these waveguide structures is not possible. Although numerical solutions can be obtained by various methods, there are several approximate analytical approaches. Here, we will introduce one of the simplest approaches, the *effective index theory*.

Referring to Fig. 11.26(c), we consider the guiding of electromagnetic radiation in a ridge waveguide. The rectangular waveguide shown in Fig. 11.26(b) can be considered as a special case of this ridge waveguide by

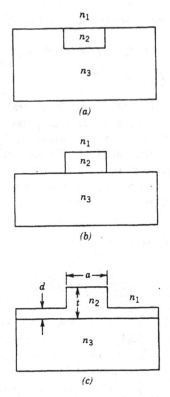

Figure 11.26 Schematic drawing of several two-dimensional waveguides: (*a*) a rectangular strip of dielectric medium embedded in another dielectric medium of lower index of refraction; (*b*) ridge waveguide structure; and (*c*) another ridge structure.

taking $d = 0$, where d is the thickness of the guiding layer on both sides of the ridge. The thickness of the guiding layer at the ridge is t, which is chosen to be greater than d because we are interested in the confinement of electromagnetic radiation at the ridge. The width of the ridge is a. The index of refraction n_2 is greater than either n_1 or n_3.

We now divide the structure into three segments for the purpose of introducing the concept of effective index theory. These segments are $y < -\frac{1}{2}a$, $-\frac{1}{2}a < y < \frac{1}{2}a$, and $\frac{1}{2}a < y$. In each of the three segments, the confinement along the vertical direction (x) is exactly identical to that of the asymmetric slab waveguide. For simplicity, we assume that only one TE mode is supported by these segments of the slab waveguide. The propagation constant of the confined mode inside the ridge is, however, different from those on both sides of the ridge because of the difference in layer thickness. The effective index of refraction (or the normalized propagation constant) defined in Eq. (11.1-20) for these three segments is

$$
n_{\text{eff}}(y) = \begin{cases} n_1, & y < -\tfrac{1}{2}a, \\ n_{\text{II}}, & -\tfrac{1}{2}a < y < \tfrac{1}{2}a, \\ n_1, & \tfrac{1}{2}a < y, \end{cases} \tag{11.9-1}
$$

where n_1, n_{II} are the effective indices of refraction.

In the theory of effective index, lateral waveguiding is treated by taking Eq. (11.9-1) as a slab waveguide structure along the y direction. Thus, confinement of electromagnetic radiation requires that $n_{\text{II}} > n_1$. This is always true when the thickness t is greater than d because, according to Fig. 11.7, the effective index of refraction of any confined mode is an increasing function of thickness. To illustrate further the lateral guiding, we will consider the following example.

11.9.1 Example: GaAs Ridge Waveguide

We consider a ridge waveguide made of GaAs layers and AlGaAs substrate. Let the indices of refraction be $n_1 = 1$, $n_2 = 3.5$, and $n_3 = 3.2$. The thicknesses are $t = 0.40\lambda$ and $d = 0.25\lambda$. According to a numerical solution of Eq. (11.2-4) or Fig. 11.17, the effective indices are

$$
n_1 = 3.301, \qquad n_{\text{II}} = 3.388.
$$

We note that a step height of 0.15λ at the ridge gives rise to a difference in the effective index of 0.087. Such an index difference is sufficient to provide lateral waveguiding. In fact, if we take $a = 0.5\lambda$ as the width of the ridge and solve for the confined TE modes of such a symmetric waveguide, we obtain a single TE mode with a normalized propagation constant of 3.348. The wavefunction of such a mode is plotted in Fig. 11.27. This wavefunction shows the lateral confinement.

11.10 COUPLING OF *N* IDENTICAL WAVEGUIDES

The modes for the periodic multichannel waveguides, which consist of N identical waveguides, are often called *supermodes* in laser diode arrays.

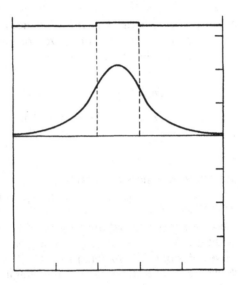

Figure 11.27 Lateral confinement of a ridge waveguide with $n_1 = 1$, $n_2 = 3.5$, $n_3 = 3.2$, $t = 0.4\lambda$, $d = 0.25\lambda$, and $a = 0.5\lambda$. The upper curve shows the effective index profile.

Although closed-form solutions for the mode dispersion relations, Eqs. (11.3-16) and (11.3-21), can be obtained by using the matrix method, numerical analysis is still needed to find the roots of these transcendental equations.

In Section 11.8, the perturbation approach was used to obtain an approximate solution for the propagation constants and the wavefunctions of two parallel waveguide. We now extend this approach to the case of N identical waveguides. For the sake of simplicity in illustrating the concept, we assume that each of the individual waveguides supports only one confined mode. In this approach, the wavefunctions of the supermodes are written as linear combinations of the unperturbed wavefunctions of the individual waveguides,

$$\mathbf{E}(x, y) = \sum_{n=1}^{N} C_n \mathbf{E}_n(x, y), \qquad (11.10\text{-}1)$$

where $\mathbf{E}_n(x, y)$ is the unperturbed wavefunction of the mode supported by the nth waveguide when the separation between waveguides is infinite. In this expansion, we neglect the coupling with other modes of different propagation constants (e.g., radiation modes).

We now follow the approach that leads to Eq. (11.8-19) and assume that only nearest-neighbor coupling is present. This leads to

$$\begin{pmatrix} K-\Delta & J-I\Delta & 0 & 0 & \cdots & 0 & 0 \\ J-I\Delta & K-\Delta & J-I\Delta & 0 & \cdots & 0 & 0 \\ 0 & J-I\Delta & K-\Delta & J-I\Delta & \cdots & 0 & 0 \\ 0 & 0 & J-I\Delta & K-\Delta & \cdots & 0 & 0 \\ \vdots & \vdots & \vdots & \vdots & \cdots & \vdots & \vdots \\ 0 & 0 & 0 & 0 & \cdots & J-I\Delta & 0 \\ 0 & 0 & 0 & 0 & \cdots & K-\Delta & J-I\Delta \\ 0 & 0 & 0 & 0 & \cdots & J-I\Delta & K-\Delta \end{pmatrix} \begin{pmatrix} C_1 \\ C_2 \\ C_3 \\ C_4 \\ \vdots \\ C_{N-2} \\ C_{N-1} \\ C_N \end{pmatrix} = 0,$$

$$(11.10\text{-}2)$$

where I, J, and K are the integrals defined in Eq. (11.8-20) for a pair of neighboring waveguides, and Δ is given by

$$\Delta = \beta^2 - \beta_0^2, \qquad (11.10\text{-}3)$$

where β_0 is the propagation constant of the unperturbed waveguides. In arriving at Eq. (11.10-2) we used $I^* = I$, $J_a = J_b$ and $K_a = K_b$.

Notice that all the elements of the matrix in Eq. (11.10-2) are zero except the diagonal elements and those immediately next to the diagonal elements. This is a result of nearest-neighbor coupling. In addition, all the diagonal elements are identical $(K - \Delta)$, and all the nonzero off-diagonal elements are also identical $(J - I\Delta)$. This is a result of the fact that all the individual waveguides are identical and equally spaced. Closed-form solutions exist for such a simple and symmetric matrix equation.

The matrix equation (11.10-2) can be rewritten as

$$(J - I\Delta)C_n + (K - \Delta)C_{n+1} + (J - I\Delta)C_{n+2} = 0, \quad (11.10\text{-}4)$$

with the boundary condition

$$C_0 = C_{N+1} = 0. \qquad (11.10\text{-}5)$$

The solution to Eq. (11.10-4) subject to the preceding boundary condition [Eq. (11.10-5)] can be written as

$$C_n = \sin \frac{ns\pi}{N+1}, \qquad (11.10\text{-}6)$$

where s is an arbitrary integer given by

$$s = 1, 2, 3, \ldots, N. \tag{11.10-7}$$

Substituting Eq. (11.10-6) for C_n in Eq. (11.10-4), we obtain the following closed-form solution for the propagation constants of the supermodes:

$$\beta^2 - \beta_0^2 = \Delta = \frac{K + 2J \cos[s\pi/(N + 1)]}{1 + 2I \cos[s\pi/(N + 1)]}, \tag{11.10-8}$$

where we recall that β_0 is the propagation constants of the unperturbed waveguide and $s = 1, 2, 3, \ldots, N$ and that I, J, and K are the integrals defined in Eq. (11.8-20) for a pair of neighboring waveguides. If the overlap integral I is much less than 1 ($I \ll 1$), Eq. (11.10-8) can be written approximately as

$$\beta^2 - \beta_0^2 = \Delta = K + 2J \cos \frac{s\pi}{N + 1}. \tag{11.10-9}$$

Notice that there are N independent supermodes. The propagation constants Δ span over a region of $K \pm 2J$. In terms of the Bloch modes and band theory, Eq. (11.10-9) represents an allowed band with a width of $4J$, when N is infinite.

To compare with earlier results for the case of two coupled waveguides, let us put $N = 2$ in Eq. (11.10-8). The propagation constants according to Eq. (11.10-8) and the wavefunctions according to Eqs. (11.10-1) and (11.10-6) are identical to the results obtained in Section 11.8. To illustrate further the use of the perturbation approach, we consider the following example.

11.10.1 Example: A Five-Channel Waveguide

Consider the coupling of five identical waveguides that are equally spaced. Let us assume that the individual waveguide supports only one mode. Find the propagation constants and the wavefunctions of the supermodes. Taking $N = 5$ and putting $s = 1, 2, 3, 4, 5$ in Eq. (11.10-9), we obtain the following values of Δ:

$$K + \sqrt{3}\,J, \quad K + J, \quad K, \quad K - J, \quad K - \sqrt{3}\,J.$$

The propagation constants are given by $\beta^2 = \beta_0^2 + \Delta$.

The wavefunctions can be obtained by using Eqs. (11.10-1) and (11.10-6) with the substitution of $N = 5$ and $s = 1, 2, 3, 4, 5$. The results are

$$E(s = 1) = \tfrac{1}{2}E_1 + \tfrac{1}{2}\sqrt{3}\,E_2 + E_3 + \tfrac{1}{2}\sqrt{3}\,E_4 + \tfrac{1}{2}E_5,$$
$$E(s = 2) = E_1 + E_2 - E_4 - E_5,$$

$$E(s = 3) = E_1 - E_3 + E_5,$$

$$E(s = 4) = E_1 - E_2 + E_4 - E_5,$$

$$E(s = 5) = \tfrac{1}{2}E_1 - \tfrac{1}{2}\sqrt{3}\,E_2 + E_3 - \tfrac{1}{2}\sqrt{3}\,E_4 + \tfrac{1}{2}E_5.$$

Notice that these wavefunctions are not normalized. They are, however, mutually orthogonal. Comparing these wavefunctions with the numerical results shown in Fig. 11.15, we find that the perturbation approach is quite good.

REFERENCES AND SUGGESTED READINGS

1. R. E. Collin, *Field Theory of Guided Waves*, McGraw-Hill, New York, 1960.
2. D. Marcuse, *Theory of Dielectric Optical Waveguides*, Academic, New York, 1974.
3. A. Yariv and P. Yeh, *Optical Waves in Crystals*, Wiley, New York, 1984.

PROBLEMS

11.1 Show that in the layers where $\beta^2 > (\omega/c)^2 n^2$, the wavefunction $E(x)$ can have, at most, one zero crossing. *Hint:* Use the continuous nature of $E(x)$ and $(1/E)(\partial^2 E/\partial x^2) > 0$.

11.2 Show that local maxima of $|E(x)|$ can only occur in the layers where $\beta^2 < (n\omega/c)^2$.

11.3 *Orthonormality relation for TE modes.* Let $(\mathbf{E}_l, \mathbf{H}_l)$ and $(\mathbf{E}_m, \mathbf{H}_m)$ be two arbitrary modes of the forms (11.1-3), and let I_{lm} be the integral

$$I_{lm} = \tfrac{1}{2} \iint (\mathbf{E}_l \times \mathbf{H}_m^*) \cdot \mathbf{z}\, da = \delta_{lm}.$$

(a) Show that

$$I_{lm} = \frac{\beta_m}{2\omega\mu} \int \mathbf{E}_l \cdot \mathbf{E}_m^*\, da + \frac{i}{2\omega\mu} \iint (\mathbf{E}_l \cdot \nabla)E_{mz}^*\, da$$

by using Maxwell's equations (11.1-2). This shows that for TE waves with $E_z = 0$, the orthornormality relation (11.6-13) reduces to (11.6-14). *Hint:* Use the vector identity

$$[\mathbf{A} \times (\nabla \times \mathbf{B})] \cdot \mathbf{C} = \mathbf{A} \cdot (\mathbf{C} \cdot \nabla)\mathbf{B} - \mathbf{C} \cdot (\mathbf{A} \cdot \nabla)\mathbf{B}.$$

(b) Show that, from (a) and integrating by parts,

$$2\delta_{lm} = I_{lm} + I_{ml}^* = \frac{\beta_l + \beta_m}{2\omega\mu} \iint \mathbf{E}_l \cdot \mathbf{E}_m^* \, da$$

$$+ \frac{i}{2\omega\mu} \iint (\mathbf{E}_{lz} \nabla \cdot \mathbf{E}_m^* + \mathbf{E}_{mz}^* \nabla \cdot \mathbf{E}_l) \, da.$$

Thus, either $\nabla \cdot \mathbf{E} = 0$ or $\mathbf{z} \cdot \mathbf{E} = 0$ is sufficient for the reduction of the orthonormality relations from Eqs. (11.6-13) and (11.6-14).

11.4 *Orthonormality relation for TM waves.* Follow the procedures in Problem 11.3 and use the same definition for I_{lm}.

(a) Show that

$$I_{lm} = \frac{\beta_m}{2\omega} \iint \mathbf{H}_l \cdot \frac{1}{\varepsilon} \mathbf{H}_m^* \, da + \iint \frac{i}{2\omega\varepsilon} (\mathbf{H}_l \cdot \nabla) \mathbf{H}_{mz}^* \, da.$$

Thus, for TM waves, $H_z = 0$, and the orthonormality relation (11.6-13) reduces to (11.6-14).

(b) Derive a relation similar to that of Problem 11.3(b). Show that $\mathbf{H} \cdot \nabla\varepsilon = 0$ or $\mathbf{z} \cdot \mathbf{H} = 0$ is sufficient for the orthonormality relation (11.6-14).

11.5 *Mode confinement factor.* Let Γ_i be the fraction of power flowing in medium i ($i = 1, 2, 3$) of the slab waveguide. In particular, Γ_2 is the fraction of mode power flowing in the guiding layer n_2 and is often called the mode confinement factor. If the mode is normalized to a power of 1 W, the Γ's are defined as

$$\Gamma_i = \tfrac{1}{2} \, \text{Re} \int_i (\mathbf{E} \times \mathbf{H}^*) \cdot \mathbf{z} \, dx, \qquad i = 1, 2, 3,$$

where the integral is over the region occupied by medium i where the index of refraction is n_i.

(a) Show that, for TE modes,

$$\Gamma_1 = \frac{1/q}{t + 1/q + 1/p} \frac{h^2}{h^2 + q^2},$$

$$\Gamma_2 = \frac{t}{t + 1/q + 1/p} + \frac{1/p}{t + 1/q + 1/p} \frac{p^2}{h^2 + p^2}$$

$$+ \frac{1/q}{t + 1/q + 1/p} \frac{q^2}{h^2 + q^2},$$

$$\Gamma_3 = \frac{1/p}{t + 1/q + 1/p} \frac{h^2}{h^2 + p^2} .$$

Note that $\Gamma_1 + \Gamma_2 + \Gamma_3 = 1$. This proves Eq. (11.2-7).

(b) Show that, for TM modes,

$$\Gamma_1 = \frac{1/q'}{t' + 1/q' + 1/p'} \frac{h^2}{h^2 + q^2} ,$$

$$\Gamma_2 = \frac{t'}{t' + 1/q' + 1/p'} + \frac{1/q'}{t' + 1/q' + 1/p'} \frac{q^2}{h^2 + q^2}$$

$$+ \frac{1/p'}{t' + 1/q' + 1/p'} \frac{p^2}{h^2 + p^2} ,$$

$$\Gamma_3 = \frac{1/p'}{t' + 1/q' + 1/p'} \cdot \frac{h^2}{h^2 + p^2} ,$$

where

$$t' = \frac{t}{n_2^2} ,$$

$$\frac{1}{q'} = \frac{q^2 + h^2}{\bar{q}^2 + h^2} \frac{1}{n_1^2 q} ,$$

$$\frac{1}{p'} = \frac{p^2 + h^2}{\bar{p}^2 + h^2} \frac{1}{n_3^2 p} ,$$

(c) Show that Γ_2 increases as β increases, and

$$\lim \Gamma_2 = 1, \qquad \beta \to \frac{n_2 \omega}{c} .$$

Thus, lower order modes are more confined than higher order modes because lower order modes have a larger propagation constant. *Hint*: Use $h^2 + p^2 = (n_2^2 - n_3^2)(\omega/c)^2$ and $h^2 + q^2 = (n_2^2 - n_1^2)(\omega/c)^2$.

(d) Compare the mode confinement factor Γ_2 for TE_m and TM_m modes.

(e) Show that in the limit of well confinement (i.e., $h \to 0$),

$$\frac{\Gamma_3^{TM}}{\Gamma_3^{TE}} = \frac{1 + p/q + pt}{1 + (n_1^2/n_3^2)(p/q) + (n_2^2/n_3^2)pt} ,$$

where $p = (\omega/c)(n_2^2 - n_3^2)^{1/2}$ and $q = (\omega/c)(n_2^2 - n_1^2)^{1/2}$.

(f) Show that for a metal-clad waveguide ($n_3^2 < 0$), $\Gamma_3^{TM} > \Gamma_3^{TE}$. In other words, TM modes penetrate more into the metal than TE modes and are thus more lossy.

11.6 *Perturbed dielectric slab waveguide.* Let the dielectric perturbation to a slab waveguide be

$$
\Delta n^2(x) = \begin{cases} \Delta n_1^2, & 0 \leqslant x, \\ \Delta n_2^2, & -t \leqslant x \leqslant 0, \\ \Delta n_3^2, & x \leqslant -t, \end{cases}
$$

where Δn_1^2, Δn_2^2, and Δn_3^2 are constants (can be complex to include gain or loss).

(a) Show that the correction to the propagation constant is

$$
\delta\beta_m^2 = \left(\frac{\omega}{c}\right)^2 (\Gamma_1 \, \Delta n_1^2 + \Gamma_2 \, \Delta n_2^2 + \Gamma_3 \, \Delta n_3^2),
$$

where

$$
\Gamma_i = \frac{\int_i \mathbf{E}_m^* \cdot \mathbf{E}_m \, dx}{\int \mathbf{E}_m^* \cdot \mathbf{E}_m \, dx}, \qquad i = 1, 2, 3,
$$

with the integration $\int_i$ performed over the medium i. These Γ's represent the fraction of power flowing in the corresponding medium.

(b) For a slab waveguide with a fixed thickness t, the propagation constants β_m^2 may be considered as functions of n_1^2, n_2^2, and n_3^2. Thus, $\delta\beta_m^2$ can be written as

$$
\delta\beta_m^2 = \frac{\partial\beta_m^2}{\partial n_1^2} \Delta n_1^2 + \frac{\partial\beta_m^2}{\partial n_2^2} \Delta n_2^2 + \frac{\partial\beta_m^2}{\partial n_3^2} \Delta n_3^2.
$$

The function $\beta_m^2 = \beta_m^2(n_1^2, n_2^2, n_3^2)$ is given implicitly in the mode conditions (11.2-5) and (11.2-11). Show directly from (11.2-5) that

$$
\frac{\partial\beta_m^2}{\partial n_i^2} = \left(\frac{\omega}{c}\right)^2 \Gamma_i, \qquad i = 1, 2, 3.
$$

11.7 Let the new mode of propagation be

$$
\mathbf{E}_m' = \mathbf{E}_m + a_{mm}\mathbf{E}_m + \delta\mathbf{E}_m,
$$

where $a_{mm}\mathbf{E}_m$ is taken out of the expansion $\delta\mathbf{E}_m$ so that $\delta\mathbf{E}_m$ is now orthogonal to $\mathbf{E}_m$. If $\mathbf{E}'_m$ is to be normalized as

$$\int \mathbf{E}'^*_m \cdot \mathbf{E}'_m \, dx \, dy = \frac{2\omega\mu}{\beta'_m},$$

where

$$\beta'^2_m = \beta^2_m + \delta\beta^2_m,$$

show that a_{mm} is given by

$$a_{mm} = -\frac{1}{4}\frac{\delta\beta^2_m}{\beta^2_m} = -\frac{1}{2}\frac{\delta\beta_m}{\beta_m}.$$

11.8 *Surface plasmons.*

(a) If the metal is viewed as a semi-infinite free electron gas with a dielectric constant given by

$$\varepsilon = \varepsilon_0\left(1 - \frac{\omega^2_p}{\omega^2}\right),$$

where ω_p is the plasma frequency (see Problem 2.2), show that surface plasmon modes exist only when $\omega^2 < \frac{1}{2}\omega^2_p$. Here, we assumed that $n^2_1 = 1$.

(b) Show that the propagation constant β derived in Eq. (11.4-4) is always greater than $n_1 k_0$, i.e., $\beta > n_1 k_0$, provided $n^2_1 > 0$ and $n^2_1 + n^2_3 < 0$. Here, $k_0 = \omega/c$.

(c) Show that $\beta^2 = pq$ and that the polarization states of the $\mathbf{E}$ vector in the two media are mutually orthogonal.

(d) Obtain an expression for the z component of the Poynting vector. Show that the Poynting power flow in the positive and negative dielectric media are opposite in direction. Thus, a surface plasmon wave propagating along the silver surface along the $+z$ direction will have a negative Poynting power flow in the silver and a positive Poynting power flow in the air.

(e) The complex refractive index of gold at $\lambda = 10\,\mu$m is $n - i\kappa = 7.4 - i53.4$. Find the propagation constant $\beta^{(0)}$ and the attenuation coefficient α of the surface plasmon wave.

11.9 *Solution of coupled-mode equations.* The coupled-mode equations in Eq. (11.8-11) can be solved by using a similar approach as used in

Section 8.5. In other words, we multiply both sides of the first equation by $\exp(-i\,\Delta\beta\,z)$ and then differentiate both sides with respect to z. The second equation is then substituted into the first.

(a) Show that the solution is

$$A(z) = [C_1 \sin sz + C_2 \cos sz]\exp(\tfrac{1}{2}i\,\Delta\beta\,z),$$

$$B(z) = \frac{iA'}{\kappa_{ab}}\exp(-i\,\Delta\beta\,z),$$

where C_1 and C_2 are constants and s is given by

$$s^2 = \kappa_{ab}\kappa_{ba} + (\tfrac{1}{2}\Delta\beta)^2.$$

(b) Derive Eq. (11.8-13).

12

Optics of Semiconductor Quantum Well and Superlattice Structures

Semiconductor quantum well and superlattice structures are ultrafine layered media whose layer thicknesses are in the range of one or a few atomic layers. These ultrastructures can be fabricated by several new material growth technologies such as molecular beam epitaxy (MBE) and metal–organic chemical vapour deposition (MOCVD). The presence of these ultrafine layers may affect the motion of electrons and thus leads to "quantum size effect" when the physical dimensions of the layers are comparable with the characteristic lengths that determine electron behavior: de Broglie wavelength, Bohr radii, and mean free path. These lengths range between 10 and 1000 Å. As a result, the refractive index of the medium can be different from that of the bulk material. In addition, the layer thicknesses of these ultrastructures are in the range of soft x rays. Thus, ultrastructures may also find applications in these spectral regimes. In this chapter, we will first discuss the confinement of charge carriers in quantum wells. We then discuss the optical properties, such as absorption and refractive indices, of some superlattices. In the last part of this chapter, we will discuss the possibility of using these ultrastructures for soft x-ray media.

12.1 QUANTUM WELLS

One of the most well-known problems in quantum mechanics is that of the confinement of a particle in a one-dimensional rectangular potential well. The quantization of states becomes significant when the de Broglie wavelength of the particle is comparable to the dimension of the well. For electrons with 1 eV kinetic energy, the de Broglie wavelength is 12.3 Å. Thus, these potential wells are best exemplified by the superlattice structures whose dimensions are in the range of 10–100 Å. In fact, quantum states of confined carriers in a very thin GaAs layer sandwiched between two $Al_x Ga_{1-x} As$ layers have been experimentally observed [1]. Gallium arsenide (GaAs) is a semiconductor with a bandgap of approximately 1.4 eV; AlAs has a larger

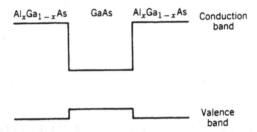

Figure 12.1   Schematic drawing of a quantum well formed by a thin GaAs layer sandwiched between two $Al_xGa_{1-x}As$ layers.

bandgap of approximately 2.7 eV [2]. The bandgap of the alloy $Al_xGa_{1-x}As$ varies almost linearly between these two extremes. The difference in bandgaps leads to a discontinuity in the band structure (see Fig. 12.1). Thus, carriers such as electrons or holes may be trapped by the GaAs layer.

To familarize ourselves with the quantum states in these structures, we briefly review the elementary quantum theory of a particle in a rectangular potential well. Let the potential energy of the particle be

$$V(z) = \begin{cases} 0, & |z| < \tfrac{1}{2}a, \\ V_0 & \text{otherwise}, \end{cases} \tag{12.1-1}$$

where V_0 is well height and a is width. The motion of this particle along the z axis is described by the following Schrödinger equation:

$$-\frac{\hbar^2}{2m}\frac{d^2}{dz^2}\psi + V(z)\psi = E\psi, \tag{12.1-2}$$

where $\hbar$ is Planck's constant, m is the mass of the particle, and E is the kinetic energy of the particle. This equation may be rewritten in the form

$$\frac{d^2}{dz^2}\psi + \frac{2m}{\hbar^2}[E - V(z)]\psi = 0, \tag{12.1-3}$$

which is formally identical to the wave equation for guided modes in a symmetric dielectric slab [see Eqs. (11.1-4)]. In addition, the boundary conditions that require that the wavefunction and its derivative be continuous are identical to those of the TE modes. Thus, the wavefunction can be written as

$$\psi(z) = \begin{cases} A\sin kz + B\cos kz, \\ C\exp(-qz), \\ D\exp(qz), \end{cases} \tag{12.1-4}$$

where A, B, C, and D are constants and the parameters k and q are related to the kinetic energy E by

$$k = \frac{1}{\hbar}(2mE)^{1/2}, \qquad q = \frac{1}{\hbar}[2m(V_0 - E)]^{1/2}. \qquad (12.1\text{-}5)$$

The parameter k is the wave number of the particle along the z axis in the guide layer. To be acceptable solutions, the wavefunction and its derivative must be continuous at the interfaces. This leads to a finite set of discrete quantum states. The energy of these quantum states is obtained by matching the wavefunction and its derivative at the boundaries. In a manner similar to that used in Section 11.1, we obtain the following transcendental equation:

$$\tan ka = \frac{2kq}{k^2 - q^2}, \qquad (12.1\text{-}6)$$

which can be solved for the quantized energy levels. Notice that this equation is also formally identical to that of the TE mode equation. The number of eigenstates is limited, being only one, or two, or three, and so on, depending on the depth V_0 and the width a of the potential well. According to the graphic method described in Section 11.1, which may also be applied to Eq. (12.1-6) for the case in which

$$V_0 a^2 < \frac{\pi^2 \hbar^2}{2m}, \qquad (12.1\text{-}7)$$

there is only one possible energy level; for

$$\frac{\pi^2 \hbar^2}{2m} < V_0 a^2 < \frac{4\pi^2 \hbar^2}{2m}, \qquad (12.1\text{-}8)$$

there are two energy levels; for

$$\frac{4\pi^2 \hbar^2}{2m} < V_0 a^2 < \frac{9\pi^2 \hbar^2}{2m}, \qquad (12.1\text{-}9)$$

there are three energy levels; and so on. For electrons, $\pi^2 \hbar^2/(2m) = 37.6 \text{ Å}^2 \text{ eV}$. Figure 12.2 plots the energy of these quantum states as a function of the parameter $V_0 a^2$.

The wavefunctions for the ground state and the first two excited states are shown in Fig. 12.3 for a rectangular potential well in which three energy levels are allowed. Note the similarity to the wavefunction of the first three

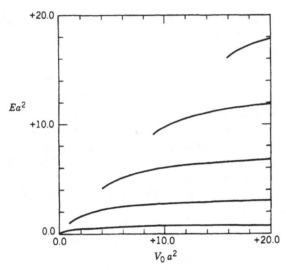

Figure 12.2 Energy of quantum states of a rectangular potential well as a function of the parameter $V_0 a^2$. The coordinates are in units of $\pi^2 \hbar^2/(2m)$.

confined TE modes of a symmetric slab waveguide (see Fig. 11.3). In the example shown in Fig. 12.1, the potential well is such that $V_0 a^2$ is $8\pi^2 \hbar^2/(2m)$, which, according to Fig. 12.2 or Eq. (12.1-9), can support three energy levels. If we take $a = 40$ Å, which corresponds to a potential well depth of 188 meV, the energy levels are $E_1 = 16$ meV, $E_2 = 61$ meV, and $E_3 = 132$ meV, respectively.

12.1.1   Example

In this example, we look at the quantum states of confined layers in a very thin GaAs layer sandwiched between two $Al_{0.2}Ga_{0.8}As$ layers.

Referring to Fig. 12.4, we consider a layer of 140-Å GaAs sandwiched between two layers of $Al_{0.2}Ga_{0.8}As$. The heterostructure behaves as a simple rectangular potential well with a depth of $0.88\,\Delta E_g$ for confining electrons and $0.12\,\Delta E_g$ for confining holes, where ΔE_g is the difference in the semiconductor energy gap. For such a heterostructure, $\Delta E_g = 250$ meV. Thus, the depth of the potential well is 220 meV for electrons and 30 meV for holes [1].

For confining electrons in GaAs, the effective mass $m_c = 0.0665\,m$. This leads to $V_0 a^2/[\pi^2 \hbar^2/(2m_c)] = 7.6$. According to Eq. (12.1-9), there are three quantum states for confining electrons. Similarly, there are also quantum

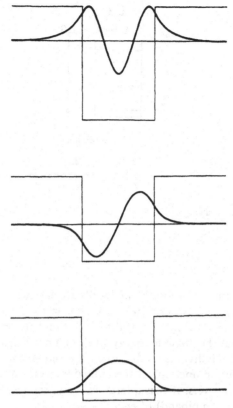

Figure 12.3 Wavefunction and energy level of the three quantum states of a rectangular potential well with $V_0 a^2 = 8\pi^2 h^2/(2m)$. The horizontal lines indicate the energy levels.

states for the confining holes. We note that the carriers (holes and electrons) are confined in the GaAs layer.

12.2 MULTIPLE QUANTUM WELLS

We now consider the energy levels of electrons in a periodic potential that consists of a series of equally spaced rectangular wells. The motion of an electron in a periodic rectangular potential well is also an old problem and had been used by Kronig and Penney [3] to model the energy band in solids. Again, such a model is best exemplified by multiple quantum wells (MQWs).

Referring to Fig. 12.5, we consider a MQW that is essentially a one-dimensional array of rectangular potential wells. Let N be the number of

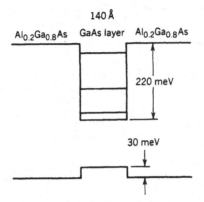

Figure 12.4   Schematic drawing of the band diagram of an $Al_{0.2}Ga_{0.8}As–GaAs–Al_{0.2}Ga_{0.8}As$ heterostructure.

wells. We assume that all the individual wells are identical. The energy levels (or quantum states) of MQWs are very similar to the guided modes in periodic multichannel waveguides. If the separation between the wells is very large, it is clear that the possible energy levels of a particle (e.g., electron) are simply those of an individual well. When the separation between wells is small so that the wavefunctions of the individual wells overlap, each energy level is split into N levels. If N is a large number, the result of this splitting leads to a subband (miniband) of energy levels.

The mathematics involved in the derivation of the energy levels and the wavefunctions is formally identical to those used in the periodic multichannel waveguides. Let the potential energy of the particle be

$$V(z) = \begin{cases} 0, & (n-1)\Lambda < z < (n-1)\Lambda + a \\ V_0, & \text{otherwise,} \end{cases} \tag{12.2-1}$$

Figure 12.5   Schematic drawing of the potential energy function of a multiple quantum well consisting of a one-dimensional array of N identical wells.

where V_0 is the well height, $n = 1, 2, \ldots, N$, Λ is the period and a is the width of the wells. The motion of the particle is again described by Eq. (12.1-2). Let the wavefunction of the particle be

$$
\psi(z) = \begin{cases}
C_0 \exp(-qz) + D_0 \exp(qz), & z < 0, \\
A_1 \exp(-ikz) + B_1 \exp(ikz), & 0 < z < a, \\
C_1 \exp[-q(z-a)] + D_1 \exp[q(z-a)], & a < z < \Lambda, \\
A_2 \exp[-ik(z-\Lambda)] + B_2 \exp[ik(z-\Lambda)], & \Lambda < z < \Lambda + a, \\
\vdots \\
A_N \exp\{-ik[z-(N-1)\Lambda]\} + B_N \exp\{ik[z-(N-1)\Lambda]\}, \\
\qquad\qquad (N-1)\Lambda < z < (N-1)\Lambda + a, \\
C_N \exp\{-q[z-(N-1)\Lambda-a]\} + D_N \exp\{q[z-(N-1)\Lambda-a]\}, \\
\qquad\qquad (N-1)\Lambda + a < z,
\end{cases}
$$

$$(12.2\text{-}2)$$

where A_n, B_n, C_n, D_n ($n = 0, 1, 2, \ldots, N$) are constants and the parameters k and q are related to the kinetic energy E by

$$
k = \frac{1}{\hbar}(2mE)^{1/2}, \qquad q = \frac{1}{\hbar}[2m(V_0 - E)]^{1/2}.
$$

The parameter k is the wave number of the particle along the z axis in the quantum well. To be acceptable solutions, the wavefunction and its derivative must be continuous at the interfaces. In addition, the wavefunction must vanish at infinity (i.e., $C_0 = D_N = 0$). These boundary conditions lead to a finite set of discrete quantum states. The energy of these quantum states is obtained by matching the wavefunction and its derivative at the boundaries. In a manner similar to that used in Section 11.3, we obtain the following trancendental equation:

$$
A \frac{\sin NK\Lambda}{\sin K\Lambda} - \frac{\sin(N-1)K\Lambda}{\sin K\Lambda} = 0, \qquad (12.2\text{-}4)
$$

with

$$
A = e^{qb}\left[\cos ka - \frac{1}{2}\left(\frac{k}{q} - \frac{q}{k}\right)\sin ka\right], \qquad (12.2\text{-}5)
$$

$$
\cos K\Lambda = \cosh qb \cos ka - \frac{1}{2}\left(\frac{k}{q} - \frac{q}{k}\right)\sinh qb \sin ka. \quad (12.2\text{-}6)
$$

where b is the thickness of the potential barriers ($b = \Lambda - a$).

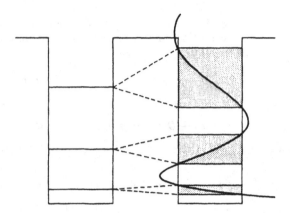

Figure 12.6 Plot of the energy bands of an array of potential wells. The energy levels of a single well are shown on the left. The height and the width of the wells are such that $V_0 a^2 = 8\pi^2 \hbar^2/(2m)$. Separation between wells is $b = 0.1a$. The curve on the right is $\cos K\Lambda$ versus E. The scale is chosen such that $\cos K\Lambda = 0$ is at the center and $\cos K\Lambda = \pm 1$ are at the walls of the well. The shaded regions are energy bands.

Equation (12.2-4) can be used to solve for the energy E of the quantum states. As shown in Appendix A, there are exactly N quantum states in each region where $K\Lambda$ varies from $m\pi$ to $(m + 1)\pi$ and none elsewhere, provided that V_0 is large enough. Physically, the MQWs can be considered as a system of N interacting potential wells. Each well can support a number of quantum states when they are far apart from each other. When these wells are brought together, the presence of other wells affects the quantum states of each individual well. Such interaction leads to the splitting of each of the N-fold degenerate quantum states. Each of the quantum states of an individual well is thus split into a band of N nondegenerate states. When N becomes very large, these N nondegenerate energy levels fill the region of $K\Lambda$ between $m\pi$ and $(m + 1)\pi$. This leads to the so-called allowed energy band in an infinite structure. In the case of superlattice structures with MQWs, the interaction leads to the existence of a subband (or miniband) of energy levels. If we plot $\cos K\Lambda$ as a function of the kinetic energy, the allowed energy bands are regions when $|\cos K\Lambda| < 1$. Energy gaps (or forbidden bands) are regions where $|\cos K\Lambda| > 1$. Figure 12.6 shows the energy levels of a single well and the energy bands of an array of wells.

Although the preceding discussion is limited to the case when $E < V_0$, energy band also exists in the energy regime when $E > V_0$. In the case of a single well, the energy levels for the range $E > V_0$ is in the continuum. This is no longer true in the case of a periodic array of wells. Both energy bands and energy gaps exist in the range $E > V_0$.

The perturbation approach used in Section 11.0 for the coupling of N identical waveguides can also be applied here to the treatment of N identical quantum wells. If we assume that coupling exists only between nearest neighbors and only between eigenstates with the same quantum number, the energy of the quantum states of the MQWs can be written as

$$E = E_0 - \frac{K + 2J\cos[s\pi/(N+1)]}{1 + 2I\cos[s\pi/(N+1)]}, \tag{12.2-7}$$

where E_0 is the energy of the unperturbed quantum states, $s = 1, 2, 3, \ldots, N$, and I, J, K are integrals defined as

$$I = \int_{-\infty}^{\infty} \psi(z)\psi(z+\Lambda)\, dz, \tag{12.2-8}$$

$$J = V_0 \int_0^a \psi(z)\psi(z+\Lambda)\, dc, \tag{12.2-9}$$

$$K = V_0 \int_0^a |\psi(z+\Lambda)|^2\, dc, \tag{12.2-10}$$

where we recall that $\psi(z)$ is the wavefunction of the quantum state of the first well (between $z = 0$ and $z = a$), a is the well width, V_0 is the well height, and Λ is the period of the MQWs. If the overlap integral I is much less than 1 ($I \ll 1$), Eq. (12.2-7) becomes

$$E = E_0 - K - 2J\cos\frac{s\pi}{N+1}. \tag{12.2-11}$$

Notice that there are N quantum states associated with each of the unperturbed quantum states of a single cell. The energy of the quantum states spans over a region of $K \pm 2J$. When N becomes infinite, the energy levels become an energy band. The width of this energy band is proportional to the integral J. For higher quantized states, the overlap integrals J are bigger. This leads to bigger energy bands. The situation is depicted in Fig. 12.6.

12.3   OPTICAL PROPERTIES OF SUPERLATTICES AND QUANTUM WELLS

In addition to the form birefringence that arises from the stratification, superlattices also exhibit some anomalous optical properties. Most of these optical anomalies are due to the confinement of charge carriers in the quantum wells.

In semiconductors such as GaAs, we consider an electron in the conduction band and a hole in the valence band. The electron and the hole are attracted to each other by the Coulomb interaction. The interaction of such an electron–hole pair leads to stable bound states of the two particles. The bound electron–hole pair is known as an *exciton*. The photon energy required to create a free electron–hole pair in the semiconductor must be equal to or greater than the bandgap E_g. However, the photon energy required to create an exciton is slightly less than the gap energy E_g as a result of the binding. The binding energy is given by (in MKS units)

$$E_b = \frac{32\pi^2 \mu e^4}{h^2 \varepsilon^2 n^2}, \tag{12.3-1}$$

where ε is the dielectric constant, n is the principal quantum number ($n = 1$, 2, 3, . . .), and μ is the reduced mass,

$$\frac{1}{\mu} = \frac{1}{m_e} + \frac{1}{m_h}, \tag{12.3-2}$$

formed from the effective mass m_e, m_h of the electron and hole, respectively. The Bohr radius of the exciton is given by (in cgs units)

$$a_0 = \frac{4\pi\varepsilon\hbar^2}{\mu e^2} n^2. \tag{12.3-3}$$

If we take $m_e = 0.067m$, $m_h = 0.08m$, and $\varepsilon = 13\varepsilon_0$ for GaAs, Eq. (12.3-1) leads to a binding energy of 3 meV for $n = 1$. The Bohr radius for $n = 1$ is approximately 200 Å. These are the typical binding energies and Bohr radii for excitons. Absorptions due to resonance excitation of such excitons are difficult to observe at room temperature because of the broadening due to many perturbations.

When the carriers are confined by the quantum wells, the exciton binding energy is greatly increased provided the layer thickness is less than the exciton Bohr radius. This may be explained by the argument that the quantum well confinement decreases the mean distance between the electron and the hole and thus effectively increases the Coulomb interaction. It may appear that the exciton is squeezed by the quantum well and becomes distorted in shape. In fact, when the layer thickness becomes so small, the motion of electrons and holes is dominated by the presence of the potential well. This can be seen from the energy of the first quantum state as calculated in Section 12.1. The binding energy of 3 meV for the bulk exciton is too small compared with the energy of the first quantum state in the well.

The increase in the exciton binding energy leads to sharper peaks in the optical absorption spectroscopy. In addition, such absorption peaks become observable at room temperature. Distinct exciton peaks at room temperature have been observed in GaAs–Al$_x$Ga$_{1-x}$As MQW samples [4]. It is known that the absorption spectra and the refractive index of semiconductors near the band edge are dominated by the contribution from the exciton lines. Thus, sharp exciton lines lead to strong dispersion of the index of refraction.

In addition, because of the anisotropy due to the layered structure, the superlattice structures also exhibit optical birefringence. Such birefringence is a result of the split of the exciton lines [5].

12.4 SUPERLATTICES AS SOFT X RAY MEDIA

As the growth technology progresses, it becomes possible to fabricate ultrathin-layered structures that are smooth on an atomic scale. Since the thickness of these layers is in the range of 10–1000 Å, these structures are immediate candidates for soft x-ray and far-ultraviolet media.

The interest in soft x rays arises from its applications in high-resolution lithography, surface science, contact microscopy, mammorgraphy, spectroscopy, plasma diagnosis, and so on. These applications require mirrors. Optics in this spectral regime has been very difficult because of the lack of mirrors. The index of refraction of most materials is very close to unity in this spectral regime, with $1 - n$ being on the order of 10^{-2}–10^{-3}. Thus, the Fresnel reflection of a single surface at normal incidence is on the order of 10^{-5}–10^{-7}, which is obviously too small for practical applications. Table 12.1 lists such indices of refraction for some materials [6].

In a manner similar to that of Bragg reflection, high-reflectance mirrors in this spectral regime can be achieved by the deposition of alternating layers of two different materials (e.g., quarter-wave stacks). Since the indices of refraction are all nearly unity, many layers may be needed to achieve high reflectance. In addition, the two materials must be very different in their complex refractive indices. To avoid material absorption, it is desirable to use materials that have low extinction coefficients.

Even though the indices of refraction are complex, the matrix formulation derived in Chapters 5 and 6 can be used to calculate the reflectance of such layered structures. Equation (6.3-12) can be used to calculate the reflectance of a stack of alternating layers. In the spectral regime of soft x rays, the term r_{al} (reflection coefficient at air–metal interface) is virtually zero. Thus, the reflection coefficient for a Bragg reflector of N periods can be written approximately as

$$r = r_N, \qquad (12.4\text{-}1)$$

Table 12.1. Index of Refraction of Some Materials in the Soft x-Ray Spectral Regime

E (eV)	Å	C		Al		Ag		Bi	
		$1-n$	k	$1-n$	k	$1-n$	k	$1-n$	k
10	1239.8	-0.811	4.49×10^{-1}	-0.572	1.05	-0.462	5.57×10^{-1}	0.290	9.68×10^{-1}
20	619.9	0.293	4.51×10^{-1}	-0.584	1.07	-0.290	7.14×10^{-1}	0.144	1.03×10^{-1}
30	413.3	0.163	1.40×10^{-1}	0.132	8.04×10^{-1}	0.069	5.41×10^{-1}	0.013	1.27×10^{-1}
40	309.9	0.107	7.30×10^{-2}	0.246	4.38×10^{-1}	0.104	3.68×10^{-1}	-0.051	1.31×10^{-1}
50	248.0	0.078	4.50×10^{-2}	0.205	2.72×10^{-1}	0.166	2.90×10^{-1}	-0.027	1.60×10^{-1}
60	206.6	0.060	2.70×10^{-2}	0.162	2.19×10^{-1}	0.127	2.21×10^{-1}	0.034	1.51×10^{-1}
70	177.1	0.045	1.50×10^{-2}	0.172	1.55×10^{-1}	0.141	1.97×10^{-1}	0.058	1.20×10^{-1}
80	155.0	0.034	1.02×10^{-2}	0.147	1.09×10^{-1}	0.152	1.39×10^{-1}	0.068	8.07×10^{-2}
90	137.8	0.027	7.25×10^{-3}	0.134	7.63×10^{-2}	0.151	8.24×10^{-2}	0.065	5.36×10^{-2}
100	124.0	0.022	5.19×10^{-3}	0.111	4.54×10^{-2}	0.124	3.82×10^{-2}	0.055	3.59×10^{-2}
110	112.7	0.018	3.77×10^{-3}	0.089	2.83×10^{-2}	0.098	1.72×10^{-2}	0.047	2.51×10^{-2}
120	103.3	0.015	2.82×10^{-3}	0.071	1.78×10^{-2}	0.074	6.22×10^{-3}	0.040	1.70×10^{-2}
130	95.4	0.013	2.35×10^{-3}	0.056	1.33×10^{-2}	0.057	3.66×10^{-3}	0.034	1.22×10^{-2}
140	88.6	0.011	1.93×10^{-3}	0.045	1.09×10^{-2}	0.045	3.54×10^{-3}	0.028	8.67×10^{-3}
160	77.5	0.008	1.21×10^{-3}	0.030	1.01×10^{-2}	0.032	4.38×10^{-3}	0.020	4.48×10^{-3}
180	68.9	0.006	8.48×10^{-4}	0.021	1.07×10^{-2}	0.022	4.31×10^{-3}	0.014	3.31×10^{-3}
200	62.0	0.005	5.85×10^{-4}	0.017	1.15×10^{-2}	0.018	3.77×10^{-3}	0.010	2.83×10^{-3}
300	41.3	-0.002	3.54×10^{-3}	0.009	8.95×10^{-3}	0.004	2.32×10^{-3}	0.003	4.18×10^{-3}

where r_N is given by Eq. (6.3-5). Because of the material absorption, the reflectance does not continue to increase by adding more layers, although many layers are needed to achieve high reflectance. Thus, one obtains a saturation of reflectance with a number of layers.

The design of high reflectance for soft x-ray radiation can also be achieved by using coupled-mode analysis as used in Section 8.5. Such a method has been used by Vinogradov and Zeldovich [7] for the design of such mirrors. In the coupled-mode analysis, the square-well index profile is approximated by the zeroth and ± 1 spatial harmonics, whereas the matrix method of Eq. (6.3-12) is an exact approach. To appreciate the effect of the number of layers on the reflectance, we consider the following two examples.

Example: Chromium–Carbon Bragg Reflector at $\lambda = 113\,\text{Å}$. Consider a Bragg reflector that consists of alternating layers of chromium and carbon [7]. The indices of refraction of these two materials at $\lambda = 113\,\text{Å}$ are given by [8]

$$n_1 = 0.9603 - 0.01790i \quad \text{for Cr,}$$

$$n_2 = 0.9792 - 0.00375i \quad \text{for C.}$$

Using layer thicknesses of $a = 21\,\text{Å}$ for chromium, and $b = 37\,\text{Å}$ for carbon, the calculated reflectances at $\lambda = 113\,\text{Å}$, according to Eq. (12.4-1), are 13% for $N = 40$, 14.2% for $N = 80$, and 14.3% for $N = 120$. Note that the Fresnel reflectance of a single boundary is less than 0.02%. We also note that peak reflectance saturates at $N = 120$. In other words, further increasing the number of layers will not increase the reflectance. Figure 12.7 plots the reflectance spectrum of such Bragg reflectors.

Example: Titanium–Nickel Bragg Reflector at $\lambda = 31.36\,\text{Å}$. Consider a Bragg reflector that consists of alternating layers of titanium and nickel [7]. The indices of refraction of these two materials at $\lambda = 31.36\,\text{Å}$ are given by [9]

$$n_1 = 0.9940 - 0.0015i \quad \text{for Ni,}$$

$$n_2 = 0.9981 - 0.0004i \quad \text{for Ti.}$$

Using layer thicknesses of $a = 5.8\,\text{Å}$ for nickel and $b = 9.9\,\text{Å}$ for titanium, the calculated reflectances at $\lambda = 31.36\,\text{Å}$, according to Eq. (12.4-1), are 18% for $N = 100$, 21% for $N = 200$, 30% for $N = 400$, and 33% for $N = 800$. Note again that the Fresnel reflectance for titanium and nickel at the air–metal boundary is less than 10^{-5} for the noted wavelength. Again, we note

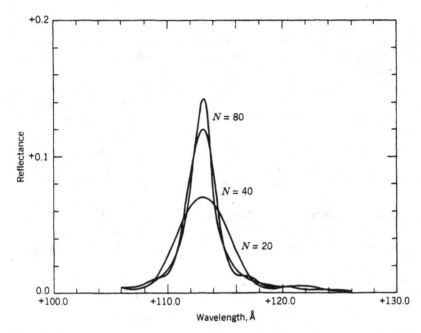

Figure 12.7   Reflectance spectrum of Cr–C Bragg reflectors at various number of periods.

that reflectance saturates at $N = 800$. Figure 12.8 plots the reflectance spectrum of such Bragg reflectors.

In these two example, we note that the layer thicknesses are chosen in such a way that the period $\Lambda = a + b$ is approximately one-half of the wavelength. This so-called *Bragg condition* is to ensure high reflectance at the desired wavelength for normal incidence. In addition, we also note that the layer thicknesses are chosen such that the lossier layer is thinner than the other layer. This is necessary to minimize the material absorption loss and to optimize the peak reflectance. In fact, if one of the layers is nonabsorbing, a high-reflectance mirror can be achieved by using a thin but highly absorbing layer and a relatively thicker nonabsorbing layer as the pair [10]. In a manner analogous to the *Borrmann effect* observed in x-ray diffraction [11], the electromagnetic wave positions itself so that the nodes of the standing wave coincide with the absorbing layers. Thus, absorption loss is minimized and high reflectance is achieved. Mirrors using such a design principle for the near ultraviolet have been demonstrated [10].

A similar principle can be used for the design of x-ray laser media that consist of alternating gain and loss layers ($\alpha_1 > 0$, $\alpha_2 < 0$). We recall

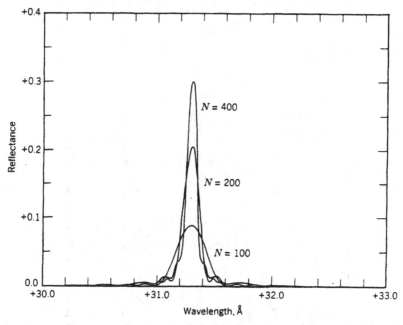

Figure 12.8 Reflectance spectrum of Ti–Ni Bragg reflectors at various number of periods.

that $\alpha = 2k\omega/c$, where k is the extinction coefficient. As an illustrative example, we shall consider a structure consisting of alternating beryllium and aluminum layers with a period $\Lambda = 57\,\text{Å}$. We are looking for x-ray oscillation at the beryllium K_α line (114 Å). If the whole structure is pumped for some external incoherent x-ray source, beryllium layers become gain layers as far as beryllium K_α emission is concerned, while the loss constant α_1 of the aluminum layers at 114 Å is about $1.0\,\Lambda^{-1}$. The real parts of the refractive indices are taken as unity ($n_{1,2} \approx 1.0$).

We investigate the reflectance of a 10-period slab ($N = 10$) as a function of the normalized frequency $\omega\Lambda/c$ and the gain α_2. The contour plot of $|r_N|$ in the $\alpha_2\omega$ plane is shown in Fig. 12.9. A series of poles where $|r_N| = \infty$ are found in the lower half plane ($\alpha_2 < 0$). The coordinates of these poles correspond to the threshold gains and the oscillation frequencies of the laser. The number of poles is exactly N, which is the number of periods. The pole trajectory in the $\alpha_2\omega$ plane indicates that the pole nearest the bandgap ($\omega\Lambda/c = \pi$) has the lowest threshold gain. The threshold gain α_{2_t} is approximately equal to the loss α_1 for modes whose frequency is far away from the bandgap. However, it is much less than the loss when the oscillation is near

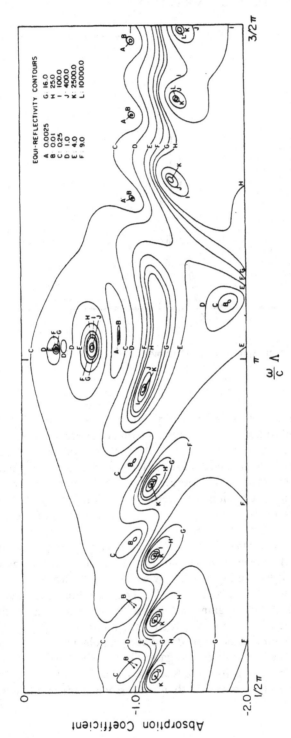

Figure 12.9 Contour plot of the reflectance $|r_N|$ of a 10-period Bragg reflector consisting of alternating gain and loss layers of beryllium and aluminum.

the bandgap. In our example, $|\alpha_{2l}| \approx \frac{1}{3}\alpha_1$. This theoretical result can be explained as follows. The generated power per unit area is proportional to

$$J = -\int \alpha(x)E^2(x) \, dx, \qquad (12.4\text{-}2)$$

where

$$\alpha(x) = \begin{cases} \alpha_1 > 0, & \text{layer 1,} \\ \alpha_2 < 0, & \text{layer 2.} \end{cases} \qquad (12.4\text{-}3)$$

If the lasing mode intensity distribution has its maxima in the gain layers and minima in the loss layers, power generation ($J > 0$) is possible even when the integrated loss is positive, or in other words, when

$$\int \alpha(x) \, dx > 0. \qquad (12.4\text{-}4)$$

In the conventional Fabry–Perot laser where $\alpha(x) = \text{const}$, power generation requires a net positive gain (negative loss),

$$\alpha L < 0. \qquad (12.4\text{-}5)$$

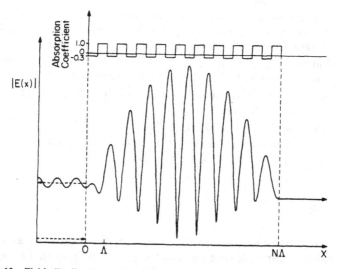

Figure 12.10 Field distribution near oscillation. The dashed arrows indicate incident and reflected waves respectively. The solid arrow at the right-hand side indicates the transmitted wave. The upper curve is the gain-loss profile.

That means the medium of a conventional laser has to be pumped until the gain exceeds the loss. In a periodic multilayer laser, however, the gain constant of the gain layer does not have to be larger than the loss constant of the loss layer, assuming the same layer thickness.

The field distribution near oscillation of a typical multilayer x ray is shown in Fig. 12.10. Notice that the local maxima of the field amplitude are all located in the gain layers. The parameters correspond to the low-threshold pole of Fig. 12.9.

A similar principle can be used for the design of x-ray laser media that consist of alternating gain and loss layers.

REFERENCES

1. See, for example, R. Dingle, W. Wiegmann, and C. H. Henry, *Phys. Rev. Lett.* **33**, 827 (1974).

2. M. A. Afromowitz, *Solid State Commun.* **15**, 59 (1974).

3. R. de L. Kronig and W. G. Penney, *Proc. Roy. Soc. (London)* **130**, 499 (1931).

4. D. A. B. Miller, D. S. Chemla, D. J. Eilenberger, P. W. Smith, A. C. Gossard, and W. T. Tsang, *Appl. Phys. Lett.* **41**, 679 (1982).

5. J. P. van der Ziel and A. C. Gossard, *J. Appl. Phys.* **48**, 3018 (1977).

6. H.-J. Hagemann, W. Gudat, and C. Kunz, *J. Opt. Soc. Am.* **65**, 742 (1975), and the references therein.

7. A. V. Vinogradov and B. Ya. Zeldovich, *Appl. Opt.* **16**, 89 (1977). (*Note:* dielectric constants quoted in this paper are in error by a factor of 2 as compared with the complex indices of refraction reported in Refs. 8 and 9.)

8. A. P. Lukirsky, E. P. Savinov, O. A. Ershov, and Yu. F. Shepelev, *Opt. Spektrosk.* **16**, 310 (1964) [*Opt. Spectrosc.* **16**, 168 (1964)].

9. O. A. Ershov, I. A. Brytov, and A. P. Lukirsky, *Opt. Spektrosk.* **22**, 127 (1967) [*Opt. Spectrosc.* **22**, 66 (1967)].

10. E. Spiller, *Appl. Phys. Lett.* **20**, 365 (1972).

11. B. W. Batterman and H. Cole, *Rev. Mod. Phys.* **36**, 681 (1964).

12. A. Yariv and P. Yeh, *Opt. Comm.* **22**, 5 (1977).

PROBLEMS

12.1 *Infinite square-well potentials.* Consider the motion of an electron inside an infinite square potential well. Let the potential energy be

$$V(x) = \begin{cases} 0, & |x| < \tfrac{1}{2}a, \\ \infty, & \text{otherwise.} \end{cases}$$

(a) Show that the wavefunctions can be written as

$$\psi(x) = \begin{cases} A \cos kx, \\ A \sin kx, \end{cases}$$

where A is a normalization constant and k is the wave number related to energy by

$$E = \frac{\hbar^2 k^2}{2m}.$$

(b) Show that the allowed energy levels are

$$E = \frac{\hbar^2 \pi^2}{2ma^2} n^2 \quad (n = 1, 2, 3, \ldots).$$

(c) Show that the ground state agrees with the uncertainty principle.

12.2 *Coupled-mode analysis for complex periodic media.* The superlattice structures described in Section 12.4 can be approximated by a sinusoidal layer. Wave propagation in such layers has been discussed in Section 8.5. The expansion equation (8.5-1) is not adequate for metallic layers whose indices of refraction are complex. Let the index of refraction be

$$n(x) = \begin{cases} n_1, & 0 < x < b, \\ n_2, & b < x < \Lambda, \end{cases}$$

where n_1 and n_2 are complex indices of refraction.

(a) Show that the index profile can be written as a Fourier expansion and is given by

$$n(x) = \sum n_p \exp[-ipKx],$$

where p is a dummy index ($p = 0, \pm 1, \pm 2, \ldots$) and n_0 is given by

$$n_0 = \frac{an_1 + bn_2}{\Lambda},$$

where we recall that $a + b = \Lambda$ and $K = 2\pi/\Lambda$.

(b) Neglecting the higher order terms, show that the index-of-refraction profile is

$$n(x) = n_0 + p_1 \exp[-iKx] + p_{-1} \exp[iKx],$$

where $p_1 = n_1$ and $p_{-1} = n_{-1}$ to avoid confusion with the layer indices. Also show that p_1 and p_{-1} are complex conjugates of each other provided that both n_1 and n_2 are real.

(c) Substitute the index profile in (b) into the wave equation (8.1-1) and show that the coupled equations become

$$\frac{d}{dx} A = -i\kappa B e^{i\Delta kx}, \qquad \frac{d}{dx} B = i\kappa' A e^{-i\Delta kx},$$

where κ and κ' are given by

$$\kappa = \frac{\omega}{2c} p_1, \qquad \kappa' = \frac{\omega}{2c} p_{-1},$$

and Δk is given by

$$\Delta k = 2\frac{\omega}{c} n_0 - K.$$

We note that Δk is now also a complex quantity.

(d) Show that the reflection coefficient can be written

$$r = \frac{-i\kappa' \sinh sL}{s \cosh sL + i(\Delta k/2)\sinh sL},$$

where s is given by $s^2 = \kappa\kappa' - (\Delta k/2)^2$. Notice the slight difference between these formula and Eq. (8.5-25).

(e) Show that the saturated reflection coefficient can be written

$$r = \frac{-i\kappa'}{s + i\Delta k/2}.$$

Hint: Let sL be infinity and $\cosh sL = \sinh sL$ in (d).

Appendix A

Zeros of Mode Dispersion Relation

To prove that Eq. (11.3-16) has N zeros in each allowed band where $K\Lambda$ varies from $m\pi$ to $(m + 1)\pi$, we need to show that the left side of the same equation changes sign N times in each allowed band. We define

$$F(K\Lambda) = A \frac{\sin NK\Lambda}{\sin K\Lambda} - \frac{\sin (N - 1)K\Lambda}{\sin K\Lambda}. \qquad (A.1)$$

where A can also be considered a function of $K\Lambda$. Let us now examine the sign of F at $K\Lambda = n\pi/N$, where $n = 0, 1, 2, 3, \ldots, N$. From Eq. (A.1), we obtain

$$F\left(\frac{n\pi}{N}\right) = \begin{cases} N[A(0) - 1] + 1, & n = 0, \\ (-1)^n, & n = 1, 2, \ldots, N - 1, \\ (-1)^{N-1}\{N[A(\pi) + 1] - 1\}, & n = N. \end{cases}$$

$$(A.2)$$

To show that F changes sign N times, we need to show that

$$N[A(0) - 1] + 1 > 0 \qquad (A.3)$$

and

$$N[A(\pi) + 1] - 1 < 0 \qquad (A.4)$$

for all N.

Since N is arbitrary, we need to show that

$$A(0) > 1 \qquad (A.5)$$

and

$$A(\pi) < -1. \qquad (A.6)$$

Using Eqs. (11.3-9) and (11.3-17), we can eliminate the factor $p/q - q/p$ and obtain, after a few algebraic steps,

$$A = \cos K\Lambda + \frac{\cosh qa \cos K\Lambda - \cos pb}{\sinh qa}. \tag{A.7}$$

Since $\cosh qa > 1$ and $\sinh qa > 0$, the second term on the right side of Eq. (A.7) is always of the same sign as the first term. Thus, at $K\Lambda = 0$ and $K\Lambda = \pi$, Eq. (A.7) yields $A(0) > 1$ and $A(\pi) < -1$, respectively. This proves that F changes sign N times in the region where $K\Lambda$ varies from 0 to π. The proof for the region where $K\Lambda$ varies from $m\pi$ to $(m + 1)\pi$ is similar.

To prove that Eq. (11.3-16) has no roots elsewhere, we need to show that the left side of the same equation does not change sign in the region where $|\cos K\Lambda| > 1$. Let $K\Lambda = 0 + i\chi$, where χ is a real number. Then A can be written, according to Eq. (A.7), as

$$A = \cosh \chi + \frac{\cosh qa \cosh \chi - \cos pb}{\sinh qa}. \tag{A.8}$$

Notice that $A > 1$ and is an increasing function of $|\chi|$. By substituting $0 + i\chi$ for $K\Lambda$ in Eq. (A.1), F becomes

$$F(0 + i\chi) = A\frac{\sinh N\chi}{\sinh \chi} - \frac{\sinh (N - 1)\chi}{\sinh \chi}. \tag{A.9}$$

where A is given by Eq. (A8). Since $A > 1$ and $(\sinh N\chi/\sinh \chi) > [\sinh (N - 1)\chi/\sinh \chi]$ for any χ, we conclude that F has no zero in the 0th forbidden band, where $K\Lambda = 0 + i\chi$. The proof for the region where $K\Lambda = m\pi + i\chi$ is similar. This proves that $F(K\Lambda)$ has no zeros in all the forbidden band, where $|\cos K\Lambda| > 1$.

Author Index

Subject Index